Designers' Guide to EN 1994-1-1: 2026

Designers' Guide to EN 1994-1-1: 2026

Eurocode 4 Design of composite steel and concrete structures

Part 1.1: General rules and rules for buildings

Third edition

Roger P Johnson and Stephen J Hicks

Published by Emerald Publishing Limited, Floor 5, Northspring, 21-23 Wellington Street, Leeds LS1 4DL.

ICE Publishing is an imprint of Emerald Publishing Limited

Other ICE Publishing titles:
Concise Guide to Reinforced Concrete Design to Eurocode 2
Patrick Purcell. ISBN 9780727765727
ICE Handbook of Concrete Durability
Edited by Marious Soutsos. ISBN 9780727763754
Fibre-reinforced Concretes for High-performance Structures
Andreas Lampropoulos. ISBN 9780727765567

A catalogue record for this book is available from the British Library

ISBN 978-1-83662-921-4

Commissioning Editor: Michael Fenton
Content Development Editor: Ryan Molyneux

Books Production Lead: Benn Linfield
Typeset by KnowledgeWorks Global Limited
Printed and bound by CPI Group (UK) Ltd, Croydon, CR0 4YY

Preface to the third edition

EN 1994-1-1, 'General rules and rules for buildings' is a part of Eurocode 4, 'Design of composite steel and concrete structures' which is one of the standards that comprise the current system of Eurocodes. Details are given in Chapter 0, 'Introduction', which corresponds to *Clause 0* of EN 1994-1-1.

Objectives of this guide

This book provides guidance on the interpretation and use of EN 1994-1-1, with comment and worked examples. It explains the relationship with the other Eurocode parts to which Eurocode 4 refers and with other standards and regulations. It also provides references and background information on the origin of the provisions of EN 1994-1-1.

Layout of this guide

EN 1994-1-1 has a foreword, twelve main Clauses, numbered 0 to 11, and ten Annexes, numbered A to J. This guide has twelve chapters, numbered 0 to 11, which correspond to the Clauses; chapters numbered 12 to 20, which correspond to Annexes A, B and D to J; Appendices A and B which provide background theory for lateral-torsional buckling and columns, respectively, and Appendix C on the new technical specification for composite dowel shear connectors. Annex C does not need comment and so has no chapter here.

The numbering and titles of the sections in this guide also correspond to those of the level 2 clauses (e.g. 1.1) of EN 1994-1-1. Some sub-sections are also numbered (e.g. 1.1.2). This implies correspondence with the sub-clause in EN 1994-1-1 of the same number. Their titles also correspond. There are extensive references to lower-level clause and paragraph numbers. The first significant reference to a clause is in ***bold italic*** type (e.g. ***clause 1.1.1(2)***). These are in strict numerical sequence throughout the book, and appear in the adjacent outer margin to help readers to find comments on particular provisions of the code. Some comments on clauses are necessarily out of sequence, but use of the index should enable these to be found.

In worked examples, results are calculated that may appear as numbers in arithmetical work pages later. To enable readers to find the source of the number, it and its symbol are printed in **bold** type at first appearance; for example: f_{ctm}= 0.89 × 2.6 = **2.31 MPa.**

All cross-references in this guide to clauses, paragraphs, annexes, figures, tables and formulae of EN 1994-1-1 are in *italic* type, which is also used where text from EN 1994-1-1 has been directly reproduced (conversely, cross-references to and quotations from other sources, including other Eurocodes, are in roman type). Figures, formulae and tables repeated from EN 1994-1-1 retain their number. They are distinguished from new material by the use of *italic* type when referring to them. The word 'formula' in Eurocode 4 includes equations and inequalities.

References to EN 1994, EN 1994-1-1 and so on should be deemed to refer to BS EN 1994-1-1:2026.

Formulae not in Eurocode 4 are referred to as 'equations' and have numbers prefixed by X (e.g. Equation X6.1 in Chapter 6).

The following abbreviations are widely used in this book.

BS EN, British and European standard

CEN, European Committee for Standardization

EAD, European Assessment Document

ETA, European Technical Assessment

ETPS, European Technical Product Specification

IABSE, International Association for Bridge and Structural Engineering

NCCI, non-contradictory complementary information

NDP, nationally determined parameter

TS, technical specification

Acknowledgements

The authors are deeply indebted to the members of the many working groups of CEN and BSI with whom they have worked on drafts of Eurocode 4, and to others who have prepared national comments. They thank the University of Warwick for facilities provided for Eurocode work.

They are grateful to the Steel Construction Institute and its Chairman and CEO, Graham Couchman, for information provided over many years. Roger Johnson gives thanks for the unselfish support from his late wife Diana over 65 years. Stephen Hicks acknowledges, with thanks, his wife Mandy for her unwavering support.

Roger P. Johnson and *Stephen J. Hicks*

Contents

Roger P Johnson and Stephen J Hicks
ISBN 978-1-83662-921-4
https://doi.org/10.1108/978-1-83662-920-720253002

CHAPTER 0
Introduction

This clause is preceded by a short 'European foreword', which refers to the preparation of this revision of EN 1994-1-1, gives some reminders to users and refers to the purpose and use of national annexes.

0.1 Introduction to the Eurocodes

Of the Eurocodes listed in ***clause 0.1***, EN 1994-1-1 cross-refers extensively only to EN 1992, 'Design of concrete structures' and EN 1993, 'Design of steel structures'. There are a few references to EN 1990, 'Basis of structural and geotechnical design', with which it has to be consistent. *Clause 0.1*

The way in which Eurocodes are used in practice depends to some extent on the laws and regulations of the country concerned, as explained in this clause. The Eurocodes recognise the responsibility of relevant authorities and have safeguarded their right to determine values related to safety matters at a national level.

The governments of the UK and its devolved regions publish documents that provide guidance on the means by which compliance with the Building Regulations in force can be achieved. For the structure and its safety these are: Approved Document A (England); Approved Document A (Wales); Scottish Technical Handbook; and Technical Booklet D in Northern Ireland.

Approved Document A (England) currently references the Eurocodes. It states: '...*British Standards for structural design based on the Eurocodes were...implemented by the British Standards Institution on 1st April 2010 and it is these standards with their corresponding UK National Annexes which are now referenced in this Approved Document as practical guidance on meeting Part A requirements*.' It further states: '*There may be alternative ways of achieving compliance with the requirements and there may be cases where it can be demonstrated that the use of withdrawn standards no longer maintained by the British Standards Institution continues to meet Part A requirements*.' The Scottish Technical Handbook, the Welsh Approved Document and the Northern Ireland Technical Booklet have been updated in a similar way.

0.2 and 0.3 Introduction to EN 1994 (all parts) and to EN 1994-1-1

No definition is given of the 'steel and concrete composite structures' to which EN 1994 applies. Composite members, beams, columns and slabs are defined in *clause 3.1*. The titles of its Parts include the well-understood words 'building' and 'bridge'. It will usually be obvious where the code applies; it is where composite behaviour, defined in *clause 3.1.4*, is being relied on in design.

EN 1994-2 has a new title, 'Bridges'. The previous EN 1994-2 included all the general rules of EN 1994-1-1. There were some provisions that were in theory 'general', but given only in EN 1994-2 because they were in practice not used for buildings, where simpler methods are preferred. They have now been moved into EN 1994-1-1, which unfortunately increases its length and complexity. EN 1994-2 is, of course, much shorter than before.

EN 1994-1-2 covers 'Structural fire design'. It is outside the scope of this guide.

The following further Parts of Eurocode 4 are also outside the scope, except for Appendix C on composite dowels.

- CEN/TS 1994-1-101: Double and single skin steel concrete composite (SC) structures (in preparation).
- CEN/TS 1994-1-102: Composite dowels (publication expected in 2026).
- CEN/TS 1994-1-103: Composite columns comprising high performance columns (in preparation).

0.4 Verbal forms used in the Eurocodes

Provisions are of two main types: Principles, which use the verb 'shall', and Application Rules, which use 'should'. In the first set of Eurocodes, the clause number of a principle ended in P. This is no longer done. If a design method deviates from a clause in some way, the verb used is important, because if 'shall' is used, that design does not comply with the Eurocodes, which may have legal implications. The situation with deviation from a 'should' clause is more complex, as explained below.

0.5 National Annex for EN 1994-1-1

Each national standards body is required to publish each Eurocode part as a national standard. It comprises, without any alterations, the full text of the Eurocode and its annexes as published by the Comité Européen de Normalisation (CEN), usually preceded by a National Title Page and a National Foreword, and followed by a National Annex.

Each Eurocode recognises the right of national regulatory authorities to determine values related to safety matters. Values, classes or methods to be chosen or determined at national level are referred to as Nationally Determined Parameters (NDPs). A recommended value for each one is given in a Note that follows the relevant clause. Those clauses are listed in this clause. The values are usually those assumed during drafting and used for calibration work.

In EN 1994-1-1 the NDPs are principally the partial factors for material or product properties peculiar to this standard – for example, relating to shear connectors or profiled sheeting. A new method for the resistance of composite slabs to vertical shear was sufficiently controversial to need three NDPs. Each national annex gives or cross-refers to the values to be used for the NDPs in its country. For the five annexes that are Informative (optional), each country has these options: to permit its use; to mandate its use (i.e. treat it as Normative, with the same status as the rest of the Eurocode); to prohibit its use in full or in part; or to provide alternative provisions in full or in part.

Complementary information (formerly referred to as non-contradictory complementary information or 'NCCI' in the first-generation Eurocodes) may be directly included within the national annex (NA) or by reference, provided that it does not alter or contradict any provisions of the Eurocode. In practice, questions on interpretation of code clauses always arise. Any organisation can publish material claimed to be 'complementary information', and particular industries may have a vested interest in doing so.

Each national annex will have been approved by the relevant national standards body (BSI for the UK), which in effect gives complementary information to which it refers a status close to that of a national standard. However, much complementary information will appear after the national annex has been published. Before using such material for work claimed to be in accordance with Eurocodes, the designer should be satisfied that it does not alter or contradict any provisions of the Eurocode.

The UK's National Annex to EN 1994-1-1 was not complete at the time of writing (April 2025), so this book uses the 'recommended' data given in Notes in Eurocode 4. The comments given on the 'expected' content of the Annex are mainly based on the NA to the 2004 Eurocode 4, and should not be relied on.

Drafting errors in codes and some questions of interpretation are resolved by the official corrigenda and amendments that appear in its lifetime. Proposals for these are classified as 'editorial' or 'technical'. As they would apply in all Member States, technical changes have to be approved by the CEN Committee TC 250/SC 4.

Roger P Johnson and Stephen J Hicks
ISBN 978-1-83662-921-4
https://doi.org/10.1108/978-1-83662-920-720253003

CHAPTER 1
Scope

- Scope of EN 1994-1-1 — *Clause 1.1*
- Assumptions — *Clause 1.2*

1.1 Scope of EN 1994-1-1

Clause 1.1 refers to 'basic rules for the design of composite structures' for 'buildings' and 'bridges'. The Eurocodes are concerned with structural behaviour. Other requirements – for example, thermal and acoustic insulation, are not considered. *Clause 1.1*

In EN 1994-1-1, all rules that are for buildings only are preceded by a heading that includes the word 'buildings' or, if an isolated paragraph, are placed at the end of the relevant clause – for example, *clause 7.4.2.3(5)*. All other provisions are 'general' – for example, most of the rules for resistance to bending apply for a composite beam in a building as in a bridge, so those rules are 'general' and appear in EN 1994-1-1, even for rules where most applications occur in bridges.

The coverage in this guide of the 'general' clauses of Part 1-1 is relevant to both buildings and bridges, except where noted otherwise. However, guidance provided by or related to the worked examples may be relevant only to applications in buildings.

The provisions of Part 1-1 cover the design of the common composite members

- beams in which a steel section acts compositely with concrete
- composite slabs formed with profiled steel sheeting
- concrete-encased and concrete-filled composite columns
- joints between composite beams and steel or composite columns.

Clauses 7 and *11* concern connected members. *Clause 7*, 'Structural analysis', is needed particularly for framed structures. Unbraced frames and sway frames are within its scope. The provisions include definitions of imperfections and the use of second-order global analysis and prestress by imposed deformations.

The scope of Part 1-1 extends to steel I-sections where the web is encased by reinforced concrete and shear connection is provided between the concrete and the steel ('partially encased' composite members). The primary reason for its choice is improved resistance in fire. Fully-encased composite beams are not included because

- no satisfactory model has been found for the ultimate strength in longitudinal shear of a beam without shear connectors
- it is not known to what extent some design rules (e.g. for moment-shear interaction and redistribution of moments) are applicable.

A fully encased beam with shear connectors can usually be designed as if partly encased or uncased, provided that care is taken to prevent premature spalling of encasement in compression.

Part 2, *Bridges*, includes further application rules that may on occasion be useful for buildings, such as those on

- composite plates (where the steel member is a flat steel plate, not a profiled section)
- composite plate girders and box girders
- tapered or non-uniform composite members.

The omission of application rules for a type of member or structure should not prevent its use, where appropriate. Some omissions are deliberate, to encourage the use of innovative design based on specialised literature, the properties of materials and the fundamentals of equilibrium and compatibility, and follow the principles given in the relevant Eurocodes. This applies, for example, to

- base plates beneath composite columns
- shear heads in reinforced concrete framed structures (Piel and Hanswille, 2006)
- many aspects of 'mixed' structures, as used in tall buildings (Henderson, 2017).

1.2 Assumptions

Clause 1.2

Clause 1.2 consists only of cross-references to assumptions made in other European standards. It does not repeat the further assumptions found in appropriate clauses of all Parts of EN 1994. Assumptions, in general, relate to the types of structures and the level of technical management of their design and execution that the authors of the standard had in mind. Their applicability to an unusual structure or type of technical management should always be considered.

Clause 1.2(3)

The Eurocodes are concerned with design and not execution, but minimum standards of workmanship are required to ensure that the design assumptions are valid. For this reason, EN 1090 Parts 1, 2 and 4 on execution of steel structures (BS EN 1090) and EN 13670 for concrete structures (BS EN 13670) are included in ***clause 1.2(3)***. All Eurocodes prefer 'execution' to the familiar term 'construction'.

Roger P Johnson and Stephen J Hicks
ISBN 978-1-83662-921-4
https://doi.org/10.1108/978-1-83662-920-720253004

CHAPTER 2
Normative references

This chapter corresponds to *clause 2* of EN 1994-1-1, 'Normative references'. References in Eurocode 4 to clauses in the documents listed here have the same status, concerning requirement, as if they were in Eurocode 4.

The Bibliography at the end of EN 1994-1-1 provides a longer list of standards that are not 'Normative', and explains the several types of reference to them that are used in the Parts of EN 1994. It includes designated standards that businesses can use to show their construction products, services or processes comply with essential requirements of legislation (e.g. EN 10025 for hot rolled structural steel products).

References in this *clause 2* are given only to other European standards, all of which are intended to be used as a package. Formally, the standards of the International Organization for Standardization (ISO) apply only if given an EN ISO designation. However, in cases when suitable EN standards do not exist, reference may be made to European Technical Specifications (TS) or assessments that comply with all of the requirements given within European Assessment Documents (EAD).

National standards for design and for products that conflict with a relevant EN standard have been (or are being) 'withdrawn' by the British Standards Institution (BSI). They will become increasingly out-of-date, but remain important references for work on existing structures that were designed using them.

In its application to buildings, EN 1994-1-1 is based on the concept of the initial erection of a steel frame, which may include prefabricated concrete or composite members. The placing of precast concrete or profiled steel sheeting for floors, reinforcement and formwork for columns and so on then follow. The addition of in situ concrete completes the composite structure. The presentation and content of EN 1994-1-1 therefore relate more closely to EN 1993-1-1 than to EN 1992-1-1.

Designers' Guide to EN 1994-1-1: 2026: Eurocode 4 Design of composite steel and concrete structures

Roger P Johnson and Stephen J Hicks
ISBN 978-1-83662-921-4
https://doi.org/10.1108/978-1-83662-920-720253005

CHAPTER 3
Terms, definitions and symbols

3.1 Terms and definitions

Where a term is used in another Eurocode to which EN 1994-1-1 refers, its definition will be found in that code, not in *clause 3.1*. The care taken during drafting to avoid defining the same term in more than one code has, generally, been successful. Reference to the definitions in other codes can be helpful. For example, 'non-linear resistance' is in EN 1994-1-1, and 'non-linear analysis' is in EN 1993-1-1, neither giving the other term. Words like 'action' are not found in *clause 3.1* because it is defined in EN 1990. It means a load and/or an imposed deformation.

Where there is doubt, the definition in the code being used takes priority. For example, the text of EN 1994-1-1 refers to 'solid slabs' and EN 1992-1-1 has a definition 'slab, solid'. The definition 'solid composite slab' in *clause 3.1.38* is obviously applicable here.

In pre-Eurocode practice in the UK, and continuing in verbal usage, the meanings of 'resistance', 'capacity' and 'strength' lacked clear definition. In Eurocode 4, 'resistance' is used for a property of a cross-section or member that includes the dimension of force (e.g. buckling resistance), and 'capacity' is used where it does not (e.g. rotation capacity, slip capacity). 'Strength' is used for a property of a material or connector that includes the dimension of force. The term 'capacity design', defined in EN 1993-1-1, does not follow this convention. It is not used in Eurocode 4.

Many of the definitions in ***clause 3.1*** include the word 'composite', which implies shear connection, defined in *clause 3.1.2*. Its main function (*clause 3.1.1*) is to *limit* longitudinal slip and separation between the steel and concrete components of a member. They cannot be eliminated. Separation is always assumed to be negligible but explicit allowance may need to be made for effects of slip – for example, as in *clauses 5.4.2.1, 8.6.1(12)* and *A.4(1)*. *Clause 3.1*

The definition 'composite frame' (***clause 3.1.12***) does not refer to concrete members. Where the behaviour is essentially that of a reinforced or prestressed concrete structure, with only a few composite members, global analysis should be generally in accordance with Eurocode 2 and specialised guidance (Plumier and Degée, 2023). *Clause 3.1.12*

The definition of 'prestress' in ***clause 3.1.18*** refers to the use of prestressing by 'controlled imposed deformations', usually by jacking at supports, which is within the scope of Eurocode 4. Clause 3 of EN 1992-1-1, which defines prestress as an action caused by the stressing of tendons, applies to EN 1994-2 but not to EN 1994-1-1, as this is outside its scope. *Clause 3.1.18*

The term 'plastic neutral axis' is not included in the definitions in *clause 3* of EN 1994-1-1. EN 1992-1-1 does not use it. Its rules for stress distribution in compression zones (clause 8.1.2) include a rectangular ('plastic') distribution of concrete stress, but the neutral axis is *not* at one edge of the stress block. In Eurocode 4 it is at one edge, so care is needed in verifications where the code refers to Eurocode 2 (e.g. for bending resistance of a concrete flange).

3.2 Symbols

This clause lists over 500 symbols, about twice the number in the previous Eurocode 4, some of which had two meanings. The increase is mostly in the subscripts used, partly to cover the seven additional annexes and to avoid double meanings. Most of the annexes have a separate list, giving the symbols that are not in the main list. Hence, it may be necessary to look at both lists to find a symbol in an annex. There is no list of subscripts. A few can have more than one meaning – for example, $_{m}$ can mean either 'measured' or 'mean'.

This book uses the Eurocode symbols. They are defined near the start of each section in which they appear, except in some worked examples.

Some subscripts are separated by commas, which indicate a different meaning – for example, $A_{c,eff}$ (concrete, effective). This does not always apply, notably in the important group of subscripts $_{Rd}$, $_{Rk}$, $_{Ed}$ and $_{Ek}$, with R for resistance, E for effect of action, d for design and k for characteristic. For a complex design method, such as that for beams with web openings, the distinction between say, M_{Ed} and M_{Rd}, is important.

The subscripts for the three main materials used in composite structures in symbols for strength are $_c$ (concrete), $_y$ (structural steel) and $_s$ (reinforcing steel). For structural steel, $_a$ is also used, as in $M_{a,Ed}$, from the French 'acier'.

There are a few changes from the previous Eurocode 4 in the definitions of symbols, including the following.

- f_{ck} is now the characteristic cylinder strength of concrete at age t_{ref}, with t_{ref} a 'reference time', given in EN 1992-1-1: 5.1.3(2). The previous 28 days remains the 'general' value, and an option to specify other times has been added.
- For profiled sheeting, the words 'excluding longitudinal stiffener' have been added to the definition of h_p, the 'overall depth of profiled sheeting'. There is a new symbol, h_{pg}, for the depth including a stiffener.

Roger P Johnson and Stephen J Hicks
ISBN 978-1-83662-921-4
https://doi.org/10.1108/978-1-83662-920-720253006

CHAPTER 4
Basis of design

This chapter corresponds to *clause 4* of EN 1994-1-1. It was clause 2 in the previous Eurocode 4. Its main clauses are as follows.

- General rules — *Clause 4.1*
- Principles of limit states design — *Clause 4.2*
- Basic variables — *Clause 4.3*
- Verification by the partial factor method — *Clause 4.4*

4.1 General rules

Clause 4.1

Clause 4.1 lists the provisions of other EN standards that are required to be followed. EN 1994 is to be used with EN 1990, 'Basis of structural and geotechnical design', which recommends values for load factors and gives various possibilities for combinations of actions (BS EN 1990:2023). The values and choice of combinations are given in the national annex for the country in which the structure is to be constructed.

Eurocode 4 is also used in conjunction with relevant Parts of EN 1991, 'Actions on structures' (BS EN 1991-1-1:2025) and their national annexes, to determine characteristic or nominal loads. Where a composite structure is to be designed for a seismic region, account needs to be taken of EN 1998, 'Design of structures for earthquake resistance', (BS EN 1998-1:2004).

Clause 4.1.1(2)

Clause 4.1.1(2) reminds the user that design is based on actions and combinations of actions in accordance with EN 1991 and EN 1990, respectively. The use of partial safety factors for actions and resistances (the 'partial factor method') is expected but is not a requirement of Eurocodes. The method is presented in clause 8 of EN 1990 as one way of satisfying the basic requirements set out in clause 4 of that standard. This is why use of the partial factor method is given 'deemed to satisfy' status in *clause 4.1.1(2)*. It is the only method within the scope of this book. To establish that a design was in accordance with the Eurocodes, the user of any other method would normally have to demonstrate, to the satisfaction of the regulatory authority and/or the client, that the method satisfied the basic requirements of EN 1990.

4.2 Principles of limit states design

The clause provides a strong reminder, saying that the influence of sequence of construction on action effects 'shall' be considered. It does not affect the bending resistance of beams that are in Class 1 or 2 (as defined in *clause 5.5*) or the resistance of a composite column, as these are determined by rigid-plastic theory, but it does affect the resistances of beams in Class 3 or 4.

4.3 Basic variables

Clause 4.3.3

The classification of effects of shrinkage and temperature in ***clause 4.3.3*** into 'primary' and 'secondary' will be familiar to designers of continuous beams, especially for bridges.

Secondary effects occur only in statically indeterminate structures. They are to be treated as 'indirect actions', which are 'sets of imposed deformations or accelerations' (EN 1990: 3.1.3.3), not as action effects. This distinction appears to have no consequences in practice for the use of EN 1994-1-1.

4.4 Verification by the partial factor method
4.4.1 Design values

Clause 4.4.1.1
Clause 4.4.1.2

Clauses 4.4.1.1 and ***4.4.1.2*** illustrate the treatment of partial factors. Recommended values are given in Notes, which are Informative, not Normative (i.e. not part of the preceding provision), so that no numerical values are given in the Normative clauses. For the partial factors for concrete, reinforcing steel and structural steel the 'Notes' give national standards bodies no choice other than to use the values of γ_C, γ_S and γ_M given in their national annexes to ENs 1992-1-1 and 1993-1-1. This was assumed in all calibration work for EN 1994. Any alternative would be impracticable.

For the design compressive strength of concrete, f_{cd}, *clause 4.4.1.2(2)* refers to EN 1992-1-1: 5.1.6, with a reference in Note 2 to *clause 5.1*. That is where important differences from Eurocode 2 are stated. The rules involve a coefficient k_{tc}, which is a Nationally Determined Parameter (NDP) in both codes. This is discussed in Chapter 5.

The partial factors for shear connection and for longitudinal shear on composite slabs are also NDPs. It is expected that the UK's NA will use the recommended values.

Clause 4.4.1.3

Clause 4.4.1.3 refers to a European Technical Product Specification for important geometrical data, such as initial imperfections. Dimensions from drawings are treated as 'nominal' values. Where the word 'nominal', usually meaning that the value has no statistical basis, is used for some properties of structural steel, its meaning is equivalent to 'characteristic' (i.e. the design value is obtained by dividing it by an appropriate partial factor). In relation to dimensions, it usually means an estimated mean value, assumed to lie within any specified tolerance interval.

Clause 4.4.2

Clause 4.4.2, on design resistances, refers to Formulae 8.20, 8.21 and 8.23 given in clause 8.3 of EN 1990, of which only Formula 8.21 allows for more than one material. The symbols need explanation, which is simplified here.

Formula 8.23 is $X_d = \eta X_k/\gamma_M$, where

X is the design or characteristic property of a material (e.g. $X_d = f_{cd}$ for concrete)

η is a conversion factor (often 0.85 for concrete).

Factor γ_M ($= \gamma_m \gamma_{Rd}$) usually appears with a subscript denoting the material, γ_C, γ_S and so on (e.g. $\gamma_C = 1.5$ for concrete), and is the product of the two factors

γ_m is a partial factor for material properties and geometric deviations

γ_{Rd} is a partial factor for the uncertainty of the resistance model.

Formula 8.37 for the plastic resistance of a cross-section is a typical resistance function in Eurocode 4

$$N_{pl,Rd} = A_a f_{yd} + A_c f_{cd} + A_s f_{sd} \qquad (8.37)$$

In the notation above, with the resistance function denoted R{.}, it is

$$R_d = R\{(\eta_i X_{k,i} / \gamma_{M,i}); a_d; F_{Ed}\} \qquad i \geq 1 \qquad (X4.1)$$

where a_d denotes the design value of geometric properties and F_{Ed} denotes design values of actions.

There are no separate terms a_d and F_{Ed} in *Formula 8.37*. Uncertainties in areas of cross-sections are allowed for in the factors γ_M.

Formula 8.21 of EN 1990 is

$$R_d = \frac{1}{\gamma_{R,1}} R\left\{\eta_1 X_{k,1}; \frac{\eta_i X_{k,i}}{\gamma_{m,i}/\gamma_{m,1}}; a_d; \sum F_{Ed}\right\} \qquad (8.21, \text{EN 1990})$$

As $\gamma_M = \gamma_m \gamma_{Rd}$, *Formula 8.37* is consistent with Formula 8.21 when the following assumptions are made

- $\gamma_{R,i}$ has the same value for all three materials
- function $R\{.\}$ is linear.

The shear resistance of stud shear connectors is the lower of two resistances, from *Formulae 8.23* and *8.24*, each of which depends on the properties of only one material. They correspond to Formula 8.19 of EN 1990, which is not referred to in this clause.

The Eurocode 4 methods above use factors γ_M without the need to evaluate γ_m or γ_{Rd}. These values may be needed for some advanced methods such as finite-element analysis, or the future calibration of new design models, and then Formula 8.21 is applicable.

4.4.3 Combination of actions

Clause 4.4.3(1) refers to EN 1990, where the relevant clause is A1.6. The number of variable actions to be included in a combination depends both on the type of combination (characteristic, frequent, quasi-permanent etc.) and on the Ψ factors given in the relevant national annex. This aspect of the Eurocodes is comprehensive and can be more complex than some previous practice. *Clause 4.4.3(1)*

4.4.4 Verification of static equilibrium

This ultimate limit state may or may not involve a structural failure (e.g. of a holding-down bolt). For partial factors for actions and effects, ***clause 4.4.4(1)*** refers to Table A.1.8 of EN 1990. These factors depend on whether the action or effect is favourable (stabilising) or de-stabilising. *Clause 4.4.4(1)*

Roger P Johnson and Stephen J Hicks
ISBN 978-1-83662-921-4
https://doi.org/10.1108/978-1-83662-920-720253007

CHAPTER 5
Materials

This chapter concerns the properties of materials needed for the design of composite structures. It corresponds to *clause 5*, which has the following clauses.

- Concrete *Clause 5.1*
- Reinforcing steel *Clause 5.2*
- Structural steel *Clause 5.3*
- Connecting devices *Clause 5.4*
- Profiled steel sheeting for composite slabs in buildings *Clause 5.5*

Clause 5 consists mainly of cross-references to other Eurocodes and EN standards. The following comments relate to provisions of particular significance for composite structures.

5.1 Concrete

Clause 5.1(1)

The reference in ***clause 5.1(1)*** to EN 1992-1-1 for properties of concrete begins '*unless otherwise given in EN 1994-1-1*'. In EN 1992-1-1, clause 5.1 gives these properties for normal-aggregate concrete. Its Annex M, for lightweight-aggregate concrete, gives properties that are dependent on the oven-dry density relative to 2200 kg/m^3. They are presented as multiples of the properties of normal concrete with the same characteristic compressive strength. They include shrinkage and creep, which are higher, and the modulus of elasticity, which is lower. Hence, design is more likely to be governed by deflections than with normal concrete. In this guide, 'concrete' means normal-aggregate concrete unless stated otherwise.

Strength and stiffness

Strength classes for normal concrete are defined as Cx/y, where x and y are respectively the cylinder and cube compressive strengths in N/mm^2 units. All compressive strengths in design rules in Eurocodes are cylinder strengths, so an unsafe error occurs if a specified cube strength is used in calculations. It should be replaced at the outset by the equivalent cylinder strength, using the relationships given by the strength classes. Compressive strength is slightly influenced by the size of the test specimen. The standard cube size is 150 mm. Standard cylinders are 150 mm in diameter and 300 mm long. Where other sizes are used, conversion factors, available in the technical literature, should be applied.

Classes for lightweight-aggregate concrete are designated LC$x/y$. The relationships between cylinder and cube strengths differ from those of normal concrete.

Except where prestressing by tendons is used (which is outside the scope of EN 1994), the tensile strength of concrete is rarely used in design calculations for composite members. The mean tensile strength f_{ctm} appears in the definitions of 'cracked' global analysis in *clause 7.4.2.3(2)*, and in *clause 9.4.2(1)* on minimum reinforcement. Its value and the 5% and 95% fractile values are given in Table 5.1 of EN 1992-1. The appropriate fractile value should be used in any limit state verification that relies on either an adverse or beneficial effect of the tensile strength of concrete.

An 'approximate indicative value' for the modulus of elasticity E_{cm} is given in Formula 5.1 of EN 1992-1-1, with comments that it can vary either way (e.g. by 30% or more), depending on the type of aggregate, and that national annexes can provide further information. Use of great precision for modular ratios is, therefore, pointless. The short-term elastic modulus increases slowly for ages greater than 28 days. This small change is negligible compared with the uncertainties in the modelling of creep, and should be ignored.

Clause 5.1(2) For the design compressive strength of concrete, ***clause 5.1(2)*** gives *Formula 5.1* which is a copy of Formula 5.3 in EN 1992-1-1:

$$f_{cd} = \eta_{cc} k_{tc} f_{ck}/\gamma_C \qquad (5.1)$$

Factor η_{cc} allows for the difference between a measured cylinder or cube strength and the strength in a member. It is a function of a reference strength $f_{ck,ref}$, which is an NDP in Eurocode 2. Factor k_{tc}, also an NDP, is discussed below. In many situations they both have the value 1.0 in EN 1992-1-1. This is inconsistent with the methods of EN 1994-1-1 which, for plastic resistances, use a rectangular stress block that extends to the neutral axis, unlike the one in EN 1992-1-1.

Clause 5.1(3) This is why ***clause 5.1(3)*** gives further rules, including that the product $\eta_{cc}\, k_{tc}$ should not exceed 0.85 where plastic resistance or some non-linear methods are used. There is an exception: the limit is 1.0 for concrete-filled sections. It allows for lateral restraint from the external steel member. It should be noted that these rules are *not* NDPs.

The factor k_{tc}

This factor allows for 'high sustained loads'. It comes from test evidence, as follows. Consider a cylinder test specimen with a stress f_{cm} at failure. If that identical specimen had been subjected to a constant stress of, say, $0.9 f_{cm}$, research has found that it would eventually fail, after a period of weeks or months.

Allowing for this has been controversial, which is why k_{tc} is an NDP in EN 1992-1-1. Possible motives for making no reduction for this reason are

- the increase in concrete strength after 28 days
- so high a sustained load would probably give warning of distress well before failure
- when structures receive their design loads, they will be older than 28 days, which is the usual basis for design calculations
- the existing design models for resistance have been calibrated against results from test specimens subjected to sustained loads, so the effects are implicitly already accounted for.

Clause 5.1(4) ***Clause 5.1(4)*** limits the scope of EN 1994-1-1 to the strength range C20/25 to C70/85 for normal concrete and from LC20/22 to LC60/66 for lightweight concrete. These ranges are narrower than those given in EN 1992-1-1 because there is limited knowledge and experience of the behaviour of composite members with weak or very strong concrete. This applies to the load/slip properties of shear connectors, the redistribution of moments in continuous beams and the resistance of columns.

Shrinkage

Clause 5.1(5) *Clause 5.1(6)* *Clause 5.1(7)*

The shrinkage of concrete referred to in ***clause 5.1(5)*** is the drying shrinkage that occurs after setting. It does not include the plastic shrinkage that precedes setting, nor autogenous shrinkage. The latter develops during hardening of the concrete (clause 3.1.76 of EN 1992-1-1) and is that which occurs in enclosed or sealed concrete, as in a concrete-filled tube where no loss of moisture occurs. ***Clause 5.1(6)*** permits its effect on stresses and deflections to be neglected, but does not refer to crack widths. It has little influence on cracking due to direct loading, and the rules for initial cracking (*clause 9.4.2*) take account of its effects. For values of shrinkage strain, ***clause 5.1(7)*** refers to the comprehensive rules in EN 1992-1-1, with a simpler alternative for buildings, the Informative *Annex C*. It is expected that the UK's National Annex will recommend the use of the latter.

In typical environments in the UK, the influence of shrinkage of normal-weight concrete on design of composite structures for buildings is significant only in

- very tall structures
- very long structures without movement joints
- the prediction of deflections of beams with high span/depth ratios (*clause 9.3.1(9)*).

Rectangular stress block for concrete in compression: Eurocodes 2 and 4 compared

The stress-strain curve given in clause 5.1.6(3) of EN 1992-1-1 is complex. There is reference to simpler stress distributions in its clause 8.1.2. One of these is a rectangular stress distribution, similar to the stress block given in the former British Standard for the structural use of concrete, BS 8110 (1997). Its shape, for concrete strength classes up to C50/60, and the corresponding strain distribution are shown in Figure 5.1.

This stress block is inconvenient for use with composite cross-sections because the region near the neutral axis assumed to be unstressed is often occupied by a steel flange, and algebraic expressions for resistance to bending become complex.

In composite sections, the contribution from the steel section to the bending resistance reduces the significance of that from the concrete. It is thus possible for EN 1994 to allow the use of a rectangular stress block extending to the neutral axis, as shown in Figure 5.1.

For a member of unit width, the moment about the neutral axis of the EN 1992 stress block is 0.48 $f_{ck}x^2/\gamma_C$, if $\eta_{cc}\ k_{tc} = 1$. The value for beams in EN 1994-1-1 is 0.425 $f_{ck}x^2/\gamma_C$. Calibration studies have shown that this overestimates the bending resistance of cross-sections of columns, so a correction factor α_M is given in *clause 8.8.3.6(1)*.

Figure 5.1. Stress blocks for concrete at ultimate limit states

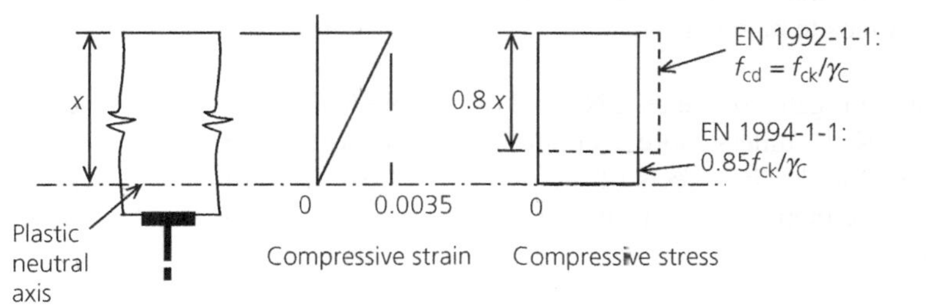

5.2 Reinforcing steel

Clause 5.2(1) refers to EN 1992-1-1, which states in clause 5.2.2(1) and Table 5.4 that its rules are valid for characteristic yield strengths f_{yk} up to 700 N/mm^2 (Class B700). ***Clause 5.2(3)*** sets the lower limit of B500 for plastic analyses and resistances but not for 'elastic' or 'strain-limited' design methods. *Clause 5.2(1)* *Clause 5.2(3)*

For simplicity, *clause 5.2(2)* permits the modulus of elasticity of reinforcement to be taken as 210 kN/mm^2, the value in EN 1993-1-1 for structural steel, rather than 200 kN/mm^2, the value in EN 1992-1-1.

The scope of clause 5.2.1 of EN 1992-1-1, and hence of EN 1994-1-1, is for reinforcement, including welded fabric ('mesh'), with bar diameters up to 50 mm for bars that are 'ribbed' (high bond) and weldable. Steel fibre reinforcement is not directly included in Eurocode 4, except for advanced design methods according to *clause 5.1(9)*.

There are three ductility classes, from A (the lowest) to C. *Clause 7.5.1(5)* excludes the use of Class A reinforcement from any composite cross-section in Class 1 or 2 with bars in tension. The minimum ductility properties for welded fabric given in Table C.2 of EN 1992-1-1 may not be sufficient to satisfy *clause 7.5.1(6)* of EN 1994-1-1, as this requires demonstration of sufficient ductility to avoid fracture when built into a concrete slab (Anderson *et al.*, 2000). It has been found in tests on continuous composite beams with fabric in tension that the cross-wires initiate cracks in concrete, so that tensile strain becomes concentrated at the locations of the welds in the fabric.

5.3 Structural steel

Clause 5.3(1) refers to EN 1993-1-1. This lists in its Table 5.1 steel grades with nominal yield strengths up to 700 N/mm^2. ***Clause 5.3(2)*** sets an upper limit of 460 N/mm^2 for use with EN 1994-1-1. There has been extensive research on the use in composite members of structural steels with yield strengths exceeding 355 N/mm^2 (Bergmann and Hanswille, 2006; Liew *et al.*, 2021; Thai *et al.*, 2021; Uy, 2003). *Clause 5.3(1)* *Clause 5.3(2)*

EN 1994-1-1 includes some modifications to design rules for use with steel grades higher than S355, to avoid premature crushing of concrete. This applies to

- redistribution of moments (*clause 7.4.4(6)*)
- rotation capacity (*clause 7.4.5(4)(a)*)
- plastic resistance moment for beams (*clause 8.2.1.2(2)*)
- lateral torsional buckling (*clause 8.4.3(1)*)
- resistance of columns (*clause 8.8.3.6(1)*).

Thermal expansion

For simplicity, the Eurocodes permit, for most situations, the use of a single value for the coefficients of thermal expansion of steel, concrete and reinforcement. For structural steel, clause 5.2.5(1) of EN 1993-1-1 gives the value 12×10^{-6} K^{-1}. (The unit K^{-1} refers to the kelvin scale, and means 'per degree Celsius'.)

The previous Eurocode 4 (but not the new one) included a note that for calculating the 'structural effects of unequal temperatures' in composite structures, the coefficient may be taken as 10×10^{-6} per °C, which is the value given for normal-weight concrete in clause 5.1.6(6)) of EN 1992-1-1, 'unless more accurate information is available'. Thermal expansion of reinforcement is presumably assumed to be the same as that of normal-weight concrete. The value 10×10^{-6} K^{-1} can normally be used for all three materials of a composite member.

For lightweight-aggregate concretes, EN 1992-1-1: Table M.2 gives the coefficient of thermal expansion as 8×10^{-6} K^{-1}, 'unless more accurate information is available'. It can range from about 4×10^{-6} to 14×10^{-6} K^{-1}. The effects of the difference from 10×10^{-6} K^{-1} should be considered in design of composite members where the temperatures of the concrete and the structural steel could differ significantly.

5.4 Connecting devices

5.4.1 General

Reference is made to EN 1993, 'Eurocode 3: Design of steel structures, Part 1-8: Design of joints' (BS EN 1993-1-8:2024) for information relating to fasteners, such as bolts, and welding consumables. Provisions for 'other types of mechanical fasteners', such as the cartridge-fired pins used for profiled sheeting are given in clause 5.3.1.2 of EN 1993-1-3, Design of steel structures. Part 1-3: Cold-formed members and sheeting (BS EN 1993-1-3:2024). Commentary on joints is given in Chapters 8 and 10.

5.4.2 Shear connectors

Clause 5.4.2

The previous EN 1994-1-1 gave detailed rules only for headed stud connectors, based on a limited range of well-known types of load-slip curves from push testing. Any other method of connection had to satisfy the general requirements of *clause 8.6.1*. This new ***clause 5.4.2*** is for shear connectors generally, with a wider range of load-slip properties, and distinguished by their ductility category. The use of adhesives on a steel flange is unlikely to be suitable.

5.4.2.1 Ductility

Clause 5.4.2.1(2)
Clause 5.4.2.1(3)

The revised Eurocode introduces ductility categories D1 to D3. The previous rules were for studs with characteristic slip capacities at ultimate limit states of 6 mm and 10 mm, which in most cases places them in category D2 (***clause 5.4.2.1(2)*** and ***(3)***), with a further condition for D3 in a Note.

The ductility categories, defined in *Table 5.1*, are based on load-slip curves (*Figure 5.1(a) to (f)*) (Figure 5.2), usually from tests in accordance with *clause B.2*. These diagrams are not actual test curves. They illustrate a type of curve, using the three characteristic values needed for a given type of connector to be classified – the shear resistance, P_{Rk}, the slip capacity, δ_{uk}, the elastic slip, δ_{ek}, and also P_{em}, defined below.

From *Table 5.1*, a connector may be in category D1 because of low initial stiffness ($\delta_{ek} > 2.5$ mm), with the risk of excessive slip at serviceability load levels, where most design rules ignore slip. Also, it may have slip capacity less than 6 mm, denoting insufficient ductility for the many methods in Eurocode 4 that rely on 'ideal-plastic behaviour'.

Figure 5.2. A load-slip curve from a push test on shear connectors

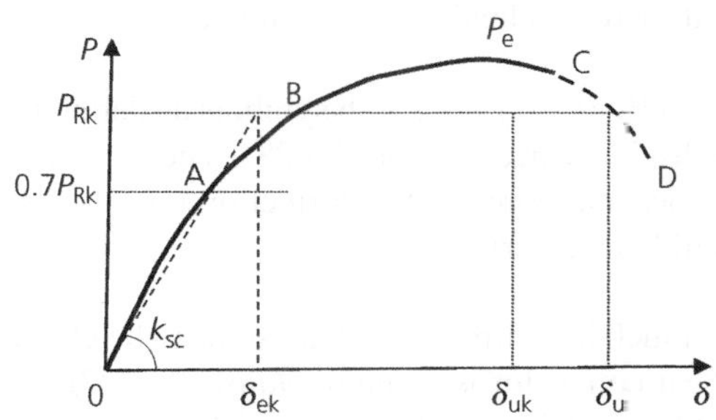

In the comments that follow it is assumed that the ductility of a type of shear connector is found by the simplest method given in *clause B.3*, that is, from push tests on three 'nominally identical' specimens. The load-slip curve of *Figure B.3*, shown here as Figure 5.2, needs explanation. Its use of symbol P_e for maximum load shows that the figure gives a curve from a particular test, with δ_{uk} less than δ_u, as expected.

From *clause B.3.5(5)*, the ultimate slip from a test, δ_u, is the maximum slip at resistance P_{Rk} and δ_{uk} is 10% below the least δ_u. If even one of the three tests were discontinued before the falling branch reached the level P_{Rk} (point C in Figure 5.2), this rule would give δ_u as the value at point B. In practice, the maximum recorded slip (point C) would be used, but it could still lead to a value of δ_{uk} lower than that required for classification as D2. Hence, tests should be continued to point D, about 25% below the peak resistance P_e.

The design of shear connection can be so much simpler for connectors in category D2 or D3 than for those in D1 that rules for upgrading connectors from D1 to D2 are given. They require the characteristic shear resistance found from tests to be reduced and are now explained. The following new symbols are used.

- P_{em} is the mean of results P_e (maximum load) from the set of tests on nominally identical push specimens that is used to determine the values P_{Rk}, δ_{uk}, δ_{ek} and k_{sc}.
- $P_{em,6}$ is similar, except that each result P_e is the maximum load before the slip exceeds 6 mm.
- The slip $\delta_{ek,lim}$ has the recommended value 2.5 mm, and is an NDP.
- α, not exceeding 1.0, is a reduction factor for the characteristic shear resistance P_{Rk}.
- β_{VL}, not exceeding α, is a further reduction factor for P_{Rk}.

'Ideal-plastic behaviour' needs to be verified (***clause 5.4.2.1(4)***) because many design methods rely on it. Reference to test records would show that most stud connectors within the scope of EN 1994 satisfy its condition. This says that if, for a test series, $P_e/P_{Rk} \leq 1.3$, there is sufficient plastic behaviour to qualify for category D2 or D3. It is based on a study of many load/slip curves from push tests. *Clause 5.4.2.1(4)*

Clause 5.4.2.1(5) is for connectors with low initial stiffness ($\delta_{ek} > \delta_{ek,lim}$), as shown in *Figures 5.1(d)* and *5.1(f)*. They can be deemed to be in category D2 if their characteristic shear resistance is reduced by the factor α from *Formula 5.2*. Two notes give values for α that 'can be taken'. The first one needs a value for P_{Rk} 'determined at a slip of $\delta_{ek,lim}$'. This could be done by finding a value from each test curve, and calculating P_{Rk} from them by one of the methods of *Annex B* (e.g. 10% below the lowest of three). *Clause 5.4.2.1(5)*

This method ($\alpha = \delta_{ek,lim}/\delta_{ek}$) is also shown by the geometry of *Figure 5.1(d)*. It is understood that recent UK practice has been to use the value 2.0 mm for $\delta_{ek,lim}$, so the 2.5 mm in this clause may be changed in the national annex.

Clause 5.4.2.1(6) is for shear connectors with insufficient plastic behaviour (i.e. $P_e/P_{Rk} > 1.3$), as shown in *Figure 5.1(d)*. If the mean maximum resistance P_{em} is much higher than P_{Rk}, the shear force resisted by some of the connectors in a shear span of a beam can be much higher than is assumed *Clause 5.4.2.1(6)* in design. The compensation to enable 'ideal-plastic behaviour' to be assumed should, in theory, be an increase to the action effect $v_{L,Ed}$. It is presented as a reduction in the resistance P_{Rk}, which has the same effect, because a rule that increases loading is unrealistic.

The reduction factor is $\beta_{VL} = \alpha\,(P_{Rk}/P_{em,6}) \leq \alpha$, with symbols as defined above, and the usual alternative of a European Technical Product Specification (ETPS). The next sub-clause makes clear that the inclusion of α in the formula does not represent the correction for initial stiffness. Both reductions are required for connectors that fail both criteria.

These rules enable connectors much less stiff than typical studs to be classed as 'ductile'. For example, the elastic stiffness k_{sc} of a shear connector is given in *clause B.3.5(7)* as P_{Rk}/δ_{ek}. A 19-mm stud may have typical values $P_{Rk} \approx 80$ kN and $k_{sc} \approx 100$ kN/mm. Then, $\delta_{ek} = P_{Rk}/k_{sc} = 80/100 = 0.8$ mm.
Clause 5.4.2.1(8) *Table 5.1* requires a ductile connector to have $\delta_{ek} \leq 2.5$ mm. ***Clause 5.4.2.1(8)*** is a reminder to consider the consequences for rotation capacity and deflection of using connectors much less stiff than studs. Usually in EN 1994-1-1, design methods that rely on rotation capacity are presented in a way that avoids calculation of rotations, with an implicit assumption that the shear connectors would be not more flexible than studs in solid slabs.

Clause 5.4.2.2 ***Clause 5.4.2.2*** refers to EN ISO 13918, 'Welding – Studs and ceramic ferrules for arc stud welding'. This gives minimum dimensions for the overall length of the stud (before welding) and the weld collars. Other methods of attaching studs, such as spinning, may not provide weld collars large enough for the resistances of studs given in *clause 8.6.8.1* to be applicable. The grades of stud referred to in EN ISO 13918 include SD1, with ultimate tensile strengths greater than 450 N/mm^2, and SD3, which is manufactured from stainless steel with a range 500 to 780 N/mm^2. The upper limit in *clause 8.6.8.1(1)* is 500 N/mm^2.

Shear connection between steel and concrete by friction is permitted only in some provisions for composite slabs and concrete-encased webs and columns. Chemical bond is not used.

5.5 Profiled steel sheeting for composite slabs in buildings

The title includes 'in buildings' as this clause and other provisions for composite slabs are not applicable to composite bridges, where durability can be uncertain. The materials for profiled steel sheet-
Clause 5.5 ing must conform to the standards listed in ***clause 5.5***. Such standards should include tolerances on embossments and indentations, which can influence resistance to longitudinal shear. Tolerances on embossments, given for test specimens in *clause B.4.3(2)*, provide guidance.

The minimum bare metal thickness ('steel core thickness' in EN 1993-1-3) was in the previous EN 1994-1-1 subject to national annexes, with a recommended minimum of 0.70 mm for use in the UK. No value for it is given here, or in *clause 6.2* (on durability). The shear resistances for studs in transverse sheeting (*clause 8.6.9.2*) are not valid for sheeting with nominal thickness less than 0.70 mm.

emerald PUBLISHING ice

Roger P Johnson and Stephen J Hicks
ISBN 978-1-83662-921-4
https://doi.org/10.1108/978-1-83662-920-720253008

CHAPTER 6
Durability

This chapter concerns the durability of composite structures. It corresponds to *clause 6* which has the following clauses.

- General *Clause 6.1*
- Profiled steel sheeting for composite slabs in buildings *Clause 6.2*

6.1 General

Almost all aspects of the durability of composite structures are covered by cross-references to EN 1990, EN 1992 and EN 1993. The material-independent provisions, in clause 4.6 of EN 1990, require the designer to take into account nine factors, which are essentially those in the previous EN 1990. These include the foreseeable use of the structure, the expected environmental conditions, the design criteria, the performance of the materials, and the quality of workmanship and level of control on site.

Clauses 6.3, 6.4 and 6.5 of EN 1992-1-1 define exposure classes, exposure resistance classes and concrete cover, respectively. Its Annex P (Informative) gives an alternative approach for minimum cover, based on structural classes to be given in national annexes. For the widely used exposure class method, the examples of the classes in Table 6.1 can be modified by national annexes. Minimum covers for durability (Tables 6.3 and 6.4) are also Nationally Determined Parameters.

Allowances for incorrect cover in practice, given in Table 6.7, are usually 5 or 10 mm. The UK's previous national annex gave a range of values that depend on the composition of the concrete.

6.2 Profiled steel sheeting for composite slabs in buildings

For profiled steel sheeting, ***clause 6.2(1)*** requires the corrosion protection to be adequate for its environment. EN 10346, referred to in ***clause 6.2(2)***, is for continuously hot-dip coated steel products for cold forming. *Clause 6.2(1)* *Clause 6.2(2)*

The Note to ***clause 6.2(2)*** says that the zinc coating specified (hot-dipped galvanising) is '*sufficient for internal elements in a non-aggressive environment*'. This implies that it may not provide sufficient durability for use in a multi-storey car park or near the sea. The total thickness of zinc coating in accordance with *clause 6.2(2)* is about 0.04 mm. *Clause 6.2(2)*

Roger P Johnson and Stephen J Hicks
ISBN 978-1-83662-921-4
https://doi.org/10.1108/978-1-83662-920-720253009

CHAPTER 7
Structural analysis

Structural analysis may be performed at three levels: global analysis, member analysis and local analysis. This chapter concerns global analysis – the calculation of deformations and internal forces and moments in beams and framed structures. It corresponds to *clause 7*, which has the following sub-clauses.

- Structural modelling for analysis — *Clause 7.1*
- Structural stability — *Clause 7.2*
- Imperfections — *Clause 7.3*
- Calculation of action effects — *Clause 7.4*
- Classification of cross-sections — *Clause 7.5*

This clause, and some others, is more complex than before, because of the new policy on division of material between Parts 1-1 and 2 of EN 1994. Previously, provisions needed for bridges but rarely, if ever, used for buildings were in Part 2. Those that could be used for buildings as well as bridges have now been moved to Part 1-1 because they are 'general' (in theory). This gives greater prominence to elastic and inelastic methods of analysis. The new presentation of partial shear connection for beams is explained in the comment on *clause 8.1.1*.

Definitions of types of global analysis

From *clause 3.1*, '*...terms and definitions given in EN 1990, EN 1992-1-1 and EN 1993-1-1 and the following apply*'. Those in *clause 3.1.23* of Eurocode 4 are for elastic, rigid-plastic and non-linear plastic global analysis. The first needs no comment and the others are rarely used for composite structures.

Other definitions are as follows:

- 'non-linear behaviour' is where action effects are not proportional to actions; a term too general to be useful (EN 1990)
- 'non-linear analysis': analysis using models that account for mechanical (i.e. material) and geometrical non-linear behaviour (EN 1992). This does not include rigid-plastic global analysis, which is treated separately in Eurocode 4 (*clause 7.4.5*)
- 'geometric non-linearity': caused by changes of geometry from the original undeformed state (EN 1990). In Eurocode 4, it is allowed for in the design rules for columns, frames and slender beams
- 'material nonlinearity': the use of a non-linear stress-strain relationship for a material (EN 1990). It occurs in Eurocode 4 in all methods using the words 'inelastic' or 'plastic', and in the design rules for shear connectors and cracking of concrete
- 'second- and first-order theories' use relationships between actions and their effects that are or are not, respectively, influenced by deformations (EN 1990)
- 'second-order effects': action effects caused by structural deformations (EN 1992).

Wherever possible, analyses for serviceability and ultimate limit states use the same methods. It is more convenient, therefore, to specify them in a single clause, rather than to include them in *clauses 8* and *9*. For composite slabs, though, all provisions, including those for global analysis, are given in *clause 10*.

The division of material between *clause 7* and *clause 8* (ultimate limit states (ULS)) is not always obvious. Calculation of vertical shear is clearly 'global analysis', but longitudinal shear is in *clause 8*. This is because its calculation for beams in buildings is dependent on the method used to determine the resistance to bending. Frames are treated in *clause 7*, but for composite columns, methods of analysis and member imperfections are considered in *clause 8.8*. This separation of imperfections in frames from those in columns requires care, and is a recurrent theme of this chapter. The flow charts for global analysis (Figure 7.1) include relevant provisions from *clause 8*.

There is more guidance on the use of inelastic (meaning non-linear) methods of analysis, such as finite-element methods (FE), than in the previous Eurocodes 3 and 4. Eurocode 3 has a new Part on FE methods (BS EN 1993-1-14:2025). In Eurocode 4, their use is implied in *clause 7.4.1.1(8)*, which begins 'Unless non-linear analysis is used…', but does not refer to EN 1993-1-14. More detail is given only in *clause 8.8.2* on its application to compression members.

The use of well-established simplified models of behaviour (e.g. 'engineers' theory of bending') is implicit in the wording of many clauses. Linear-elastic FE programs are useful for some purposes, such as finding elastic critical buckling loads, but, in other analyses, their neglect of inelastic or plastic redistribution can lead to over-conservative results.

Where non-linear FE programs are used, it can be difficult to establish whether the analysis is in accordance with the Eurocodes and to check the results. The basis and limitations of the software should be well understood (Sandberg and Hendy, 2010). This reference to the Proceedings of a conference, *Codes in Structural Engineering*, International Association for Bridge and Structural Engineering, Zurich, 2010, contains other papers on the Eurocodes as drafted at that time.

Flow charts for global and member analysis

A typical design process should be simpler than the flow charts of Figure 7.1. Their complexity arises because Eurocodes 3 and 4 include alternative methods and calculations that define boundaries – for example, to determine whether side-sway is significant or not. Often, the user's experience will enable them to make such decisions without calculation.

It is impracticable to provide comprehensive charts. For those in Figures 7.1 and 7.3 it is assumed that linear-elastic global analysis is used, and that where there are alternative methods, only the simpler one is included – for example, in the choice of modular ratios and the treatment of side-sway. The limits to their scope are as follows.

- The beams are composite and of uniform depth. None of them are concrete-encased, in slenderness class 4, or have a concrete flange below the steel section.
- The columns in the frame are composite, vertical and of uniform cross-section.
- For Figure 7.1(a), the beam-to-column joints are either simple (nominally pinned) or fully continuous (see Table 7.1 of EN 1993-1-8).
- Calculations do not need to take account of the effects of shrinkage, temperature, prestress by imposed deformations and secondary effects of creep.
- The loading is as given at the top of Figure 7.1(a).

The references to imperfections (see *clause 7.3*) in the charts answer these questions:

- When is second-order global analysis required?
- Can global imperfections be omitted from the global analysis?
- When should imperfections in column lengths be included in global analysis?
- When is second-order analysis of a column length required?
- When should column imperfections be included in analyses of column lengths?

7.1 Structural modelling for analysis

Clause 7.1.1(2)

Clause 7.1.1(2) refers to a clause in EN 1990 that says, in effect, that models shall be appropriate and based on established theory and practice and that the variables shall be relevant.

Clause 7.1.1(3)

Clause 7.1.1(3), by referring to the use of Eurocode 2 as an exception, makes clear that composite members and joints are envisaged as being used in conjunction with others of structural steel. *Clause 7* is aligned with and cross-refers to clause 7 of EN 1993-1-1 wherever possible. Any significant differences between these two clauses are referred to in this guide.

Figure 7.1. Flow chart for global and member analyses for a plane frame

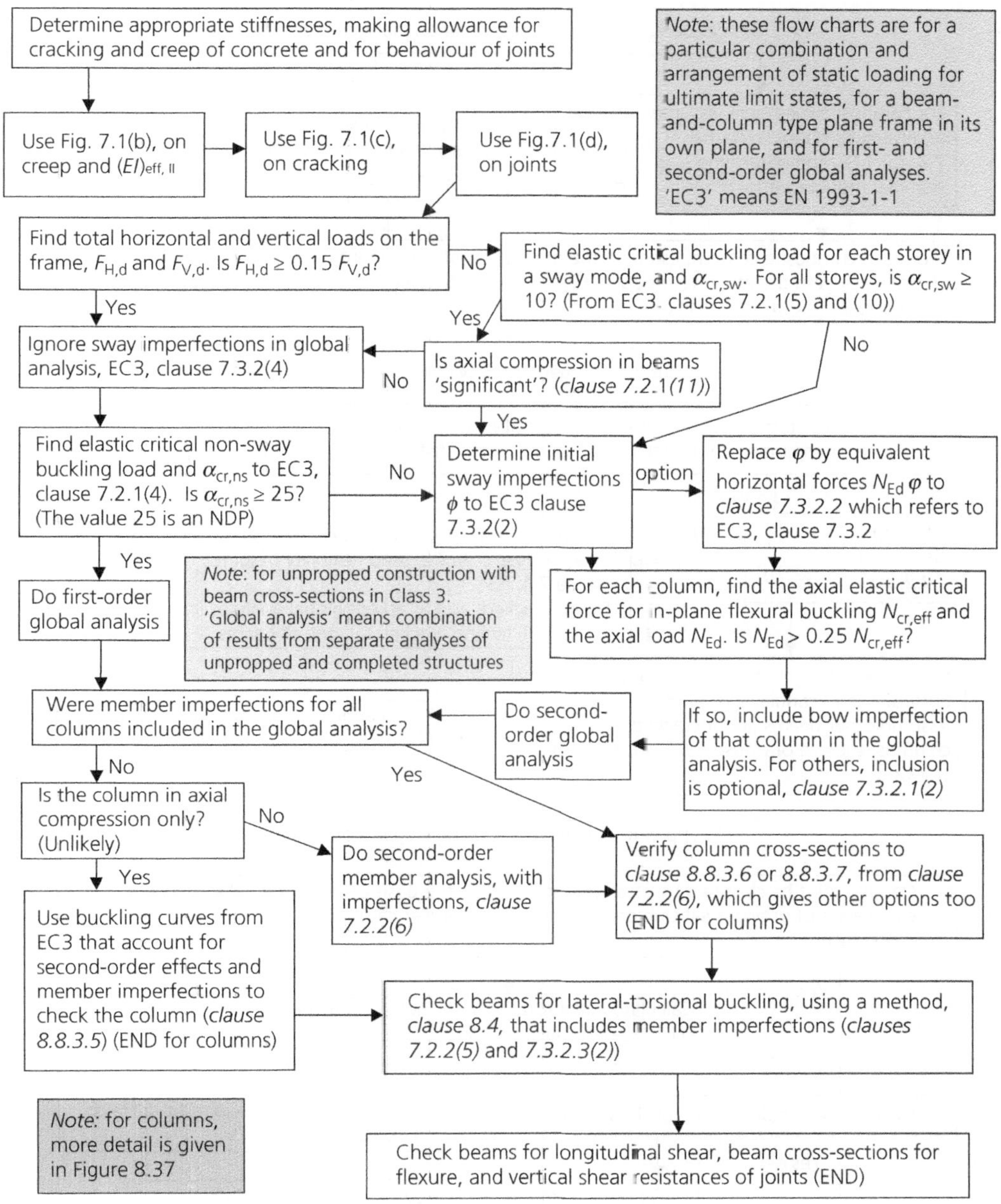

(a) Flow chart, global analysis of a plane frame with composite columns

Shrinkage of concrete and non-uniform changes of temperature affect the whole width of a concrete flange, but shear lag enables calculation of their effect in a composite beam from the use of effective widths, ***clause 7.1.1(5)***. *Clause 7.1.1(5)*

7.1.2 Joint modelling

Clause 7.1.2(2) The three simplified types of joint model listed in ***clause 7.1.2(2)*** – simple, continuous and semi-continuous – are those given in Table 7.1 of EN 1993-1-8 (BS EN 1993-1-8:2024) for joints in steelwork. For composite joints, its provisions are modified and supplemented by *clause 11* and *Annex A* of EN 1994-1-1. Joint models are further defined by one or more of the three types of joint stiffness – nominally-pinned, semi-rigid and rigid – and one of the three types of moment

Figure 7.1. Continued

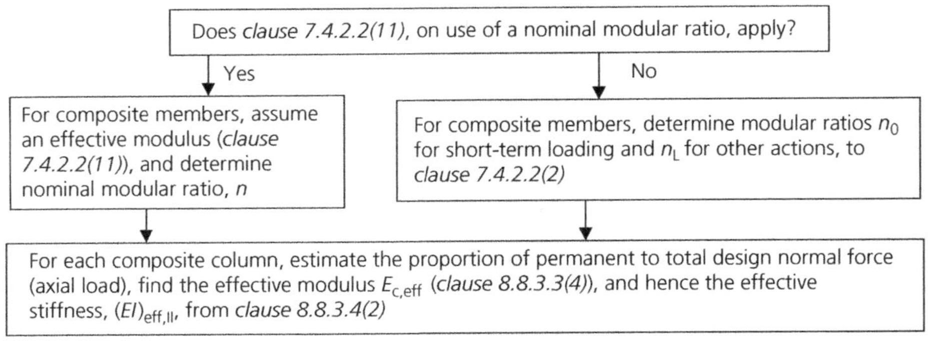

(b) Supplementary flow chart, creep

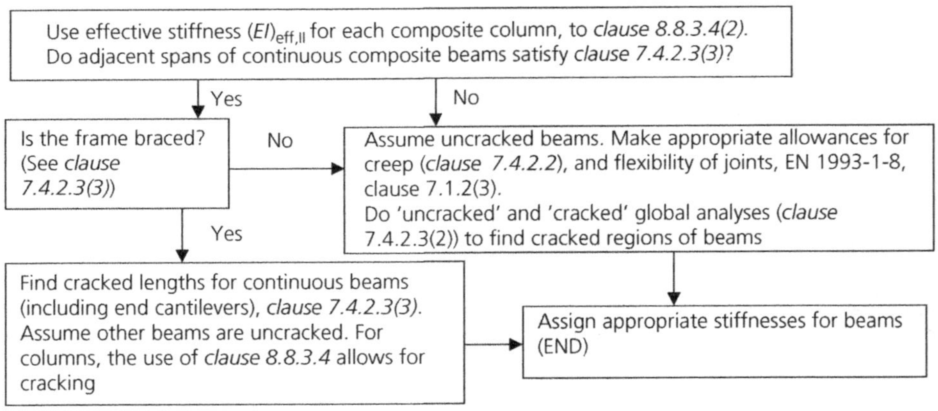

(c) Supplementary flow chart, cracking of concrete

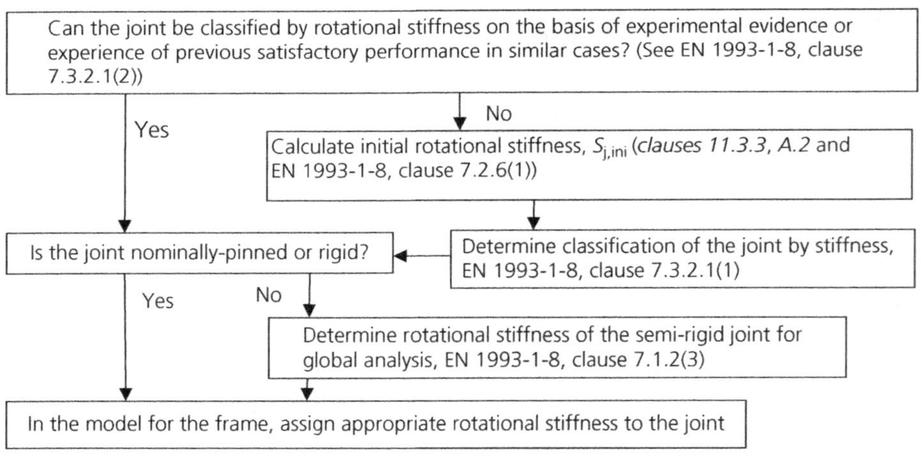

(d) Supplementary flow chart, stiffness of joints, for elastic global analysis only

resistance – nominally-pinned, partial-strength and full-strength. These are defined in clause 7.3, 'Classification of joints', of EN 1993-1-8.

The first two joint models are those commonly used for beam-to-column joints in steel frames. For each joint in the 'simple' model, the location of the nominal pin relative to the centre-line of the column, the 'nominal eccentricity', has to be chosen. It determines the effective span of each beam and the bending moments in each column. Practice varies across Europe, and neither EN 1993-1-1 nor EN 1994-1-1 gives values for nominal eccentricities. Guidance may be given in a national annex by reference to other literature ('complementary information'). The UK's National Annex does not give specific guidance on joints, the subject of *clause 11* and *Annex A*.

In reality, most joints in frames for buildings are neither 'simple' (i.e. pinned) nor 'continuous'. The third model, 'semi-continuous', is appropriate for a wide range of joints with moment-rotation behaviours intermediate between 'simple' and 'continuous'. This model is rarely applicable to bridges, so the cross-reference to EN 1993-1-8 in ***clause 7.1.2(3)*** is 'for buildings'. The provisions of EN 1993-1-8 are for joints 'subjected to predominantly static loading' (its clause 1.1(1)). They are applicable to wind loading on buildings, but not to fatigue loading, which is covered in EN 1993-1-9 and in *clause 8.7*. *Clause 7.1.2(3)*

For composite beams, the need for continuity of slab reinforcement past the columns, to control cracking, causes joints to transmit bending moment. This is assumed here to be hogging moment, though the accidental limit state of removal of a column could cause sagging bending. For the joint to 'have no effect on the global analysis' (from the definition of a 'continuous' joint in clause 7.1.1(6) of EN 1993-1-8), so much reinforcement and stiffening of steelwork would often be needed that the design would become uneconomic. Joints with some continuity often qualify as semi-continuous. Structural analysis then requires prior calculation of properties of joints, except where they can be treated as 'nominally pinned' or 'rigid' on the basis of 'experience of previous satisfactory performance in similar cases' (clause 7.3.2.1(2) of EN 1993-1-8, referred to from *clause 11.2.3(1)*), or experimental evidence. It will be found in Chapter 11 and *Annex A* that calculations for semi-rigid joints can be extensive, so this concession is important in practice.

Clause 7.1.2(2) refers to clause 7.1.1 of EN 1993-1-8, which gives the terminology for the semi-continuous joint model. For elastic analysis, the joint is 'semi-rigid'. It has a rotational stiffness, and a design resistance which may be 'partial-strength' or 'full-strength', normally meaning respectively less than or greater than the bending resistance of the connected beam. Global analyses have to take account of a concentrated rotation, which may be elastic or elastic-plastic, at the location of each connection (a 'connection' is part of a 'joint') between a beam and a supporting column. Examples are given in Chapters 11 and 12.

7.2. Structural stability

The following comments refer mainly to beam-and-column frames, and assume that the global analyses will be based on elastic theory. The exceptions, in *clauses 7.4.3* to *7.4.5*, are discussed later. All design methods must take account of errors in the initial positions of joints (global imperfections) and in the initial geometry of members (member imperfections), of the effects of cracking of concrete and of any semi-rigid joints, and of residual stresses in compression members. The stage at which each of these is considered or allowed for will depend on the software being used, which leads to some complexity in *clauses 7.2* to *7.4*.

7.2.1 Effects of deformed geometry of the structure

Clauses 7.2.1(2) and ***7.2.1(3)*** give criteria for deciding whether first-order global analysis can be used in preference to the more complex second-order analysis. That will usually be iterative but may be the more convenient method if appropriate software is available. Methods for second-order analysis are described in text books such as Trahair *et al.* (2008). *Clause 7.2.1(2)* *Clause 7.2.1(3)*

A disadvantage of second-order analysis is that the useful principle of superposition does not apply. The effects of a combination of actions cannot be calculated by applying load factors to and summing the effects of each type of characteristic loading (permanent, wind etc.), found separately, as is often done in first-order analyses. Separate second-order analyses are needed for each combination and set of load factors.

Clause 7.2.1(3) provides a basis for the use of first-order analysis with references to conditions in clauses 7.1 and 7.2 of EN 1993-1-1. They concern whether the behaviour of joints affects global analyses, and give rules related to buckling of frames and members that involve finding the relevant elastic critical load. These analyses are done for a particular load combination and arrangement.

In an elastic frame, second-order effects are dependent on the nearness of the design loads to the elastic critical load. This is the basis for both conditions, in which factors α_{cr} are defined as the '*factor by which the design value of the loading would have to be increased to cause elastic instability…*'. This load may be taken as that at which bifurcation of equilibrium occurs. For a conventional beam-and-column frame, it can be assumed that the frame has no global imperfections, and that only vertical loads are present, usually at their maximum design values. These are replaced by a set of loads which produces the same set of member axial forces without any bending. An eigenvalue analysis then gives the factor α_{cr}, applied to the whole of the loading, at which the total frame stiffness vanishes and elastic instability occurs. For first-order analysis to be used, the factor has to be large – $\alpha_{cr,sw} \geq 10$ for sway buckling of a frame and recommended as $\alpha_{cr,ns} \geq 25$ (an NDP) for non-sway buckling of members.

7.2.2 Methods of analysis

Clause 7.2.2(1) **Clause 7.2.2(1)** provides a simpler check on sway mode failure not involving either second-order or critical buckling analyses, for beam-and-column type plane frames. It requires calculation of sway deflections due to horizontal loads only, for which first-order analysis can used, and assumes that any significant second-order effects will arise only from interaction of column forces with sway deflection. It follows that the check will only be valid if axial compression in beams is 'not significant', as is usual in buildings, and that cracking and creep are allowed for. The methods of Eurocode 4 do this. Also, flexibility of joints must be allowed for. This can be done by assuming them to be nominally pinned. Then, factor α_{cr} ($\alpha_{cr,sw}$ in Eurocode 3) is found for each storey by the method of clause 7.2.1(10) of EN 1993-1-1. For many frames for buildings, it will exceed 10, and the check on sway failure is satisfied.

Clause 7.2.2(2)
Clause 7.2.2(3) **Clause 7.2.2(2)** allows second-order effects to be allowed for in a first-order analysis 'with appropriate amplification', but no further information is given. **Clause 7.2.2(3)** gives an alternative: inclusion of all these effects in global analysis, which would need appropriate software. This is not needed for beam-type members, because the methods of *clause 8* take account of local buckling, shear buckling and lateral-torsional buckling.

Clause 7.2.2(6) **Clause 7.2.2(6)** refers to 'compression members' as well as columns, to include composite members used in bracing systems and trusses. It refers to the method of *clause 7.2.2(3)*, not normally used, and gives two alternatives – the 'simplified method' of *clause 8.8.3*, and a method limited to 'members in axial compression' based on EN 1993-1-1. This means members without bending moment, which rarely occur.

Further comments on *clause 7.2.2* are made in the sections of this guide dealing with *clauses 8.4* and *8.8*. Figure 7.1(a) illustrates the use of global and member analyses for a plane frame including composite columns.

7.3. Imperfections

7.3.1 Basis

Clause 7.3.1(1) **Clause 7.3.1(1)** lists possible sources of imperfection. Subsequent clauses (and also *clause 7.2*) describe how these should be allowed for. This may be by inclusion in the global analyses or in methods of checking resistance, as explained above.

Clause 7.3.1(2) **Clause 7.3.1(2)** requires imperfections to be in the most unfavourable direction and form. The most unfavourable geometric imperfection normally has the same shape as the lowest buckling mode. This can sometimes be difficult to find; but it can be assumed that this condition is satisfied by the Eurocode methods for checking resistance that include effects of member imperfections.

7.3.2 Imperfections in structures

7.3.2.1 General

An explicit treatment of geometric imperfections is usually required for composite frames. In both
Clause 7.3.2.1(1) EN 1993-1-1 and EN 1994-1-1 the values are equivalent rather than measured values (***clause 7.3.2.1(1)***) because they allow for effects such as residual stresses, in addition to geometric imperfections. The codes define both global sway imperfections for frames and local bow imperfections of individual members (meaning a span of a beam or the length of a column between lateral restraints).

The usual aim in global analysis is to determine the action effects at the ends of members. If necessary, a member analysis is performed subsequently, as illustrated in Figure 7.1(a) – for example, to determine the local moments in a column due to transverse loading. Normally, the action effects at members' ends are affected by the global sway imperfections but not significantly by the local bow imperfections. The effect of a bow imperfection in a member on its end moments and forces should be included in the global analysis if the design normal force, N_{Ed}, exceeds 25% of the Euler buckling load for the pin-ended member (***clause 7.3.2.1(2)***). This condition is unchanged from the previous EN 1994-1-1, although its presentation is different. *Clause 7.3.2.1(2)*

Clause 7.3.2.1(3) is a reminder that explicit treatment of bow imperfections is always required for checking individual composite columns because the resistance formulae are for cross-sections only and do not allow for action effects caused by these imperfections. The reference to EN 1993-1-1 in ***clause 7.3.2.1(4)*** leads to equivalent bow imperfections for steel compression members. Where they have been included in global analysis, their inclusion in member analysis is discussed under *clause 7.2.2*. *Clause 7.3.2.1(3)* *Clause 7.3.2.1(4)*

7.3.2.2 Global imperfections

Clause 7.3.2.2(1) refers to clause 7.3.2 of EN 1993-1-1. This gives values for the global sway imperfections, describes how imperfections may be replaced by equivalent horizontal forces, and permits these to be disregarded if real horizontal forces, such as effects of wind, are significant relative to the design vertical load. For building frames, 'significant' is defined as greater than 15%. *Clause 7.3.2.2(1)*

7.3.2.3 Member imperfections

Clause 7.3.2.3(1) refers to *Table 8.7*, which gives the amplitudes of the central bow of a member designed as straight. It makes little difference whether the curve is assumed to be a half sine wave or a circular arc. These single-curvature shapes are assumed irrespective of the shape of the bending-moment diagram for the member, but the designer has to decide on which side of the member the bow is present. ***Clauses 7.3.2.3(2)*** and ***7.3.2.3(3)*** refer to member imperfections that need not be included in global analyses. If they are, only cross-section properties are required for checking resistances. *Clause 7.3.2.3(1)* *Clause 7.3.2.3(2)* *Clause 7.3.2.3(3)*

7.4. Calculation of action effects

7.4.1 Methods of global analysis

7.4.1.1 General

Types of global analysis are defined at the beginning of this chapter. ***Clause 7.4.1.1*** gives guidance on their use. For simplicity and economy, plastic (rectangular stress-block) theory is the preferred method for checking the resistance of most cross-sections. In such cases, ***clause 7.4.1.1(1)*** allows the action effects to be determined by elastic analysis. *Clause 7.4.1.1* *Clause 7.4.1.1(1)*

Clauses 7.4.1.1(2) and ***7.4.1.1(3)*** require the use of elastic global analysis for serviceability limit states (SLS) and for fatigue. Linear elastic analysis is based on linear stress/strain laws but for composite structures, cracking of concrete needs to be considered and some redistribution of bending moment is permitted (*clause 7.4.4*). Other possible non-linear effects include the flexibility of semi-continuous joints (*clause 11* and *Annex A*)). *Clause 7.4.1.1(2)* *Clause 7.4.1.1(3)*

Methods for satisfying the principle of ***clause 7.4.1.1(4)*** are given for local buckling in ***clauses 7.4.1.1(5)*** and ***7.4.1.1(6)***, and for shear lag in concrete flanges in *clause 7.4.1.2*. Reference is made to the classification of cross-sections. This is the established method of taking account of local buckling of steel elements in compression. It determines the available methods of global analysis and the basis for resistance to bending. The classification system is defined in *clause 7.5*. *Clause 7.4.1.1(4)* *Clause 7.4.1.1(5)* *Clause 7.4.1.1(6)*

There is apparent incompatibility between the methods used for global analysis and for resistance for beams with cross-sections in Class 1 or 2. It occurs also in the methods of Eurocodes 2 and 3, for concrete and steel structures respectively. It simplifies both types of analysis and creates a need for redistribution of moments. The extent to which this is permitted in Eurocode 4 for beams with cross-sections in Classes 1 to 3 is discussed under *clause 7.4.4(2)*. For Class 4 sections (those in which local buckling may occur before the attainment of yield), *clause 7.4.1.1(6)* gives a general reference to clause 4.3 of EN 1993-1-5 (BSI, 2024), which defines those situations in which the effects of shear lag and local buckling in steel plating can be ignored in global analyses.

Clause 7.4.1.1(7) reflects a general concern about slip in bolt holes and connecting devices, shared by EN 1993-1-1. It can be beneficial – for example, in reducing the stiffness of so-called 'simple' beam-to-column joints. For composite joints, *clause A.4* gives a method to account for deformation of the shear connectors adjacent to a joint. *Clause 7.4.1.1(7)*

Clause 7.4.1.1(8)

In ***clause 7.4.1.1(8)***, 'non-linear analysis' may mean global analysis to *clause 7.4.3*, which refers to clause 7.4.3 of EN 1993-1-1. That clause is headed 'Plastic global analysis', and is rarely applicable to composite structures. This is not a reference to the non-linear resistance given by *clause 8.2.1.5*.

There is reference to slip and to separation, commonly called uplift. The effect of slip resulting from use of partial shear connection in beams is offset by the conservative use of elastic global analysis where beams are in class 1 or 2. The effects of uplift with shear connection provided in accordance with *clause 8.6* are negligible.

7.4.1.2 Effective width of flanges for shear lag

Values for effective width of an uncracked concrete elastic flange can be determined by numerical analysis. However, effective width is influenced by many parameters and varies significantly along each span. It is increased both by inelasticity and by cracking of concrete. For the bending resistance of a beam, underestimates are conservative, so values in codes have often been based on elastic values. However, safe design of shear connection needs overestimates, and it may be unconservative to consider a narrower effective width in design. Resistance to bending is insensitive to effective width, so simplification is possible.

Clause 7.4.1.2(5) refers to 'continuous composite beams', which are not defined; neither is '*continuous beam*', either here or in Eurocode 3. The treatment of cantilevers is usually obvious, but the extent to which these rules apply to beams in frames is not always clear, especially concerning unbraced frames and the type of joint used. The five clauses listed below are consistent with the following definition.

> A continuous composite beam is a linear member subjected mainly to bending with at least three supports. At its internal supports it is connected to another member by a rigid and full-strength joint or is continuous into another span.

- *Clause 7.4.1.2(5)* on effective width
- *Clause 7.4.2.3(3)* on 'cracked' global analysis, which excludes unbraced frames, implying that braced frames are included
- *Clause 7.4.4(4)* on redistribution of moments, requiring joints to be either rigid full-strength or nominally pinned joints. Its restriction to beams with 'only vertical loading' does not exclude unbraced frames, but it is unclear whether they are included
- *Clause 7.4.5* on rigid-plastic global analysis appears to be applicable to unbraced frames and requires the joints to have sufficient rotation capacity
- *Clause 8.4.3(1)* on lateral bracing begins '*A continuous beam (or a beam within a frame...)*', which implies that a beam within a frame is not a 'continuous beam' ('including' would be a better word than 'or').

Clause 7.4.1.2

The simplified values for effective width given in ***clause 7.4.1.2*** of EN 1994-1-1 are similar to those in the previous BS 5950: Part 3.1 (BSI, 2010). They are based on the distance between points of contraflexure. In EN 1992-1-1, the sum of the distances for sagging and hogging regions equals the span of the beam. EN 1994-1-1 gives a larger effective width at an internal support, to be on the safe side for the provision of shear connection for that region, and to reflect the greater distance between points of contraflexure for the critical load arrangement for this cross-section. The assumed length for a sagging region is the same in the two codes.

Clause 7.4.1.2(4)

Although there are significant differences between effective widths for supports and mid-span regions, it is possible to ignore this in elastic global analysis (***clause 7.4.1.2(4)***). This is because shear lag has limited influence on the results.

Clause 7.4.1.2(6)

Clause 7.4.1.2(6) is a reminder that *Figure 7.1* is based on beams fully continuous over point supports, and gives the width at an end support, using 'may' rather than 'should'. This width can depend on the type of joint used or the detailing of the local longitudinal reinforcement, which for hogging bending should be anchored to the column.

Example 7.1: effective width of concrete flange

The notation and method used are those of *clause 7.4.1.2*. A continuous beam of uniform section consists of two spans and a cantilever, as shown in Figure 7.2. Values for b_{eff} are required for the mid-span regions AB and CD, for the support regions BC and DE, and for the support at A.

Figure 7.2. Worked example: effective width

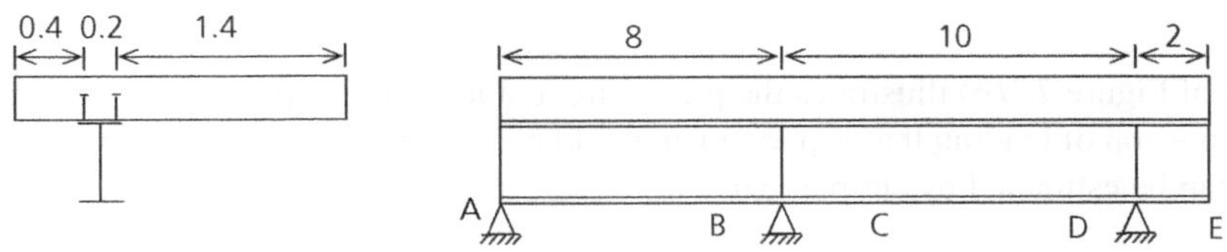

The calculated values are shown in Table 7.1. The width b_{e1} is limited by the actual width, 0.4 m; width b_{e2} is always less than the actual width, 1.4 m; and b_{0i}, the width of the shear connection, is 0.2 m. Global analyses may be based on stiffness calculated using the results for AB and CD, but the difference between them is so small that member ABCDE would be analysed as a beam of uniform section.

For support A, from *clause 7.4.1.2(6)* and *Formulae 7.2* and *7.3*, $L_e = 6.8$ m

$\beta_1 = 0.55 + 0.025 \times 6.8/0.4 = 0.975$; $\beta_2 = 0.55 + 0.025 \times 6.8/0.85 = 0.75$;

$b_{eff} = 0.2 + 0.975 \times 0.4 + 0.75 \times 0.85 = 1.23$ m

Table 7.1. Effective width of the concrete flange of a composite T-beam

	Region				
	AB	BC	CD	DE	Support A
L_e (from *Figure 7.1*): m	6.80	4.5	7.0	4.0	6.80
$L_e/8$: m	0.85	0.562	0.875	0.50	–
b_{e1}: m	0.40	0.40	0.40	0.40	0.40
b_{e2}: m	0.85	0.562	0.875	0.50	0.85
b_{eff}: m	1.45	1.162	1.475	1.10	1.23

7.4.2 Linear elastic analysis

The restrictions on the use of rigid-plastic global analysis (plastic hinge analysis), in *clause 7.4.5*, are such that linear-elastic global analysis is normally used for composite frames.

7.4.2.2 Creep and shrinkage

Clause 7.4.2.2 begins with a governing principle that applies more widely than just to creep and shrinkage: 'Appropriate allowance shall be made…'. The user of EN 1994-1-1 has to decide whether, in an unusual situation, following its rules on creep and shrinkage is sufficient and appropriate, and not to follow an application rule if it is clearly inappropriate. *Clause 7.4.2.2*

The modular ratio method of ***clause 7.4.2.2(2)*** is the simplest way of allowing for the influence of creep on action effects. From EN 1992-1-1, the elastic modulus for concrete under short-term loading, E_{cm}, is a function of the type of aggregate and grade and density of the concrete. For quartzite aggregate and normal-density concrete, it ranges from 28.8 kN/mm^2 for grade C20/25 to 40.5 kN/mm^2 for grade C70/85. With the elastic modulus for structural steel, E_a, given as 210 kN/mm^2 in EN 1993-1-1, the short-term modular ratio, given by $n_0 = E_a/E_{cm}$, ranges from 7.3 to 5.2. *Clause 7.4.2.2(2)*

The effect of type of loading is introduced by the creep multiplier ψ_L in the formula

$$n_L = n_0 (1 + \psi_L \phi_t) \qquad (7.4)$$

It is taken as zero for short-term loading. For permanent loading, a value 1.0 has been traditional for reinforced concrete. It is increased to $\psi_P = 1.1$ here because creep causes less relaxation of compressive stress in composite members as the steel component does not creep. Initial shrinkage occurs before the concrete is fully stiff, so there is less creep, hence the value 0.55. The effects of imposed deformations can be significantly reduced by creep when the concrete is young. The value $\psi_D = 1.5$ allows for this.

The flow chart of Figure 7.1(b) illustrates the procedure to allow for creep in members of a composite frame. The proportion of loading that is permanent could be obtained by a preliminary global analysis, but often this can be estimated by simpler calculations.

Clause 7.4.2.2(3) Although *clause 7.4.2.2(4)* gives the age of loading by the effects of shrinkage as one day, ***clause 7.4.2.2(3)*** allows one mean value of t_0 to be assumed. If, for example, unpropped construction is used for floor slabs, this might be taken as the age at which they could be subjected to non-trivial imposed loads. These are likely to be construction loads.

It makes quite a difference whether this age is assumed to be (for example) one week or three months. From *clause 7.4.2.2(2)*, the creep coefficients for normal-weight concrete are found from clause 5.1.5 of EN 1992-1-1. For example, suppose that normal cement is used for grade C25/30 concrete, that the building will be centrally heated so that 'dry conditions' apply, and that composite floor slabs with a mean concrete thickness of 100 mm are used. Only one side of the slabs is exposed to drying, so the notional size is 200 mm. The increase in t_0 from 7 to 91 days reduces the creep coefficient from 3.6 to about 2.2.

Clause 7.4.2.2(6) The '*time-dependent secondary effects due to creep*' referred to in ***clause 7.4.2.2(6)*** are unlikely to be found in buildings. Their calculation is quite complex, and has been explained, with an example (Johnson and Hanswille, 1998).

Clause 7.4.2.2(7) ***Clause 7.4.2.2(7)*** allows neglect of creep and shrinkage effects in beams where the bending resistance is taken as $M_{pl,Rd}$ because flexural strains are then high enough for creep and shrinkage effects to be negligible. It does not apply for fatigue or where resistance to bending is reduced by lateral-torsional buckling. This restriction can be significant.

Clause 7.4.2.2(8) ***Clause 7.4.2.2(8)*** allows the option of neglecting primary shrinkage curvature in cracked regions (Johnson, 1987). This complicates the determination of the secondary effects, because the extent of the cracked regions has to be found (*clause 7.4.2.3*), and the beam then has a non-uniform section. It may be simpler not to take the option, even though the secondary hogging bending moments at internal supports are then higher. These moments, being a permanent effect, enter into all load combinations and may influence design of what is often a critical region.

Clause 7.4.2.2(9) For creep in columns, ***clause 7.4.2.2(9)*** refers to *clause 8.8.3.4(2)*, which in turn refers to an effective modulus for concrete given in *clause 8.8.3.3(4)*. Unlike simplified values used elsewhere, this value depends on the proportion of the loading that is permanent.

Clause 7.4.2.2(10) ***Clause 7.4.2.2(10)*** excludes the use of the simplified method of *clause 7.4.2.2(11)* for certain members with both flanges composite, but as these occur mainly in bridges, no alternative is given in Part 1.1.

Clause 7.4.2.2(11) For composite beams in structures for buildings where first-order global analysis is acceptable (the majority), ***clause 7.4.2.2(11)*** allows the modular ratio to be taken as $2n_0$ for both short-term and long-term loading – an important simplification. The only exceptions are

- structures where *clause 7.2.2(1)* does not apply and second-order global analysis is required by *clause 7.2*
- structures for buildings mainly intended for storage
- structures prestressed by 'controlled imposed deformations' – this would apply, for example, to the bending of steel beams by jacking before concrete is cast around one of the flanges.

Where the conditions of *clause 7.4.2.2(11)* do not apply, the modular ratio for use in analyses for the effects of long-term loading, n_L, depends on the type of loading and the creep coefficient $\varphi(t, t_0)$. This coefficient depends on both the age of the concrete on first loading, t_0, and on the age at the time considered in the analysis, which is normally taken as 'infinity'.

These application rules are based mainly on extensive theoretical work on composite beams of many sizes and proportions (Haensel, 1975) and find application more in design of composite bridges than in buildings.

Further notes on shrinkage

For the determination of shrinkage strain, reference may be made to the comments on *clause 5.1*. The effects in columns are unimportant, except in very tall structures. In beams with the slab above the steel member, shrinkage causes sagging curvature. This is its '*primary effect*', which is negligible in regions where the concrete slab is cracked through its thickness.

In continuous beams, primary curvature from shrinkage is incompatible with the levels of the supports. It is counteracted by bending moments caused by changes in the support reactions, which increase at internal supports and reduce at end supports. The moments and the associated shear forces are the '*secondary effects*' of shrinkage. There are calculations in Example 8.6.

The long-term effects of shrinkage are significantly reduced by creep. In the example above, on creep of concrete, with $t_0 = 1$ day, clause 5.1.5 of EN 1992-1-1 gives $\varphi_t = 4.9$, and *Formula 7.4* gives the modular ratio as

$$n = n_0(1 + 0.55 \times 4.9) = 3.7n_0$$

Where it is necessary to consider shrinkage effects within the first year or so after casting, a value for the relevant free shrinkage strain can be obtained from Annex B of EN 1992-1-1.

Where shrinkage effects are considered at ultimate limit states, a partial factor is needed. Clause 4.3.1(1) of EN 1992-1-1 recommends $\gamma_{SH} = 1.0$, and this is expected to be confirmed in the UK's National Annex. For the influence of shrinkage on serviceability verifications, *clause 9.3.1(9)* and the comment on it are relevant. The partial factor is 1.0.

Effects of cracking of concrete

This ***clause 7.4.2.3*** is applicable to both serviceability and ultimate limit states. Figure 7.1(c) illustrates the procedure. In conventional composite beams with the slab above the steel section, cracking of concrete reduces the flexural stiffness in hogging moment regions but not in sagging regions. The change in relative stiffness needs to be taken into account in elastic global analysis. This is unlike analysis of reinforced concrete structures, where cracking occurs in both hogging and sagging bending, and uncracked cross-sections can usually be assumed throughout. *Clause 7.4.2.3*

EN 1994-1-1 provides alternative methods for allowing in global analysis for cracking in beams. ***Clause 7.4.2.3(2)*** provides a general method. This is followed in ***clause 7.4.2.3(3)*** by a simplified approach with limited application. For buildings, cracking is also considered in the rules of *clause 7.4.4* on redistribution of bending moments. *Clause 7.4.2.3(2)* *Clause 7.4.2.3(3)*

In the general method, the first step is to determine the expected extent of cracking in beams. The envelope of moments and shears is calculated for characteristic combinations of actions, assuming uncracked sections and including long-term effects. The contribution from reinforcement is small, and usually is neglected. The section is then assumed to be 'cracked' at cross-sections where, for normal-density concrete, the extreme-fibre tensile stress in concrete exceeds twice the mean value of the axial tensile strength given by EN 1992-1-1. The 'cracked' stiffness, with reinforcement, is then adopted for such sections, and the structure re-analysed. This requires the beams with cracked regions to be treated as members of non-uniform section. The 'uncracked' and 'cracked' flexural stiffnesses E_aI_1 and E_aI_2 are defined in *clause 3.1*.

The stress at which the stiffness is reduced to the 'cracked' value is chosen for these reasons

- the concrete is likely to be stronger than specified
- reaching f_{ctm} at the surface may not cause the slab to crack right through

- until after yielding of the reinforcement, the stiffness of a cracked region is greater that $E_a I_2$ because of tension stiffening between the cracks
- the calculation uses an envelope of moments, for which regions of slab in tension are more extensive than they are for any particular loading.

Clause 7.4.2.3(3)
Clause 7.4.2.3(4)
Clause 7.4.2.3(5)

Clauses 7.4.2.3(3) to ***7.4.2.3(5)*** provide a non-iterative method, but one that is applicable only in some situations. These include conventional continuous composite beams and beams in braced frames. The cracked regions could differ significantly from the assumed values in an unbraced frame that resists wind loading by bending. Where the conditions are not satisfied, the general method of *clause 7.4.2.3(2)* should be used.

Cracking affects the stiffness of a frame and, therefore, needs to be considered in the criteria for the use of first-order analysis (*clauses 7.2.1(4)* and *7.2.2(1)*). For braced frames within the scope of *clause 7.4.2.3(3)*, the cracked regions in beams are assumed to be of fixed extent, and the effective stiffness of the columns is given by *clause 8.8.3.4(2)*. The value of the elastic critical factor α_{cr} can, therefore, be determined prior to analysis under the design loads.

Clause 7.4.2.3(3) refers to continuous composite beams. Here, 'continuous' is understood to mean that the joints between spans qualify as 'rigid' and that flexural interaction with bending moments in supporting columns can be neglected.

The 'encasement' in *clause 7.4.2.3(5)* is a reference to the partially-encased beams treated in *clause 8.3*. Fully encased beams are outside the scope of EN 1994-1-1.

Temperature effects

Clause 7.4.2.5(2)

Clause 7.4.2.5(2) states that temperature effects, specified in EN 1991-1-5 (BSI, 2025), may normally be neglected in analyses for certain situations. Its narrow scope provides a further incentive to select steel sections for beams that are not weakened by lateral-torsional buckling.

Study of the Ψ factors of Table A.1.7 of EN 1990 (BSI, 2023) for combinations of actions for buildings will show, for many projects, that temperature effects do not influence design. This is now illustrated for the design action effects due to the combination of imposed load (Q) with temperature (T) for a building with floors in category B, office areas. Similar comments apply to other combinations of actions and types of building.

The combination factors recommended in clause A.1.5.3 of EN 1990 are given in Table 7.2 and expected to be confirmed in the UK's National Annex. For ultimate limit states, the combinations to be considered, in the usual notation and with the recommended γ_F factors, are

$$1.35G_k + 1.5(Q_k + 0.6T_k) \quad \text{and} \quad 1.35G_k + 1.5(T_k + 0.5Q_k)$$

The second one, with T leading, governs only where $T_k > 1.25Q_k$. Normally, action effects due to temperature are much smaller than those due to imposed load and additional action effects resulting from the inclusion of T in the first combination are not significant.

For serviceability limit states, much depends on the project. Clause 5.4(4) of EN 1990 states that 'Serviceability requirements should be specified individually for each project'. This is repeated, for buildings, in its clause A.1.8.1(1).

Table 7.2. Combination factors for imposed load and temperature

Action	Ψ_0	Ψ_1	Ψ_2
Imposed load, building in category B	0.7	0.5	0.3
Temperature (non-fire) in buildings	0.6	0.5	0

There are three combinations of actions given in EN 1990 for serviceability limit states – characteristic, frequent and quasi-permanent. The first of these uses the same combination factors Ψ_0 as for ultimate limit states, and the comments made above apply. The quasi-permanent combination is normally used for long-term effects and temperature is, therefore, not included.

For the frequent combination, the alternatives are

$$G_k + 0.5Q_k \quad \text{and} \quad G_k + 0.5T_k + 0.3Q_k$$

The second one governs only where $T_k > 0.4Q_k$. This example shows that unless there are members for which temperature is the most severe action, as can occur in some industrial structures, the effects T are unlikely to influence verifications for buildings.

Prestressing by controlled imposed deformations or by jacking

Clause 7.4.2.6(2) draws attention to the need to consider the effects of deviations from assumed values of stiffness and imposed deformations. If the deformations are controlled, design values of internal forces and moments arising from this form of prestressing should be calculated from the characteristic or nominal value of the deformation, which will usually be the intended or measured value. The nature of the control required is not specified. It should take account of the sensitivity of the structure to any error in the deformation. Prestressing by jacking of supports is rarely used in buildings. The subsequent loss of prestress can be high. *Clause 7.4.2.6(2)*

7.4.3 Non-linear global analysis

Clause 7.4.3 adds little to the corresponding clauses in EN 1992-1-1 and EN 1993-1-1, to which it refers. This clause gives principles that apply to any method of global analysis that does not conform to *clause 7.4.2, 7.4.4* or *7.4.5*. Its reference to EN 1993-1-14 is relevant to the use of finite-element methods. Its use of 'shall' for clauses referred to does not mean that 'should' in those clauses must be treated as 'shall'. *Clause 7.4.3*

Clause 7.3.4 of EN 1992-1-1, referred to from ***clause 7.4.3(1)***, is 'non-linear analysis'. It does not distinguish between global and member analyses, and gives many general requirements. These are normally satisfied by analyses in accordance with the other provisions of EN 1994-1-1, but designers should be aware of them. Clause 7.4.3 of EN 1993-1-1, also referred to, is 'plastic global analysis'. It includes the well-known plastic hinge method, and is concerned with material non-linearity. *Clause 7.4.3(1)*

In practice, for composite structures, elastic global analysis is normally used, because use of *clause 7.4.3*, 'non-linear global analysis' is complex and the only other alternative, 'rigid plastic global analysis' to *clause 7.4.5*, has limited applicability. The term 'non-linear', used in *clause 8.2.1.5* for a type of resistance, is confusing. It presumably refers to non-linear analysis, because almost all resistances of composite members are non-linear functions of the strengths of their materials.

7.4.4 Linear elastic analysis with limited redistribution for buildings

The concept of redistribution of moments calculated by linear-elastic theory is well established in the design of concrete and composite framed structures. It makes allowance for inelastic behaviour, and enables the size of a design envelope of bending moments (from all relevant arrangements of variable loads) to be reduced.

In composite beams it is easier to provide resistance to bending in mid-span regions than at internal supports, so reduction of design hogging moments can reduce cost. The flexural stiffness at mid-span is higher, sometimes much higher, than at internal supports, so that 'uncracked' global analyses overestimate hogging bending moments in continuous beams. A flow chart for this clause is given in Figure 7.3. It applies for verifications for ULS other than fatigue and for SLS, based on linear-elastic global analysis.

This quite complex subject is treated here in this sequence, with ultimate limit states first

- limits to the use of redistribution
- limits to the amount of redistribution
- beams with cross-sections in Class 3 or 4
- partially-encased beams
- redistribution from mid-span to supports
- redistribution for 'steel members'

- redistribution in columns
- redistribution for serviceability verifications.

Figure 7.3. Flow chart for redistribution of moments in a continuous composite beam and adjacent columns

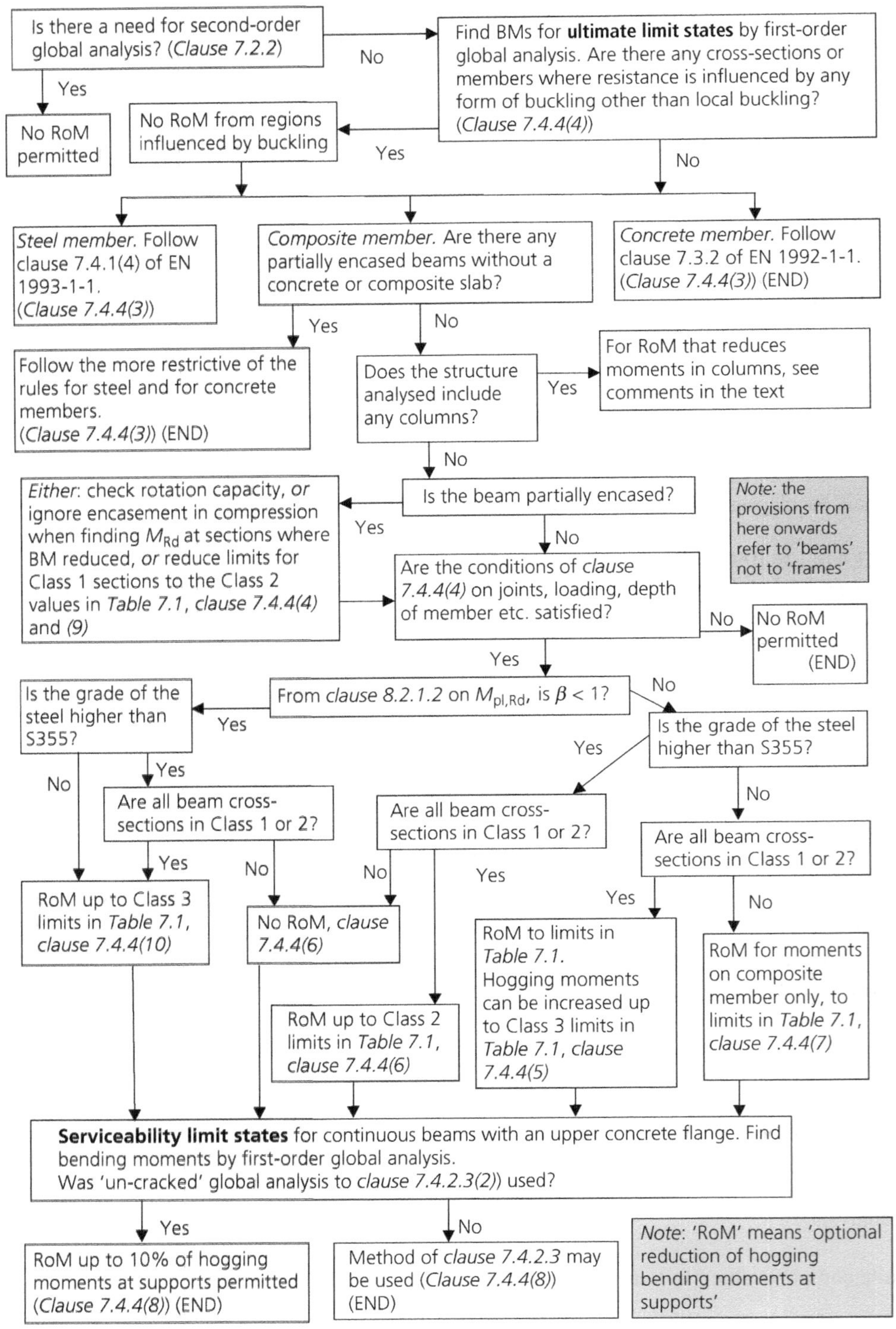

Redistribution for ultimate-limit-state verifications up to the limits of *clause 7.4.4* is not applicable where

(*a*) second-order global analysis is required
(*b*) the structure is an unbraced system

(*c*) a fatigue limit state is being verified
(*d*) semi-rigid or partial-strength joints are used
(*e*) the depth of a beam varies within a span
(*f*) the resistance of the beam is reduced to allow for lateral-torsional buckling
(g) the loading does not satisfy the condition of *clause 7.4.4(4)* for 'mainly uniform' distribution
(*h*) the beam has some non-vertical loading
(*i*) a beam with steel of grade higher than S355 has cross-sections in Class 3 or 4.

The reasons for these exclusions are now briefly explained.

(*a*) Redistribution arises from inelastic behaviour which reduces the stiffness of the structure. The rules that determine whether second-order analysis is required do not take account of this reduction.
(*b*) The amounts of redistribution given in *Table 7.1* have been established taking no account of the interaction with bending moments in the supporting columns that occurs in unbraced frames. Also, lateral loading may cause axial forces in beams in these frames.
(*c*) Fatigue verification needs reliable stress ranges. Those found by elastic analysis will be altered by redistribution.
(*d*) The limits given in *Table 7.1* allow for inelastic behaviour in composite beams, but not for the moment-rotation characteristics of semi-rigid or partial-strength joints.
(*e*) The limits given in *Table 7.1* have been established only for beams of uniform section.
(*f*) Lateral-torsional buckling may limit the rotation capacity available. This condition can be restrictive, as it may apply where $\bar{\lambda}_{LT} > 0.4$ (see a note to clause 8.3.2.3(1) of EN 1993-1-1). This is a nationally determined value, expected to be confirmed in the UK's National Annex for rolled and certain hollow sections. Welded sections require a value of 0.2.
(g) The rules do not take account of highly non-uniform design loading.
(*h*) Non-vertical loading causes axial tension or compression, not allowed for in the rules. The beam is being assumed to be horizontal. Sloping beams can resist non-vertical loading without axial force, but presumably are excluded.
(*i*) The greater strains associated with higher grade steels may increase the rotation needed to achieve redistribution.

Limits to the amount of redistribution

The limits given in *Table 7.1* apply only to beams and are for ultimate limit states. It should not be inferred that no redistribution is permitted in structures that fail to satisfy one or more of the conditions for its use. It would be necessary to show that any redistribution proposed satisfied ***clause 7.4.4(2)***. The percentages given for 'cracked' analyses of beams in Class 1 or 2 are more restrictive by 5% than in BS 5950 (BSI, 2010). *Clause 7.4.4(2)*

This reflects the finding (Johnson and Buckby, 1986) that the difference caused by cracking, between moments calculated by elastic theory in such beams, is nearer 15% than 10%.

Inelastic flexural behaviour usually redistributes bending moment from hogging to sagging regions of a beam (except of course at an end adjacent to a cantilever). The limits to this redistribution in *Table 7.1* are based on extensive experience in the use of earlier codes and on research. They have been checked by parametric studies, based on Eurocode 4, of beams in Classes 1 and 2 (Johnson and Chen, 1991) and Class 3 (Johnson and Fan, 1988).

The use of plastic resistance moments for action effects found by elastic global analysis (*clause 7.4.1.1(1)*) implies redistribution of moments, usually from internal supports to mid-span regions, additional to the degrees of redistribution permitted, but not required, by ***clause 7.4.4(4)*** (Johnson and Huang, 1995). *Clause 7.4.4(4)*

Redistribution that increases hogging bending moments

Where there are heavy point loads and, in particular, for adjacent spans of unequal length, there can be a need for redistribution from mid-span to supports. This is allowed, to a limited extent, by ***clause 7.4.4(5)***. *Clause 7.4.4(5)*

Beams with cross-sections in Class 3 or 4

The effects of sequence of construction should be considered where unpropped construction is used and the composite member is in Class 3 or 4. As *Table 7.1* makes allowance for inelastic behaviour in a composite beam, ***clause 7.4.4(7)*** limits redistribution of moments to those arising after composite action is achieved. *Clause 7.4.4(7)*

No such limitation applies to a cross-section in Class 1 or Class 2, as the moment resistance is determined by plastic analysis and is, therefore, independent of the loading sequence. The reference in *clause 7.4.4(3)(b)* to redistribution in steel members is to those in Class 3 or 4 that do not subsequently become composite.

One of the requirements of *clause 7.4.4(2)* is that redistribution should take account of 'all types of buckling'. Where the shear resistance of a web is reduced to below the plastic value $V_{pl,Rd}$ to allow for web buckling, and the cross-section is not in Class 4, it would be prudent either to design it for the higher of the vertical shears before and after redistribution, or to treat it as if in Class 4.

Partially encased beams

Crushing of the concrete encasement in compression may limit the rotation capacity needed to achieve redistribution. However, it is permitted, to *clause 7.4.4(3)(d)*.

Redistribution for 'steel members'

For these members *clause 7.4.4(3)(b)* refers to EN 1993-1-1 and presumably applies to both beams and columns. It is not applicable to composite beams constructed unpropped at the stage where the steel beam acts alone. Design with yielding of the beam at this stage is unlikely because of its effect on deflections.

Redistribution of moments in columns

Clause 7.4.4(1) refers to redistribution in 'continuous beams and braced frames', but composite columns are not mentioned. At a beam-column intersection in a frame, there are usually bending moments in the column arising from interaction with the beam. Redistribution for the beam may be done by assuming it to be continuous over simple supports. If the hogging moments are reduced, the bending moments in the column should be left unaltered. If hogging moments are increased, those in the column should be increased in proportion.

Serviceability limit states

Clause 7.4.4(8)

Clause 7.4.4(8) refers to *Clause 7.4.2.3* on the effects of cracking of concrete, the rules of which include some redistribution of bending moments and are applicable for all limit states. The alternative, up to 10% redistribution, is also to allow for cracking of concrete. It is a correction to results from an 'uncracked' global analysis, not an allowance for yielding of steel.

Clause 7.4.4(9)

Clause 7.4.4(9) is for partially encased beam sections with a concrete or composite top flange. It gives conditions under which hogging moments may be redistributed up to the limits in *Table 7.1* for Class 2 sections. This clause is not included in the flow chart of Figure 7.3.

7.4.5 Rigid-plastic global analysis for buildings

Plastic hinge analysis, the well-known method for steel portal frames, is referred to as 'rigid-plastic' analysis because it is based on the assumption that the response of a member to bending moment is either rigid (no deformation) or plastic (rotation at constant bending moment). Other types of inelastic global analysis, defined in clause 3.1.6 of EN 1990, have the general name 'non-linear analysis' (*clause 7.4.3*). No application rules are given for them because they require purpose-written computer programmes. These other methods are potentially more accurate than rigid-plastic analysis, but only if the stress-strain curves are realistic and account is taken of longitudinal slip, as required by *clause 7.4.3(2)*.

Clause 7.4.5

Taking account of the cross-references in ***clause 7.4.5***, the conditions under which use of rigid-plastic global analysis is allowed extend over two pages, which need not be summarised here. Development of a collapse mechanism in a composite structure requires a greater degree of redistribution of elastic moments than in most steel structures because beams are usually much stronger at mid-span than at supports. The purpose of the conditions is to ensure that this redistribution, and the large in-plane deformations associated with it, can occur without loss of resistance caused by buckling, fracture of steel or crushing of concrete. In practice, elastic global analysis is normally used because of these conditions and the complexity of non-linear global analysis.

Rigid-plastic analysis is applicable only if second-order effects are not significant. *Clause 7.2.1(3)* refers to the method from EN 1993-1-1 that is used for checking this. The ratio α_{cr} is calculated, which depends on the elastic critical buckling load for the frame. Care is needed if plastic hinges are expected to form in partial-strength joints (ECCS TC11, 1999). These may be substantially weaker than the connected members and the plastic hinges may form at relatively low levels of load. It would be prudent to neglect the stiffness of such joints when determining α_{cr}.

Clause 7.4.5(1) requires cross-sections of 'steel members' to satisfy clause 7.6 of EN 1993-1-1. This applies during unpropped construction, but not to steel elements of composite members, except as provided in *clause 7.4.5(6)* for members of non-uniform cross-section. The rules on rotation capacity in clause 7.6(2) of EN 1993-1-1 are replaced by the conditions of ***clause 7.4.5(4)*** of EN 1994-1-1. The rule on neutral axis depth in *clause 7.4.5(4)(g)* corresponds to a limit $z_{pl}/h \leq 0.15$. *Figure 8.3* shows that this excludes situations where $M_{pl,Rd}$ is subject to reduction by the factor β.

Clause 7.4.5(1)

Clause 7.4.5(4)

7.5 Classification of cross-sections

Typical types of cross-section are shown in *Figure 8.1*. The classification method for cross-sections of composite beams is the established method of EN 1993-1-1 for taking account in design of local buckling of steel elements in compression. It determines the available methods of global analysis and the basis for resistance to bending. For columns, local buckling is covered in *clause 8.8.1(8)*, rather than by reference to EN 1993-1-1.

Figure 7.4. Flow chart for classification of a cross-section of a composite beam

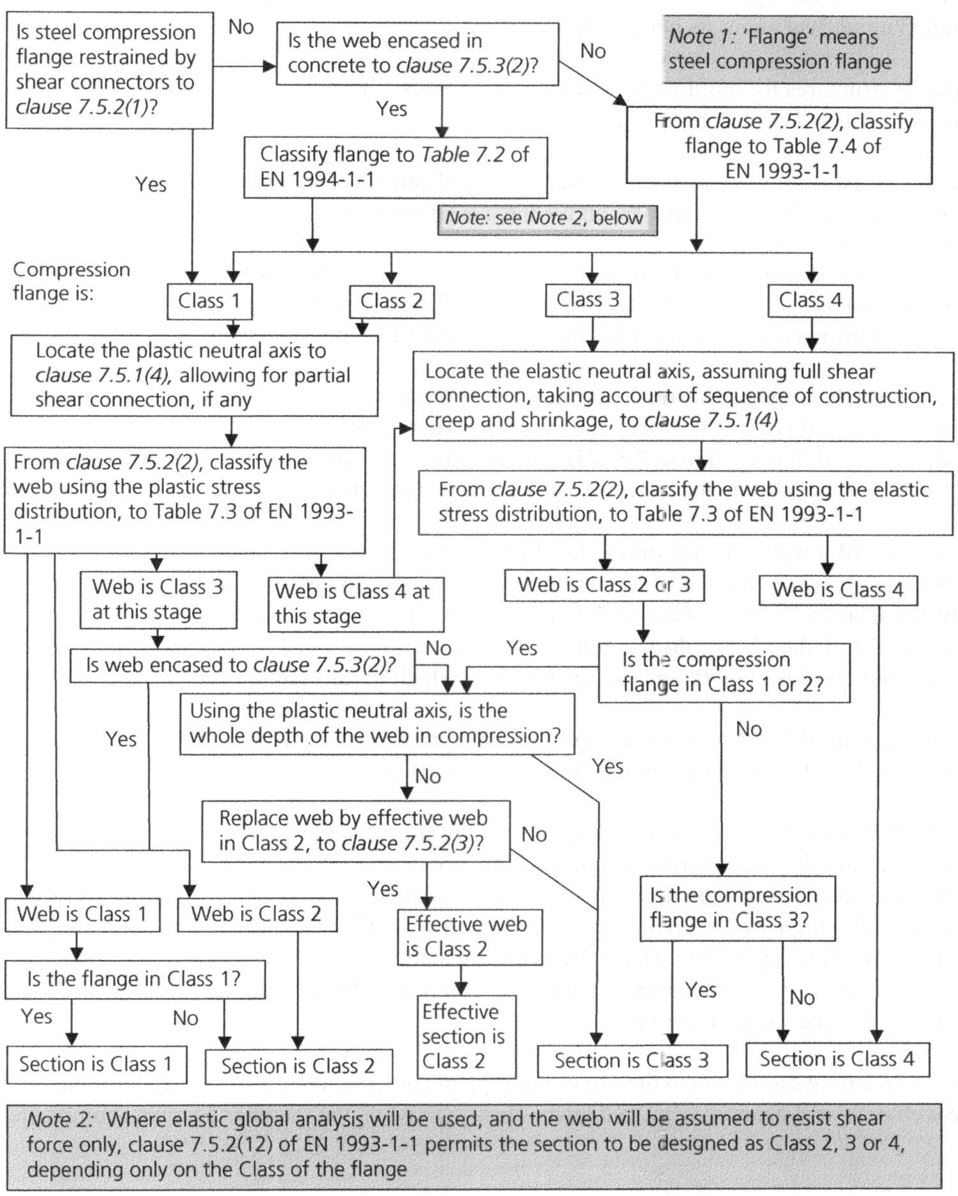

Clause 7.5 A flow diagram for the provisions of ***clause 7.5*** is given in Figure 7.4. The clause numbers there are from EN 1994-1-1, unless noted otherwise.

Clause 7.5.1 ***Clause 7.5.1*** refers to EN 1993-1-1 for definitions of the four classes and the slendernesses that define the class boundaries. Classes 1 to 4 correspond respectively to the terms 'plastic', 'compact', 'semi-compact' and 'slender' that were used in former British codes. The scope of EN 1994-1-1 includes members where the cross-section of the steel component has no plane of symmetry parallel to the plane of its web (e.g. a channel section). Asymmetry of the concrete slab or its reinforcement is also acceptable.

The classifications are done separately for steel flanges in compression and steel webs, but the methods interact as described below. The class of the cross-section is the least favourable of the classes so
Clause 7.5.1(2) found (***clause 7.5.1(2)***), with three exceptions. One is where a web is assumed to resist shear forces only (clause 7.5.2(12) of EN 1993-1-1). Then, the class of the cross-section is that of the flange in compression. The others are the 'hole-in-web' option of *clause 7.5.2(3)*, which is available only where the plastic neutral axis lies within the web, and the use of web encasement which is discussed later.

Reference is sometimes made to a beam in a certain class. This means that none of its cross-sections are in a less favourable class than the one stated. *Clause 7.5.1(2)* warns that the class of a composite section depends on the sign of the bending moment (sagging or hogging), as it does for a steel section that is not symmetrical about its neutral axis for bending.

Designers of structures for buildings normally select beams with steel sections such that the composite sections are in Class 1 or 2, for the following reasons.

- Bending resistances of beams can be found using plastic theory. For composite sections, this gives resistances from 20% to 40% above the elastic resistance, whereas the increase for steel sections is about 15%.
- The limits to redistribution of moments are more favourable than for Classes 3 and 4.
- Where composite floor slabs are used, it may be difficult to provide full shear connection. *Clause 8.6.3.2(3)* permits partial connection, but only where all beam cross-sections are in Class 1 or 2.

Simply-supported composite beams in buildings are almost always in Class 1 or 2, because the depth of web in compression (if any) is small and the connection to the concrete slab prevents local buckling
Clause 7.5.1(3) of the adjacent steel flange. ***Clause 7.5.1(3)*** refers to this, and *clause 7.5.2(1)* refers to the more useful *clause 8.6.10.5*, which limits the spacing of the shear connectors required.

Since the class of a web depends on the level of the neutral axis, and this is different for elastic and plastic bending, it is not obvious which stress distribution should be used for a section near the bound-
Clause 7.5.1(4) ary between Classes 2 and 3. ***Clause 7.5.1(4)*** provides the answer, the plastic distribution. This is because the use of the elastic distribution could place a section in Class 2, for which the bending resistance would be based on the plastic distribution, which in turn could place the section in Class 3.

The use of the partial factors for the strengths of steel and concrete affects the level of the plastic neutral axis. This should be found using the design (factored) values, from *clause 7.5.1(4)*.

Clause 7.5.1(5) ***Clause 7.5.1(5)***, on the minimum area of reinforcement for a concrete flange, appears here, rather than in *clause 8*, because it gives a further condition for a cross-section to be placed in Class 1 or 2. The reason is that these sections must maintain their bending resistance, without fracture of the reinforcement, while subjected to higher rotation than those in Class 3 or 4. This is ensured by disallowing the use of bars in ductility Class A (the lowest), and by requiring a minimum cross-sectional area, which depends
Clause 7.5.1(6) on the tensile force in the slab just before it cracks (Johnson, 2003; Kemp, 2006). ***Clause 7.5.1(6)***, on welded mesh, has the same objective.

Clause 7.5.1(7) ***Clause 7.5.1(7)*** draws attention to the use of unpropped construction, during which both the top flange and the web of a steel beam may be in a lower class than when the member becomes composite.

The hole-in-web method

This useful device is in clause 8.2.2.4 of EN 1993-1-1, which is referred to from ***clause 7.5.2(3)***. In beams subjected to hogging bending, it often happens that the bottom flange is in Class 1 or 2, and the web is in Class 3. The initial effect of local buckling of the web would be a small reduction in the bending resistance of the section. The assumption that a defined depth of web (the 'hole') is not effective in bending enables the reduced section to be upgraded from Class 3 to Class 2, with the advantages for design that are listed above. The method is analogous to the use of effective areas for Class 4 sections, to allow for local buckling. *Clause 7.5.2(3)*

The flow diagram of Figure 7.4 includes a limitation to its scope that is not evident from the wording in EN 1993-1-1

> ...the effective cross-sectional properties should be determined replacing the portion of the web in compression by an effective part of $20\varepsilon t_w$ adjacent to the compression flange and another part of $20\varepsilon t_w$ adjacent to the plastic neutral axis of the effective cross-section...

It follows that for a design yield strength f_{yd}, the compressive force in the web is limited to $40\varepsilon t_w f_{yd}$. For a composite beam in hogging bending, the tensile force in the longitudinal reinforcement in the slab can exceed this value, especially where f_{yd} is reduced to allow for vertical shear. The method is then not applicable, because the second 'part of $20\varepsilon t_w$' is not adjacent to the plastic neutral axis, which lies within the top flange. The method, and this limitation, is illustrated in Example 8.1.

Figure 7.4 shows that where an effective web in Class 2 is used, though the plastic check gives Class 3 (case P_3, say), no check need be made on the stress distribution in the elastic range of behaviour, even though more types of action are then taken into account. It is being assumed, in effect, that even if the web is Class 3 in the elastic range (E_3), plastic redistribution moves it into Class 2 before local buckling reduces the resistance of the cross-section to flexure. This implies a further assumption that the elastic check would not place the web in Class 4 (E_4). That combination, P_3 and E_4, is thought to be unlikely in buildings in practice. The alternative, to make an elastic check a precondition for the use of an effective web, would be laborious. It follows that at cross-sections where the effects of sequence of construction, shrinkage or creep are a major component of the total action effects and, especially for plate girders, effective webs should be used with caution.

Partially-encased cross-sections

Concrete encasement, as specified in ***clause 7.5.3***, refers to the steel web, as shown in *Table 7.2*. The encasement improves the resistance of both the web and the steel flange to local buckling, represented by increases in the slenderness ratios for compression flanges in Classes 2 and 3. The limit for Class 1 is unaltered. *Clause 7.5.3*

The encasement thus enables a Class 3 web to be treated as Class 2, without loss of cross-section. Conditions under which the encasement contributes to the bending and shear resistance of the member are given in *clause 8*, where relevant comments will be found

Roger P Johnson and Stephen J Hicks
ISBN 978-1-83662-921-4
https://doi.org/10.1108/978-1-83662-920-720253010

CHAPTER 8
Ultimate limit states

This chapter corresponds to *Section 8* of EN 1994-1-1, which has the following clauses.

- Beams *Clause 8.1*
- Resistances of cross-sections of beams *Clause 8.2*
- Resistance of cross-sections of beams with partial encasement for buildings *Clause 8.3*
- Lateral-torsional buckling of composite beams *Clause 8.4*
- Transverse forces on webs *Clause 8.5*
- Shear connection *Clause 8.6*
- Fatigue *Clause 8.7*
- Composite columns and composite compression members *Clause 8.8*

Clauses 8.1 to *8.6* and 8.8 define resistances of cross-sections to static loading, for comparison with action effects determined by the methods of *Section 7*. The ultimate limit state considered is defined in clause 5.3(3) of EN 1990 as

> Failure of the structure or the ground, or any part of them including supports and foundations, by rupture, excessive deformation, transformation into a mechanism, or buckling.

For lateral-torsional buckling of beams and for columns, the resistance is influenced by the properties of the whole member, and there is an implicit assumption that the member is of uniform cross-section, apart from variations arising from cracking of concrete and from detailing.

Most of the provisions of *Section 8* are applicable to both buildings and bridges, but a number of clauses are headed 'for buildings', and are replaced by other clauses in EN 1994-2. Some of these differences arise from the different treatments of shear connection for the two types of structure, which are compared in comments on ***clause 8.1.1***. *Clause 8.7*, 'Fatigue', refers to EN 1994-2 for fatigue checks and defines the situations where no fatigue checks are required for buildings. *Clause 8.1.1*

Annexes A to *J* give further rules relevant to ultimate limit states, some of which are Informative only, meaning that national standards bodies that are members of CEN, or countries that have adopted the Eurocodes as their national standards, can decide whether to give them the same 'Normative' status as the main text of EN 1994-1-1.

The main worked example in this book, a two-span continuous beam, is Example 8.6, given after the section on *clause 8.6*.

8.1 Beams
8.1.1 General

Figure 8.1 shows typical examples of beams for buildings within the scope of EN 1994-1-1, including web encasement, and profiled sheeting with spans at right angles to the span of the beam and continuous or discontinuous over the beam. The third diagram shows a haunched floor slab. Beams with openings in webs are treated in *Annex D*, shallow-floor beams in *Annex I* and beams supporting precast concrete slabs in *Annex J*.

The steel cross-section may be a rolled I- or H-section or may be a doubly symmetrical or mono-symmetrical plate girder. Other possible types include any of those shown in Tables 7.3 and 7.4 of EN 1993-1-1 – for example, rectangular hollow sections. Channel and angle sections should not be used unless the shear connection is designed to provide torsional restraint. Some types of floor construction are not within the scope of EN 1994-1-1 because their design rules are not well established – for example, beams fully encased and composite shear heads in flat slabs.

Figure 8.1. Typical cross-sections of composite beams (from BS EN 1994-1-1, by permission of the British Standards Institution)

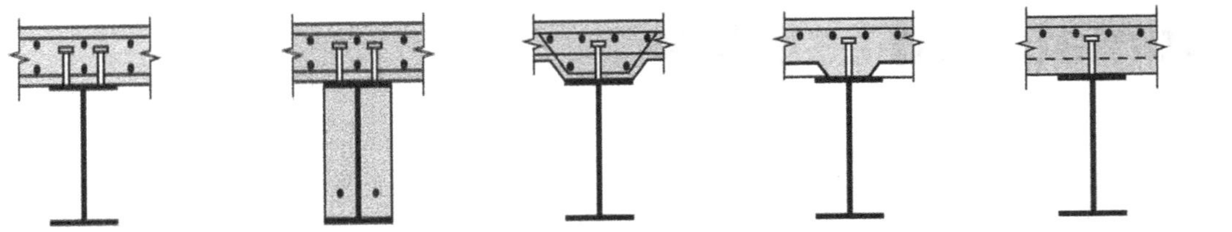

Situations commonly found in bridges, such as use of plate girders and double composite action, are covered in EN 1994-2 and its guide to the 2005 edition (Hendy and Johnson, 2006).

Shear connection

Clause 8.1.1(3)

In buildings, composite cross-sections are usually in Class 1 or 2, and the bending resistance is determined by plastic theory. At the plastic moment of resistance, the longitudinal force in a concrete flange is easily found, so design of shear connection for buildings is often based on the change in this force between two cross-sections where the force is known. This led to the concepts of critical cross-sections (***clause 8.1.1(3)***) and critical lengths (*clause 8.1.1(6)*). These concepts are not used in bridge design. Cross-sections in Class 3 or 4 are common in bridges and elastic methods are used. Then, longitudinal shear flows are found from the well-known result from elastic theory, $v_L = VA\bar{z}/I$.

Points of contraflexure are not critical cross-sections, partly because their location is different for each arrangement of variable load. A critical length in a continuous beam may, therefore, include both a sagging and a hogging region. Where connectors are uniformly spaced over this length, the number in the hogging region may not correspond to the force that has to be transferred from the longitudinal slab reinforcement. This does not matter, provided that the reinforcing bars are long enough to be anchored beyond the relevant connectors. The need for consistency between the spacing of connectors and curtailment of reinforcement is treated in *clause 8.6.3.2(4)*.

Clause 8.1.1(5)

A sudden change in the cross-section of a member changes the longitudinal force in the concrete flange, even where the vertical shear is zero. ***Clause 8.1.1(5)*** gives a criterion for deciding whether the change is sudden enough to be allowed for and will normally show that changes in reinforcement can be ignored. Where the clause is applied, the new critical section has different forces in the flange on each side of it. It may not be clear which one to use.

The wording of clauses on shear connection for buildings tends to assume that a critical length is from the cross-section of maximum sagging moment to an adjacent support. It may need interpretation where an additional critical cross-section has been inserted between these points.

One method is to use the result that gives the greater change of force over the critical length being considered. An alternative is to locate critical cross-sections on both sides of the change point, not more than about two beam depths apart. The shear connection in the short critical length between these two sections, based on the longitudinal forces at those sections, needs to take account of the change of section. The application of *clause 8.1.1(5)* is clearer for a beam that is composite for only part of its length. The end of the composite region is then a critical section.

A tapering member has a gradually changing cross-section. This can occur from variation in the thickness or effective width of the concrete flange, as well as from non-uniformity in the steel section. Where elastic theory is used, the equation for longitudinal shear force per unit length, $v_{L,Ed} = V_{Ed}A\bar{z}/I$, should be replaced by

$$v_{L,Ed} = V_{Ed}\,A\bar{z}/I + M_{Ed}\,\frac{d(A\bar{z}/I)}{dx} \tag{X8.1}$$

where x is the coordinate along the member, A is the 'excluded area' and $\bar{z}$ is the distance from its centre of area to the centroid of the cross-section.

For buildings, where resistances may be based on plastic theory, ***clause 8.1.1(6)*** enables the effect to be allowed for by using additional critical sections. It is applicable, for example, where the steel beam is haunched. The treatment of vertical shear then requires care, as part of it is resisted by the sloping steel flange. *Clause 8.1.1(6)*

Provisions for composite floor slabs, using profiled steel sheeting, are given only for buildings. The space within the troughs available for the shear connection is often insufficient for the connectors needed to develop the ultimate compressive force in the concrete flange, and the resistance moment corresponding to that force is often more than is required because of other constraints on design. This has led to the use of partial shear connection, which is defined in *clause 3.1.27*. It is applicable only where the critical cross-sections are in Class 1 or 2. Thus, in buildings, bending resistances are often limited to what is needed – that is, to $M_{\rm Ed}$, with shear connection based on bending resistances (see *clause 8.6.3.2(1)*).

Where bending resistances of cross-sections are based on an elastic model and limiting stresses, longitudinal shear flows can be found from $v_{\rm L,Ed} = V_{\rm Ed}A\bar{z} / I$. They are related to action effects, not to resistances. Shear connection designed in this way, which is usual in bridges, is technically 'partial' according to the definition in *clause 3.1.26* because increasing it would increase the bending resistances in the vicinity – although not in a way that is easily calculated, because inelastic behaviour and partial interaction are involved, and for partial interaction, slip or separation at the steel-concrete interface needs to be accounted for. For these reasons, the concept 'partial shear connection' is confusing in bridge design, and not applicable (***clause 8.1.1(7)***). Clauses in EN 1994-1-1 that refer to it are, therefore, labelled 'for buildings'. *Clause 8.1.1(7)*

8.1.2 Effective width for verification of cross-sections

The variation of effective width along a span, as given by *clause 7.4.1.2*, is too complex for verification of cross-sections in beams for buildings. The simplification in ***clause 8.1.2(2)*** often enables checks on the bending resistance of continuous beams to be limited to the supports and the mid-span regions. This paragraph should not be confused with the slightly different *clause 7.4.1.2(4)*, which applies to global analysis. *Clause 8.1.2(2)*

8.2 Resistances of cross-sections of beams

This clause is for beams with full or partial shear connection. Most of it, other than partial shear connection, is applicable to bridges as well as to buildings. Partial encasement in concrete, *clause 8.3*, is for beams in buildings only. Full encasement can improve resistance to fire but is uncommon in UK practice. Its rational design of shear connection and for crack control are difficult. It is outside the scope of EN 1994.

Application rules are given for resistance to bending, vertical shear, and combined bending and shear. Guidance on combinations of action effects, including torsion, is available (Tan and Uy, 2011). Design rules for bending combined with axial tension or compression are available (Vasdravellis *et al.*, 2015), and have been summarised (Johnson, 2019).

Beams with openings in steel webs are treated in *Annex D*. The much smaller holes needed for fasteners may be neglected (EN 1993-1-1, 8.2.2.1(1)).

8.2.1 Bending resistance

This clause covers three different methods for resistance of cross-sections, based on plastic theory, elastic analysis and non-linear theory, which should not be confused with non-linear global analysis.

Clause 8.2.1.1(1) refers to 'plastic design bending resistance'. Its calculation, *clause 8.2.1.2(1)*, involves a rectangular block of compressive stress, Figure 5.1, that is different from the rectangular stress block defined in clause 8.1.2 of EN 1992-1-1. That stress block is inconvenient for use with composite cross-sections, because the region near the neutral axis assumed to be unstressed is often occupied by a steel flange, and algebraic expressions for resistance to bending become complex (Figure 8.2). *Clause 8.2.1.1(1)*

From EN 1992-1-1 clause 5.1.6, $f_{\rm cd} = \eta_{\rm cc}k_{\rm tc}f_{\rm ck} / \gamma_{\rm C}$ (5.3) in EN 1992-1-1

The same expression is used in Eurocode 4, but the definition of the coefficient $\eta_{\rm cc}k_{\rm tc}$ in *clause 5.1(3)* is different. There is discussion under *clause 5.1(3)*.

Figure 8.2. Inconvenient stress distribution for $M_{pl,Rd}$ with a Eurocode 2 stress block for concrete

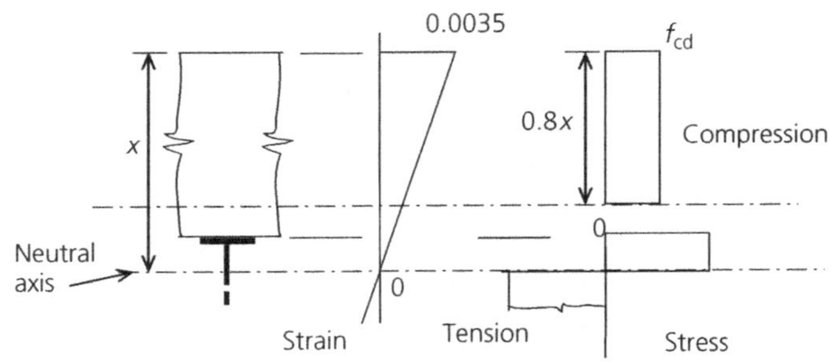

Clause 8.2.1.1(2) ***Clause 8.2.1.1(2)*** requires that the resistance of hogging moment regions in Class 3 or 4 is determined by elastic theory. These occur mainly in bridges because, in buildings, most Class 3 sections can be replaced by an effective section in Class 2, and Class 4 is rarely used.

Clause 8.2.1.1(3) For elastic and non-linear theory, the assumption that composite cross-sections remain plane is always permitted (***clause 8.2.1.1(3)***) because the methods are designed to ensure that the effects of longitudinal slip are either negligible or have been allowed for. There is no requirement for longitudinal slip to be determined. This would be difficult because the stiffness of shear connectors is not known accurately, especially where the slab is cracked.

Clause 8.2.1.1(5) For beams with curvature in plan sufficiently sharp for torsional moments not to be negligible, ***clause 8.2.1.1(5)*** gives no guidance on how to allow for the effects of curvature. In analysis from first principles, checks for beams in buildings can be made by assuming that the changing direction of the longitudinal force in a flange (and a web, if significant) creates a transverse load on that flange, which is then designed as a horizontal beam to resist that load. A steel bottom flange may require horizontal restraint at points within the span of the beam, and the shear connection should be designed for both longitudinal and transverse forces. Transverse bending of steel flanges, to which
Clause 8.2.1.1(7) ***clause 8.2.1.1(7)*** refers, occurs in some types of beams for shallow floors. This subject is also treated in Tan and Uy (2011).

Plastic moment of resistance

Clause 8.2.1.2 ***Clause 8.2.1.2*** gives the most widely used method for beams in buildings. It includes bending resistance with the slab in tension, referred to here as 'hogging bending', as well as resistance to sagging bending. Both here and in Eurocode 4, wording is often based on an implied assumption that the concrete in beams is mostly above the steel section, rather than below it. The method is available for cross-sections in Classes 1 to 3 because of the hole-in-web method (*clause 7.5.2(3)*), which is discussed in Example 8.1.

This clause is for beams with 'full interaction' (*paragraph (1)(a)*), which is a different assumption from 'full shear connection' (*clause 3.1.26*). It means that no account need be taken of slip or separation at the steel-concrete interface. Partial shear connection, for regions in sagging bending only, is the subject of *clause 8.2.1.3*.

Clause 8.2.1.2(1)(c) It is usual to neglect slab reinforcement in compression (***clause 8.2.1.2(1)(c)***). If it is included, and the concrete cover is little greater than the bar diameter, consideration should be given to possible buckling of the bars. Detailing rules for reinforcement in concrete walls given in clause 12.7(3) of EN 1992-1-1 may be relevant.

Cross-sections with composite slabs

Clause 8.2.1.2(1)(d) ***Clause 8.2.1.2(1)(d)*** says that 'the effective area of concrete in compression' resists a stress f_{cd} over the whole depth to the plastic neutral axis. The definition of f_{cd} is different from that in the previous Eurocode 4, as explained under *clause 5.1(3)*.

For sagging bending, the effective thickness of the concrete flange can depend on the angle θ between the directions of span of the beam and of the slab. For a flange that is a composite slab of total thickness h_{cs}, with profiled sheeting that has small top rib above the main top flange, the depths of the sheeting are labelled h_p and h_{pg} (shown in *Figure 8.21*).

Where the ribs run transverse to the beam, as in *Figure 8.1(d)* and *(e)* ($\theta = 90°$), the effective thickness of concrete $h_{eff,c}$ is $h_{cs} - h_{pg}$. For parallel sheeting, $\theta = 0$ and the troughs between the ribs can be included within the effective area if the plastic neutral axis is sufficiently low in the cross-section, but it is simpler to assume $h_{eff,c} = h_{cs} - h_p$. For ribs at other angles, θ, no evidence from research is known to the authors. The effective area of concrete within an effective width of flange may be taken as the full area above the ribs plus $\cos^2\theta$ times the area of concrete within the ribs, with $\cos^2\theta$ taken as zero where $\theta > 60°$. This method was given in former British standard BS 5950-3.1. Similar situations can arise where the concrete flange is partly or wholly composed of hollow-core precast slabs (see Annex J of this guide).

Plastic neutral axis
This clause includes the first of the many uses of the term 'plastic neutral axis' in Eurocode 4. It is not defined in *clause 3* or in Eurocode 2. It lies either at one edge of the rectangular stress block used for concrete in compression or within the steel section, as in *Figure 8.3*. With partial shear connection there will be two plastic neutral axes and it is not always clear from the text which one is being referred to.

Small concrete flanges
The method of *clause 8.2.1.2* is based on the assumption that where the concrete slab is in compression, the whole effective areas of steel and concrete can reach their design strengths before the concrete begins to crush. This may not be so if the concrete flange is small compared with the steel section. This lowers the plastic neutral axis and so increases the maximum compressive strain at the top of the slab for a given tensile strain in the steel bottom flange. Premature crushing may occur. This effect increases with the yield strength of the steel section.

Clause 8.2.1.2(2)(b)

The method of ***clause 8.2.1.2(2)(b)*** is simpler than the use of *clause 8.2.1.5* and should be used for beams within its scope. For steel grades S235 to S460, it applies a reduction factor β to the full-interaction bending resistance $M_{pl,Rd}$, subject to the conditions given. The factor depends on the ratio of z_{pl}, the depth of the plastic neutral axis, to the total depth of the beam, h. The greatest reduction is for grade S460 steel. For concrete flanges with effective width of at least 1.0 m, it begins at $z_{pl}/h = 0.15$, with a maximum reduction of 10% at $z_{pl}/h = 0.4$, and higher ratios of z_{pl}/h are not allowed. Most ratios z_{pl}/h for conventional T-beams are below 0.15, but this reduction can affect slim-floor design. The basis for these rules is given by Schäfer *et al.* (2021).

Clause 8.2.1.2(4)

In comparison with the previous Eurocode 4, this rule is more severe for steel grades lower than S420, but there are new exemptions, in ***clause 8.2.1.2(4)***. The simplest of these allows β to be taken as 1.0 for cross-sections where $A_c f_{cd}/(A_a f_{yd}) \geq 2$. This is unlikely in practice unless the concrete is very strong – for example, for C30/35 concrete and S355 steel, the area of concrete would have to be more than 40 times the area of steel, but for concrete grade C70/85 the ratio is 17.

For a cross-section where β applies, the compressive force in the concrete slab when $M_{Ed} = \beta M_{pl,Rd}$ is likely to be less than that required for resistance $M_{pl,Rd}$. However, the number of shear connectors needed for $M_{pl,Rd}$ should not be reduced. The design of the shear connection is discussed below.

The risk of premature crushing also led to the rules for the factor α_M, given in *clause 8.8.3.6(1)* for composite columns, and to concern over redistribution of elastic moments in partially encased beams (*clause 7.4.4(3)(d)*). Its effect on the rotation capacity of plastic hinges led to a rule given in *clause 7.4.5(4)(g)* which is applicable where rigid-plastic global analysis is used.

Ductility of reinforcement
Clause 8.2.1.2(5)

Reinforcement with insufficient ductility to satisfy *clause 7.5.1(5)* and welded mesh that is in ductility Class A should not be included within the effective section for plastic resistance of beams in Class 1 or 2 (***clause 8.2.1.2(5)***). Tests have shown that the bars can fracture, as explained in comments on *clause 5.2(1)*. Steel fibre reinforcement is outside the scope of EN 1994-1-1, but its use is not excluded.

Profiled steel sheeting

Clause 8.2.1.2(6)

The contribution to the plastic moment of resistance of a beam from profiled steel sheeting in compression is ignored (***clause 8.2.1.2(6)***) because at large strains its resistance can be much reduced by local buckling. Profiled sheeting with troughs that are not parallel to the span of a beam is ineffective in tension. This is because deformation could arise from change in shape of the profile rather than strain resulting from stress. Where the troughs are parallel, resistance to tension may still be difficult to achieve.

Beams with partial shear connection in buildings

Clause 8.2.1.3(1)
Clause 8.2.1.3(2)

The background to the use of partial shear connection is explained in comments on *clause 8.1.1*. It is permitted only for concrete flanges in compression (***clause 8.2.1.3(1)***). Where the slab is in tension the shear connection must be sufficient to 'ensure yielding' (***clause 8.2.1.3(2)***) of the reinforcement within the effective section. This applies even where there is more reinforcement than that required for the design hogging moment. Full shear connection is required in hogging regions of composite beams for several reasons.

- The bending moment may be larger than predicted because the concrete has not cracked or, if it has, because of tension stiffening.
- The yield strength of the reinforcement is likely to exceed the design value f_{sd}.
- Tests show that at high curvatures, strain hardening occurs in the reinforcement.
- The design rules for lateral-torsional buckling do not allow for the effects of partial interaction.

The words 'hogging bending' in this clause imply that the concrete slab is above the steel beam, as is usual in buildings. A better wording is 'regions where the slab is in tension' to allow for the use of double composite action.

Ductility and initial stiffness of shear connectors

The basic condition for the use of partial shear connection is that the bending resistance must not fall below the design value until after the curvature has reached the minimum value relied on in the method of global analysis used – for example, the use of redistribution of moments relies on curvatures beyond the elastic range. A condition for this is

$$\text{longitudinal slip required (i.e. relied on in design)} \leq \text{slip available} \quad \text{(X8.2)}$$

It has been shown by extensive numerical analyses, checked against test results (Aribert, 1990; Johnson, 1991), that the slip required increases with the span of the beam and, of course, as the number of shear connectors is reduced. The latter parameter is represented by the ratio of the number of connectors provided within a critical length n to the number n_f required for 'full shear connection' (defined in *clause 3.1.26*) – that is, the number that will transmit the force $N_{c,f}$ to the slab (see *Figure 8.2*). A reduced number, n, will transmit a reduced force, N_c. Thus, for connectors of a given shear resistance, the 'degree of shear connection' is

$$\eta = N_c / N_{c,f} = n / n_f \quad \text{(X8.3)}$$

Subsequent studies of slip required have generally confirmed this aspect of the Eurocode rules. It has been found (Banfi, 2006) that slip under point loads is less than under the distributed loading used in earlier work. For tapering members, it was found that in comparison with a uniform beam with the same shear connection, loading and maximum depth, the maximum slip was significantly greater. The rule in *clause 8.1.1(6)* is intended to allow for this and, in practice, other checks would cause the design loading for the tapered beam to be lower.

The new 'ductility categories' for available slip of shear connectors are discussed under *clause 5.4.2.1*. Design rules for studs were initially based on a slip capacity of 6 mm (now in category D2, 'ductile'). Then, rules for studs with a slip capacity of 10 mm were introduced (Couchman, 2015) as complementary information. These are now in category D3, 'high ductility'. There is a new category D1, which may be 'brittle' or 'flexible'. It enables Eurocode 4 to be used for a wider range of connectors.

Calculation models

Clause 8.2.1.3(3) limits the use of partial shear connection to categories D2 and D3 and defines the 'equilibrium' method of calculation for partial-interaction resistance M_{Rd}. For $\eta < 1$, the concrete stress block has a reduced depth, and a neutral axis at its lower edge. For longitudinal equilibrium, part of the steel beam must also be in compression, so it too has a neutral axis. The model assumes no separation of the slab from the beam, so their curvatures must be the same. The strain distribution is thus as shown in Figure 8.3, which is for the situation shown in *Figure 8.4* in EN 1994-1-1. At the interface between steel and concrete there is *slip strain* – that is, a rate of change of longitudinal slip. Neither the slip at this point nor the slip strain need be calculated in practice. *Clause 8.2.1.3(3)*

Figure 8.3. Plastic stress and strain distributions under sagging bending for partial shear connection

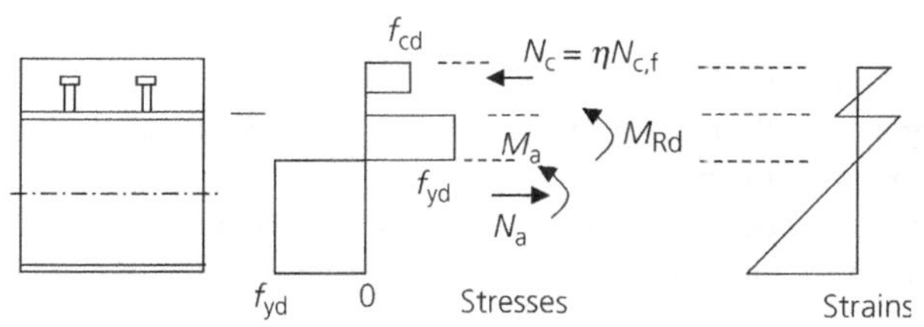

For non-ductile connectors, the rules of *clauses 8.2.1.4* and *8.2.1.5* may be used, where further comment is given. The rules for their spacing, in *clause 8.6.3.2*, are more restrictive than for ductile connectors.

Clauses 8.2.1.3(4) and ***8.2.1.3(5)*** give two relationships between resistance moment M_{Rd} and degree of shear connection, assuming that reduction factor β (*clause 8.2.1.2(2)*) is not applicable. Calculations using the equilibrium method above give curve AHC in the upper part of Figure 8.4(a), in which $M_{pl,a,Rd}$ is the plastic resistance of the steel section. The line AC, the 'interpolation' method, is a simpler and conservative approximation to it. Their use is now illustrated *Clause 8.2.1.3(4)* *Clause 8.2.1.3(5)*

The lower half of Figure 8.4(a) gives some of the limits to the use of partial shear connection from *clause 8.6.3.3*. It shows the variation of minimum degree of shear connection with yield strength of the structural steel for a cross-section with $M_{Ed} = M_{Rd}$, and the much higher degree of shear connection needed where the steel beam has a relatively small top flange.

Figure 8.4. Design methods for partial shear connection

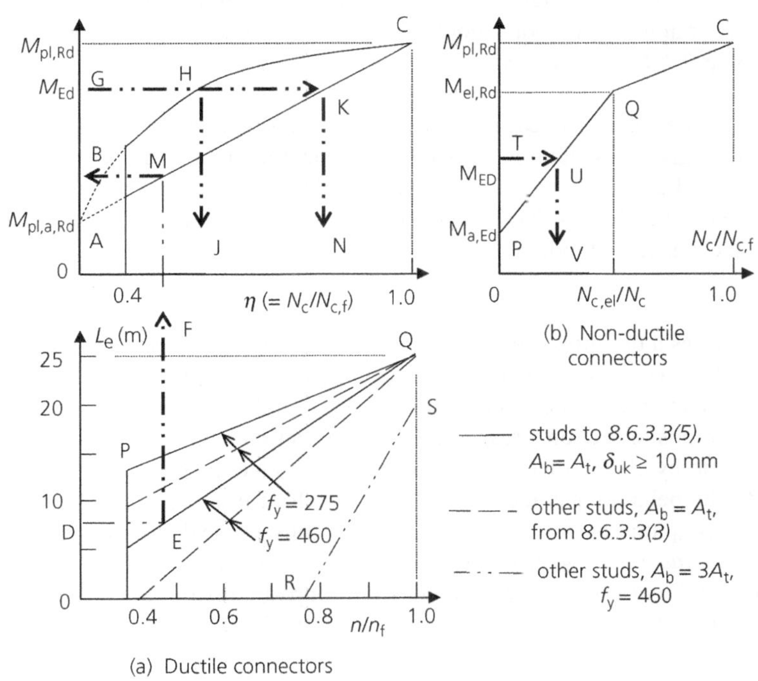

The upper part of Figure 8.4(a) is from *Figure 8.5* in EN 1994-1-1. *Figure 8.8* has the same axes, and is compared with *Figure 8.5* in Figure 8.5, which is discussed under *clause 8.2.1.6*.

Clause 8.2.1.3(6)

Clause 8.2.1.3(6) applies where the reduction factor β is applicable – for example, in slim-floor construction. It refers to the method in *paragraph (7)* shown in *Figure 8.6*.

Outline of a typical design procedure

The example is a simply supported beam of span L, with a Class 1 section at mid-span. The troughs of profiled sheeting, if used, constrain the layout of shear connectors and lead to a different design process (see Example 8.2). Here, it is assumed that the concrete flange is a solid slab. With full shear connection the resistance is $M_{pl,Rd}$. The steel section has flanges of equal area, and the loading is uniformly distributed. Stud connectors are to be used that satisfy the conditions of *clause 8.6.3.3(2)*, so that its rules for partial shear connection apply.

(*a*) Find the minimum shear connection, $(n/n_f)_{min}$, for which the connectors are ductile, from *clause 8.6.3.3(2)* (e.g. route DE in Figure 8.4(a), with $L_e = L$). Route FMB then gives a resistance to bending, by the interpolation method.

(*b*) If that resistance exceeds the design moment M_{Ed}, the specified minimum degree of shear connection is sufficient. Calculate the number of connectors for full shear connection and then, from $(n/n_f)_{min}$, the number required.

(*c*) If that resistance, point B, is below M_{Ed}, as shown in Figure 8.4(a), use of η_{min} is too low. Route GHKN gives the required degree of shear connection by the interpolation method.

(*d*) The degree of shear connection required can be reduced by using the equilibrium curve AHC instead of the interpolation curve for both of steps (*a*) and (*c*), though the two curves are often closer together than in Figure 8.4. The calculation to find point H is best done by trial and error: guess η and find M_{Rd}, and repeat until M_{Rd} just exceeds M_{Ed}.

(*e*) The spacing of the n connectors is now considered, along the length L_{cr} between the two relevant critical cross-sections (here, mid-span and a support). If the conditions of *clause 8.6.3.2(5)* are satisfied, which include $M_{pl,Rd} \leq 2.5\ M_{a,pl,Rd}$, uniform spacing may be used. Otherwise, an intermediate critical section must be chosen (*clause 8.6.3.2(6)*) or the spacing must be related to the elastic distribution of longitudinal shear (*clause 8.6.3.2(9)*).

There is much interaction between *clause 8.2.1.3* and *clauses 8.6.3.2* and *8.6.3.3*. The use of partial shear connection is illustrated in Example 8.6, which follows the comments on *clause 8.6*.

Flexible non-ductile connectors

Clause 8.2.1.4

These are connectors in ductility category D1. Plastic behaviour of the shear connection can no longer be assumed. Elastic or non-linear theory should be used to determine resistance to bending. Provisions are given in ***clause 8.2.1.4*** and *clause 8.2.1.5*.

The effect of slip at the steel-concrete interface is to increase curvature, and usually to reduce longitudinal shear, for a given distribution of bending moment along a span. Where connectors are not 'ductile', slip must be kept small, so it is rational to neglect slip when calculating longitudinal shear. This is why *clause 8.2.1.5(2)* says that composite cross-sections should be assumed to remain plane. It applies also to the elastic method.

Elastic resistance to bending

Clause 8.2.1.4(2)

Clause 8.2.1.4(6) gives a definition of $M_{el,Rd}$ that arises from the use of unpropped construction for composite beams. The elastic resistance to bending depends on the proportion of the total load that is applied before the member becomes composite. Let $M_{a,Ed}$ and $M_{i,Ed}$ be the design bending moments for the steel and composite sections, respectively, for a section in Class 3. The resulting total bending stresses are found by elastic theory. If any of them is above the relevant limiting stress given in ***clause 8.2.1.4(2)***, the behaviour is inelastic. If all are below, it is elastic. To find the elastic resistance to bending, $M_{el,Rd}$, one or both of $M_{a,Ed}$ and $M_{i,Ed}$ must be scaled up or down until one of the limiting stresses in *clause 8.2.1.4(2)* is just reached. The summation of elastic stresses for sagging bending is shown in *Figure 8.7*, which includes unpropped construction, shrinkage, and variable and permanent action effects. The effects of creep, usually allowed for in the modular ratio, are not shown separately.

To enable a unique result to be obtained, ***clause 8.2.1.4(6)*** says that $M_{i,Ed}$ is to be scaled by a factor k_{el}, and $M_{a,Ed}$ left unchanged. This is because $M_{a,Ed}$ is mainly from permanent actions, which are less uncertain than the variable actions whose effects comprise most of $M_{i,Ed}$. *Clause 8.2.1.4(6)*

Unpropped construction normally proceeds by stages, which may have to be considered individually in bridge design. For simply supported spans in buildings, it is usually sufficiently accurate to assume that the whole of the wet concrete is placed simultaneously on the bare steelwork. The weight of formwork is, in reality, applied to the steel structure and removed from the composite structure. This process leaves self-equilibrated residual stresses in composite cross-sections. For composite beams in buildings, these can usually be ignored in calculations for the final situation.

Symbols $M_{i,PT,Ed}$ and $N_{i,PT,Ed}$ are defined in *clause 8.2.1.4(6)* as action effects 'developed in time due to creep'. They are the changes found in global analyses before and after creep arising from changes in the relative stiffnesses of members. They are not considered for buildings when the simplified modular ratios are used.

One permanent action that influences $M_{i,Ed}$ is shrinkage of concrete. *Clause 8.2.1.4(5)* enables the primary stresses to be neglected in cracked concrete, but the implication is that they should be included where the slab is in compression. This provision, which affects $M_{el,Rd}$, should not be confused with *clause 7.4.2.2(8)*, which concerns global analysis to determine the secondary effects of shrinkage in statically indeterminate structures. Secondary effects are defined in *clause 4.3.3(2)*.

Non-linear resistance to bending

Unlike *clause 8.6.2(4)*, ***clause 8.2.1.5*** does not refer to load-slip properties of shear connectors. It says that cross-sections may be assumed to remain plane, with slip allowed for by the rules for partial shear connection. A note to *clause 8.2.1.5(2)* refers to 'more advanced methods'. This means unspecified finite-element methods. *Clause 8.2.1.5*

There are two methods with calculations done at the critical sections for the design bending moments. The first, given in ***clauses 8.2.1.5(1)*** to ***8.2.1.5(5)***, enables the resistance of a section to be determined iteratively from the stress-strain relationships of the materials. A strain distribution is assumed for the cross-section and the resulting stresses determined. Usually, the assumed strain distribution will have to be revised to ensure that the stresses correspond to zero axial force on the section. Once this condition is satisfied, the bending moment is calculated from the stress distribution. This may show that the design bending moment does not exceed the resistance, in which case the calculation for bending resistance may be terminated. *Clause 8.2.1.5(1)* *Clause 8.2.1.5(5)*

Otherwise, a general increase in strain should be made and the calculations repeated. For concrete, EN 1992-1-1 gives ultimate strains for concrete and reinforcement which eventually limit the moment resistance.

This procedure is best done using software. The second method, for sections in Class 1 or 2 only, is simpler, as given in ***clause 8.2.1.6.*** This is based on three points on the curve relating longitudinal force in the slab, N_c, to design bending moment M_{Ed} that are easily determined. With reference to Figure 8.4(b), which is based on *Figure 8.8* for unpropped construction, and repeated in Figure 8.5, these points are *Clause 8.2.1.6*

- P, where the composite member resists no bending moment, so $N_c = 0$
- Q, which is defined by the results of an elastic analysis of the section
- C, based on plastic analysis of the section.

Accurate calculation shows QC to be a convex-upwards curve, so the straight-line QC is a conservative approximation. *Clause 8.2.1.6* thus enables hand calculation to be used.

Figure 8.5 and *Figure 8.8* are shown combined in Figure 8.5 to illustrate the following comments on the difference between them. Both of them apply where the concrete slab is in compression, and provide approximate values for resistance moment $M_{Rd}(\eta)$ as a function of degree of shear connection.

Figure 8.5. Resistance moment as function of degree of shear connection

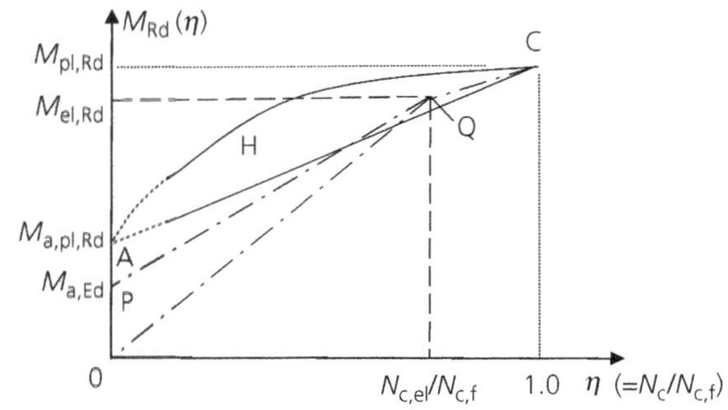

Curve AHC is based on rigid-plastic theory for ductile connectors, ignoring elastic deformation and the effects of sequence of construction. It is not applicable for degrees of shear connection below the minima given in *clause 8.6.1.2* – for example, if $\eta = 0.2$, the resistance $M_{Rd}(\eta)$ might not exceed that of the steel beam, $M_{pl,a,Rd}$, because excessive slip and failure of the shear connection could have occurred. Line AC is a conservative approximation.

The lines PQC for unpropped construction are applicable to both ductile and non-ductile connectors. For low degrees of shear connection, elastic theory is used (line PQ) until the applied moment M_{Ed} reaches resistance $M_{el,Rd}$, where the compressive force in the slab is $N_{c,el}$, which may be governed by any one of the stress limits in *clause 8.2.1.4(2)*. Slip of the shear connection is ignored. For propped construction, PQ is replaced by 0Q.

Shear connection with non-linear resistance to bending

Computations based on the stress-strain curves referred to in *clause 8.2.1.5(3)* to *8.2.1.5(5)* lead to a complete moment-curvature curve for the cross-section, including a falling branch. From the definition of partial shear connection in *clause 3.1.27*, the number of connectors for full shear connection can be found only from the stress distribution corresponding to the maximum resistance moment. This is why $N_{c,f}$ in Figure 8.4(b) is based on plastic analysis to *clause 8.2.1.2.*

For a given design moment M_{Ed}, which includes $M_{a,Ed}$, the required ratio $N/N_{c,f}$ is given by route TUV in Figure 8.4(b), with point U in either of lines PQ and QC.

Clause 8.2.1.5 does not refer to the spacing of shear connectors when moment resistance is determined by non-linear theory. The theory is based on plane cross-sections (*clause 8.2.1.5(2)*) and neglects slip. When using *clause 8.2.1.5(6)* with $M_{Rd} \leq M_{el,Rd}$, the spacing of connectors should follow the rules given in *clause 8.6.5*. Where $M_{Rd} > M_{el,Rd}$, the rules in *clause 8.6.4* apply.

For ultimate-limit-state verifications for buildings, these complications show that design methods based on elastic behaviour are best avoided.

Calculations for an effective Class 2 cross-section using the hole-in-web method

This method, given in clause 8.2.2.4 of EN 1993-1-1, enables a Class 3 web to be treated as Class 2 by reducing its area. Figure 8.3 of EN 1993-1-1 is for sagging bending. It does not give the depth of the hole and the location of the new plastic neutral axis (pna). These, and some limits to the use of the method, are now given.

The method reduces the depth of web in compression from $\alpha_c c$ to $40t_w\varepsilon$. For longitudinal equilibrium, there must be an equal reduction in the depth of the web in tension. (Here, ε is the usual correction factor for yield strength of steel, and the other symbols are shown in Figure 8.7.) Hence, the depth of the hole is

$$h_h = 2\,(\alpha_c c - 40t_w\varepsilon) \tag{X8.4}$$

and the new plastic neutral axis is $h_h/2$ higher in the cross-section than the former one, as shown in Figure 8.7(a).

Limits to the use of the hole-in-web method

From clause 8.2.2.4(1) of EN 1993-1-1, the new plastic neutral axis should be within the depth c of the web. Otherwise, the assumption that the compression zone is within a web of uniform width t_w is wrong. Hence, a condition for use of the method is $h_h + 40t_w\varepsilon \leq c$. With h_h from Equation X8.4

$$c/t_w\varepsilon \leq 40/(2\alpha_c - 1) \quad \text{(X8.5)}$$

For example, if $\alpha_c = 0.8$, then, $c/t_w\varepsilon \leq 67$. This limits the use of the method where a web is mainly in compression. A second condition is that the depth of the hole cannot be negative, so from Equation X8.4

$$c/t_w\varepsilon \geq 40/\alpha_c \quad \text{(X8.6)}$$

These two conditions limit the use of the method to web slendernesses $c/t_w\varepsilon$ within a range of values that depends on ratio α_c, as shown in Figure 8.6.

Figure 8.6. Applicability of the hole-in-web method

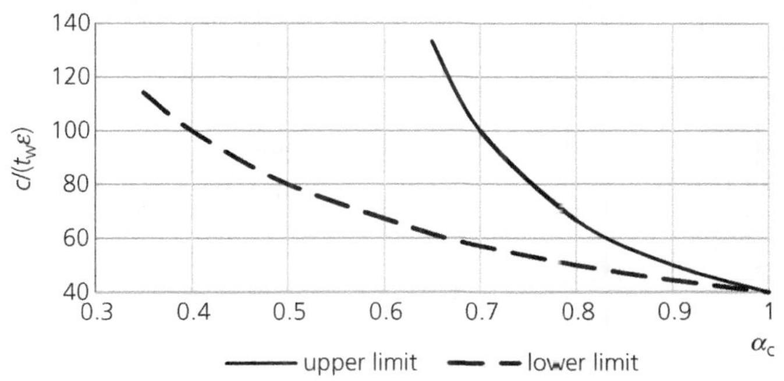

One would expect the depth of the hole to be zero for a web at the boundary between Class 2 and Class 3. Then, for a web with less than half of its depth in compression ($\alpha_c < 0.5$), Table 7.3 of EN 1993-1-1 gives the boundary as $c/t_w\varepsilon = 41.5/\alpha_c$. From Equation X8.4, $h_h = 3t_w\varepsilon$ rather than zero. This is a small error on the safe side, as it reduces the resistance of the cross-section to hogging bending. It is caused by the use of a single size of stress block, $20t_w\varepsilon$, for the whole range of Class 3 webs. Where $\alpha_c > 0.5$, the class boundary is

$$c/t_w\varepsilon = 188/(6.53\alpha_c - 1) \quad \text{(X8.7)}$$

As α_c increases, the depth of the hole for the class boundary decreases and becomes negative at $\alpha_c \approx 0.55$. Then, α_c $c/t_w\varepsilon = 39.9$ from Equation X8.7 and the condition of Equation X8.6 is not met. Hence, these rules are practicable.

Depth of the hole in the web

Where the steel section is symmetrical, a further result may be useful. The depth of web in compression, $40t\varepsilon$, includes the depth needed to balance the tensile force in the reinforcement, which is

$$h_r = A_s f_{sd} / t_w f_{yd}$$

The tension in the top flange, including web fillets, balances the compression in the bottom flange, so for longitudinal equilibrium, the depth of web in tension is

$$h_t = 40t\varepsilon - h_r \quad \text{(X8.8)}$$

The total depth of the web is c, so the depth of the hole is

$$h_h = c - 40t\varepsilon - h_t = c - 80t\varepsilon + h_r \quad \text{(X8.9)}$$

Example 8.1: resistance moment in hogging bending for an effective Class 2 cross-section

A typical cross-section near an internal support of a continuous composite beam with transverse profiled sheeting is assumed (Figure 8.7(a)). The plastic and elastic methods of calculation for the hogging moment of resistance and the use of an effective web to *clause 7.5.2(3)*, discussed above, are illustrated.

The depth of the 'web', as used here, is the depth between flange fillets (or welds), defined as c in Table 7.3 of EN 1993-1-1, not the clear depth between the flanges. Also, from EN 1993-1-1, $\alpha_c c$ is the depth of the web in compression, for a section in Class 1 or 2.

Figure 8.7. Plastic resistance moment in hogging bending, for $A_s = 296$ mm^2: (a) Class 3 cross-section with hole in web; (b) stress blocks for $M_{pl,Rd}$ for Class 2 cross-section; (c) stress blocks for web for $M_{pl,Rd}$ for Class 3 cross-section (units: mm and kN)

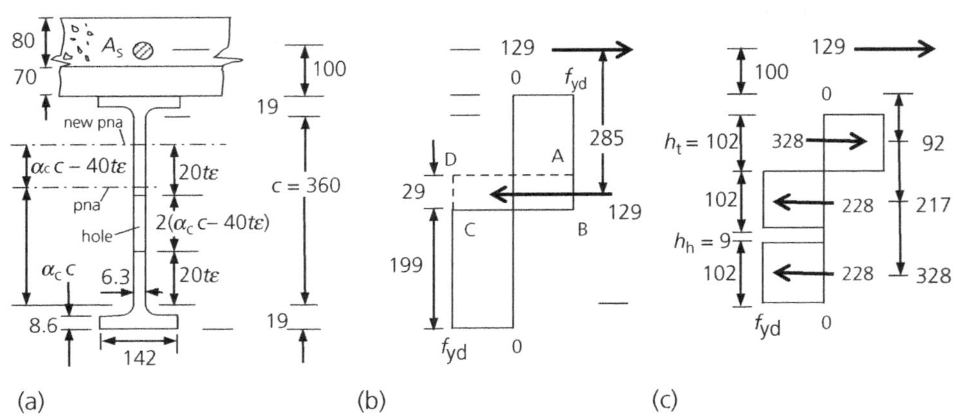

Data for the calculations, in addition to the dimensions shown in Figure 8.7 are

- structural steel: $f_y = 355$ N/mm^2, $\gamma_M = 1.0$, so $f_{yd} = 355$ N/mm^2, $\varepsilon = (235/355)^{1/2} = 0.81$
- reinforcement: $f_{sk} = 500$ N/mm^2, $\gamma_s = 1.15$, so $f_{sd} = 435$ N/mm^2
- steel section: 406 × 140 UB39 with $A_a = 4940$ mm^2, $I_a = 124.5 \times 10^6$ mm^4, $r = 10.2$ mm.

The profiled sheeting plays no part in these calculations. For simplicity, reinforcement is assumed to be in one layer, with propped construction.

Classification of the cross-section

Clause 7.5.1(1) refers to EN 1993-1-1, where Table 7.3 applies. For the bottom flange

$$c = (142 - 6.3)/2 - 10.2 = 57.6 \text{ mm}$$

$$c/t_f\varepsilon = 57.6/(8.6 \times 0.81) = 8.3$$

This is less than 9, so the flange is in Class 1 irrespective of the area of slab reinforcement.

The web depth between fillets is $c = 360$ mm, so

$$c/t_w\varepsilon = 360/(6.3 \times 0.81) = 70.5$$

This is less than 72, so when $A_s = 0$, the web is in Class 1. Addition of reinforcement increases the depth of web in compression, so its class depends on A_s. For this symmetrical steel section, $\alpha_c > 0.5$, and from Table 7.3 in EN 1993-1-1, the limit for Class 2 is

$$c/t_w\varepsilon = 188/(6.53\alpha_c - 1)$$

For $c/t_w\varepsilon = 70.5$, this gives $\alpha_c = 0.546$. Here, a cross-section that is just into Class 3 is chosen, with the assumption $\alpha_c = 0.58$. Then, the depth of web in tension is 0.42 $\alpha_c c$, and the difference between the tensile and compressive forces resisted, if the web were in Class 2, gives the area of reinforcement

$$(0.58 - 0.42) \times 360 \times 6.3 \times 355 = A_s \times 435 \text{ whence } A_s = 296 \text{ mm}^2$$

$M_{pl,Rd}$ for the section assumed to be in Class 2 (for comparison only)

The stress blocks for the web are shown in Figure 8.7(b). For a rolled section, the plastic section modulus is usually found from tables. Here

$$W_{pl} = 0.721 \times 10^6 \text{ mm}^3, \text{ so } M_{pl,a,Rd} = 0.721 \times 355 = 256 \text{ kN m}$$

F_s, the force in the reinforcement at yield, is: $F_s = 296 \times 0.435 = 129$ kN

The depth of web for this force is: $h_r = 129/(6.3 \times 0.355) = 57.6$ mm

To balance the force F_s, the stress in a depth of web $h_r/2$ changes from $+f_{yd}$ to $-f_{yd}$. This is shown as ABCD in Figure 8.7(b). The lever arm for the forces F_s is 285 mm, so taking moments

$$\mathbf{M_{pl,Rd}} = 256 + 129 \times 0.285 = \mathbf{293 \text{ kN m}}$$

$M_{pl,Rd}$ for the Class 3 section treated as Class 2 by the hole-in-web method

The results above for $M_{a,pl,Rd}$, F_s and h_r apply.

The contribution of the web is deducted from $M_{pl,a,Rd}$

$$M_{pl,a,fl,Rd} = 256 - 0.36^2 \times 6.3 \times 355/4 = 183.5 \text{ kN m}$$

This value is not calculated directly because of the complex shape of each 'flange', which includes the web fillets, or, for a plate girder, a small depth of web.

From Equation X8.9, the depth of the hole is $h_h = 360 - 80 \times 6.3 \times 0.81 + 57.6 = 9.4$ mm

From Equation X8.8, the depth of web in tension is $h_t = 40 \times 6.3 \times 0.81 - 57.6 = 146.5$ mm

and the force in it is

$$F_t = 146.5 \times 6.3 \times 0.355 = 328 \text{ kN}$$

The stress blocks in the web are shown in Figure 8.7(c). Taking moments about the bottom of the slab,

$$\mathbf{M_{pl,Rd}} = 183.5 + 129 \times 0.1 - 328 \times 0.092 + 228\ (0.217 + 0.328) = \mathbf{291 \text{ kN m}}$$

This is only slightly less than the result for the Class 2 member, because this section is near the class boundary.

Elastic resistance to bending

For simplicity, $M_{el,Rd}$ has been calculated assuming propped construction. The stress in the reinforcement governs, and $\mathbf{M_{el,Rd}} = \mathbf{231 \text{ kN m}}$. Thus, the benefit from the change from an elastic to a plastic model far exceeds the reduction in plastic resistance caused by assuming part of the web to be unstressed.

8.2.2 Resistance to vertical shear

Clause 8.2.2

Clause 8.2.2 is for beams without large web openings (covered in *Annexes D* and *E*) or web encasement (*clause 8.3*). The methods given mostly assume that the whole of the vertical shear is resisted by the steel section, as in previous codes for composite beams. This enables the design rules of EN 1993-1-1, and EN 1993-1-5 where necessary, to be used. For cross-sections in Class 1 or 2 the use of a contribution from the concrete flange is not excluded. There are conditions (***clause 8.2.2.2(2)***), and reference to 'more complex design methods'. *Clause 8.2.2.2(2)*

There is no further guidance, and in practice it may be difficult to make use of the concrete slab. Neglect of its shear resistance can be conservative where the slab is in compression. Even where it is in tension and cracked in flexure, consideration of equilibrium shows that the slab must make some contribution, except where the reinforcement has yielded.

The proportion of shear resisted by the concrete slab increases with the ratio of the thickness of the slab to the depth of the steel beam and is significant for solid slabs where the ratio exceeds about 0.5. This can occur in shallow floor beams of the type treated in *Annex I*, on which comment is given.

Background to the design model for solid slabs and composite slabs with re-entrant decking is available (Vasdravellis and Uy, 2014).

Clause 8.2.2.3
Clause 8.2.2.3(2)
Clause 8.2.2.4

These rules for plastic resistance to vertical shear are presumably for cross-sections in Class 1 or 2 only. ***Clause 8.2.2.3*** refers to EN 1993-1-1 for the elastic resistances of all classes. For cross-sections in Class 3 or 4, ***clause 8.2.2.3(2)*** shows that the sequence of construction must be considered, so design would be complex where account is taken of a contribution from the concrete flange. For resistance to shear buckling, ***clause 8.2.2.4*** also refers to Eurocode 3, with notes on its application.

Clause 8.2.2.4(2)

In composite plate girders with vertical stiffeners, the concrete slab can contribute to the anchorage of a tension field in the web (Allison *et al.*, 1982), but the shear connectors must then be designed for vertical forces (***clause 8.2.2.4(2)***). The contribution to shear resistance in *clause 8.2.2.2* is not the 'contribution' referred to here. The simpler alternative is to follow Eurocode 3, ignoring both the interaction with the slab and vertical tension across the interface.

Bending and vertical shear

Clause 8.2.2.5
Clause 8.2.2.5(1)

The methods of ***clause 8.2.2.5*** for cross-sections in Class 1 or 2 are summarised in Figure 8.8(a). Shear stress does not significantly reduce bending resistance unless the shear is quite high. For this reason, the interaction may be neglected until the shear force exceeds half the shear resistance (***clause 8.2.2.5(1)***). The value 'half' can be changed by a national annex.

Figure 8.8. Resistance to bending and vertical shear of cross-sections in Class 1 or 2 (dimensions in mm)

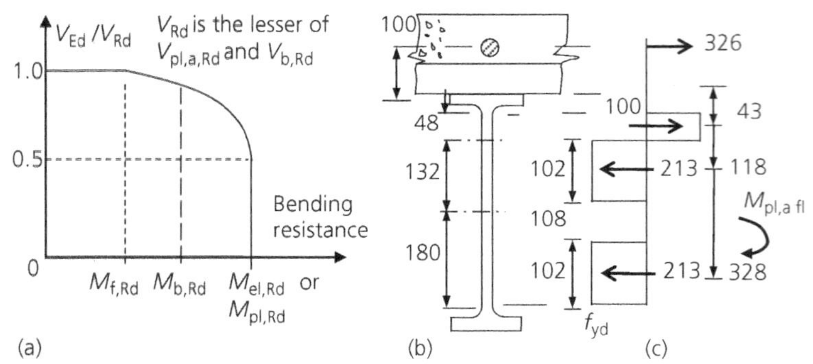

Clause 8.2.2.5(2)

Both EN 1993-1-1 and EN 1994-1-1 use a parabolic interaction curve. In ***clause 8.2.2.5(2)*** the reduction factor for the design yield strength of the web is $(1 - \rho)$, where

$$\rho = [(2V_{a,ED}/V_{pl,RD})-1]^2$$

For a design shear force equal to $V_{pl,a,Rd}$, the bending resistance is that provided by the flanges alone, denoted $M_{f,Rd}$.

Where the design yield strength of the web is reduced to allow for vertical shear, the effect on a Class 3 section in hogging bending is to increase the depth of web in compression. If the change is small, the hole-in-web model can still be used. For a higher shear force, the new plastic neutral axis may be within the top flange and the hole-in-web method is inapplicable.

The section is then treated as Class 3 or 4, and ***clause* 8.2.2.5(4)** may apply. It refers to EN 1993-1-5. Its symbols relate to the steel section only. For a composite section, longitudinal stresses are found by elastic theory. These lead to values of $M_{a,Ed}$, $N_{a,Ed}$ and $V_{a,Ed}$ acting on the steel section, which is then checked to EN 1993-1-5. It is fairly easy to check if a given combination of these action effects can be resisted, but calculation of bending resistance for a given vertical shear is difficult. *Clause 8.2.2.5(4)*

8.3 Resistance of cross-sections of beams with partial encasement for buildings

Encasement in concrete of the webs of steel beams is normally done before erection, casting one side at a time. This obviously increases the cost of fabrication, transport and erection, but when the details of the encasement satisfy *clause 7.5.3(2)*, it has several advantages for design.

- It provides complete fire resistance for the web and, with longitudinal reinforcement, compensation for the weakness of the bottom flange in fire, in accordance with EN 1994-1-2.
- It enables a Class 3 web to be upgraded to Class 2, and the slenderness limit for a Class 2 compression flange to be increased by 40% (*Table 7.2*).
- It widens significantly the range of steel sections that are not susceptible to lateral-torsional buckling (*clause 8.4.3(1)(h)*).
- It can increase resistance to vertical shear, to *clause 8.3.3(2)*.

8.3.1 General

Clause 7.5.3(2) requires web encasement to be 'capable of preventing buckling of the web'. The rules for encasement apply for webs up to the high slenderness limit given in ***clause* 8.3.1(2).** In practice, with encasement, steel sections in buildings are almost certain to be in Class 1 or 2. *Clause 8.3* is applicable only to these. *Clause 8.3.1(2)*

8.3.2 Bending resistance

The rules for resistance to bending, ***clause* 8.3.2**, correspond to those for uncased sections of the same class, except that the rules for lateral-torsional buckling (*clause 8.4.3*) are more liberal because encasement greatly improves resistance. The comments on *clause 8.4.2(7)* are relevant. However, it is possible for a beam within the scope of *clause 8.3* to be susceptible. A web-encased IPE 500A section in S420 steel is an example as its section parameter exceeds the limit of 13.9 in *Table 8.2*. *Clause 8.3.2*

Partial shear connection is permitted for a concrete flange, but not for web encasement. The resistance to longitudinal shear provided by studs within the encasement is found in the usual way, but no guidance is given on the contribution from bars that pass through holes in the web or stirrups welded to the web, in accordance with *clauses 7.5.3(2)* and *8.3.3(2)*, because their contribution may be reduced if the concrete around them is cracked. They are provided to ensure the integrity of the encased section. The load-slip properties of different types of shear connection should be compatible (*clause 8.6.1(6)*).

8.3.3 and 8.3.4 Resistance to vertical shear, and to bending and vertical shear

The rules in ***clause* 8.3.3** are based on the concept of superposition of resistances of composite and reinforced concrete members. The shear connection must be designed to ensure that the shear force is shared between the steel web and the concrete encasement. The references to EN 1992-1-1 are intended to ensure that the web encasement retains its shear resistance at a shear strain sufficient to cause yielding of the steel web. No account need be taken of web buckling in shear. Moment-shear interaction is treated in ***clause* 8.3.4** in a manner consistent with the rules for uncased sections. *Clause 8.3.3* *Clause 8.3.4*

8.4 Lateral-torsional buckling of composite beams

8.4.1 General

It is assumed in this section that in completed structures for buildings, the steel top flanges of composite beams will be stabilised laterally by connection to a solid concrete or composite slab (***clause* 8.4.1(1)**), following the rules on maximum spacing of connectors in *clause 8.6.10.5*. *Clause 8.4.1(1)*

Clause 8.4.1(2)

Any steel top flange in compression that is not so stabilised should be checked for lateral buckling (***clause 8.4.1(2)***) using EN 1993-1-1. This applies particularly during unpropped construction. Where profiled sheeting is in place, it may be possible to rely on it to restrain the top flange. Its required stiffness is given by Formula D.1 in Annex D of EN 1993-1-1.

To check a steel beam that is composite along only part of its length, the general method of *clause 8.4.2* is applicable. It needs a computed value of the elastic critical moment M_{cr}. No detailed guidance on the calculation of M_{cr} for a partly composite beam is given in either EN 1993-1-1 or EN 1994-1-1. Buckling of web-encased beams without a concrete flange has been studied (Lindner and Budassis, 2002). However, for buildings, the construction phase is rarely critical in practice, because the loading is so much less than the design total load.

Steel bottom flanges are in compression only in cantilevers and continuous beams. In beams designed as continuous, the length in compression may include most of the span, when that span is lightly loaded and one or both adjacent spans are fully loaded. Bottom flanges in compression should always be restrained laterally at supports (*clause 8.4.3(1)(f)*). It should not be assumed that a point of contraflexure is equivalent to a lateral restraint.

In a composite beam, the concrete slab provides lateral restraint to the steel member and also restrains its rotation about a longitudinal axis. Lateral buckling is always associated with distortion (change of shape) of the cross-section. Design methods for composite beams must take account of the bending of the web, Figure 8.9(b). They differ in detail from the method of clause 8.3.2 of EN 1993-1-1 but the same imperfection factors and buckling curves are used, in the absence of any better-established alternatives.

Clause 8.4.1(3)

The reference in ***clause 8.4.1(3)*** to EN 1993-1-1 provides a general method for use where neither of the methods of EN 1994-1-1 are applicable (e.g. for a Class 4 beam).

8.4.2 Verification of the resistance to lateral-torsional buckling of continuous composite beams with cross-sections in Class 1, 2 and 3

This general method of design is written with distortional buckling of bottom flanges in mind. It would not apply, for example, to a mid-span cross-section of a beam with the slab at bottom-flange level. Although not stated, it is implied that the span concerned is of uniform composite section, excluding minor changes such as reinforcement details and effects of cracking of concrete. The use of this method for a two-span beam is illustrated in Example 8.6.

Clause 8.4.2(1)
Clause 8.4.2(3)

The method is based closely on clause 8.3.2 of EN 1993-1-1. There is correspondence in the definitions of the reduction factor χ_{LT} (*clause 8.4.2(1)*) and the relative slenderness $\bar{\lambda}_{LT}$ (*clause 8.4.2(4)*). The reduction factor is applied to the design resistance moment M_{Rd}, which is defined in ***clauses 8.4.2(1)*** to ***8.4.2(3)***. Expressions for M_{Rd} include the design yield strength f_{yd}. The reference in these clauses to the use of γ_{M1} is provided because this is a check on instability. The default values for γ_{M0} and γ_{M1} are the same (1.0) in EN 1993-1-1. It is expected that the UK's National Annex will use these values. The determination of M_{Rd} for a Class 3 section differs from that of $M_{el,Rd}$ in *clause 8.2.1.4(6)* only in that the limiting stress f_{cd} for concrete in compression need not be considered. It is necessary to take account of the method of construction.

The buckling resistance moment $M_{b,Rd}$ given by *Formula 8.5* must exceed the greatest applied moment M_{Ed} within the unbraced length of compression flange considered. Its calculation involves the elastic critical bending moment M_{cr}, which depends on the distribution of bending moment in the region of hogging bending, which in turn depends on the redistribution of hogging moments, if done. That redistribution is not permitted if $M_{b,Rd} < M_{Rd}$ so an iterative calculation could be needed. Redistribution will slightly increase the calculated buckling moment. To avoid iteration, $M_{b,Rd}$ should be calculated using undistributed bending moments, as shown in the flow chart of Figure 8.11.

Lateral buckling for a Class 3 cross-section with unpropped construction

The influence of method of construction on verification of a Class 3 composite section for lateral buckling is as follows. From *Formula (8.2)*, $M_{Rd} = M_{el,Rd} = M_{a,Ed} + k_{el}M_{i,Ed}$ where the subscript i indicates the sum of the action effects on the composite member. Factor k_{el} may be less than 1 for a Class 2 section, but will exceed 1 for a Class 3 section with $M_{el,Rd} > M_{Ed}$.

Elastic critical buckling moment

Clause 8.4.2(4) requires the determination of the elastic critical buckling moment, taking account of the relevant restraints, so their stiffnesses have to be calculated. The lateral restraint from the slab can usually be assumed to be rigid. Where the structure is such that a pair of steel beams and a concrete flange attached to them can be modelled as an inverted U-frame (*Figure 8.13*), continuous along the span, the rotational restraining stiffness at top-flange level, k_s, can be found from ***clauses 8.4.2(5)*** to ***8.4.2(7)***.

Clause 8.4.2(4)

Clause 8.4.2(5)
Clause 8.4.2(7)

Clause 8.4.2(5) gives conditions that define this frame. Analysis is based on its stiffness per unit length k_s given by the ratio F/δ, where δ is the lateral displacement caused by a force F (Figure 8.9(a)). The flexibility δ/F is the sum of flexibilities due to

- bending of the slab, which may not be negligible: $1/k_1$ from *Formula 8.8*
- bending of the steel web, which predominates: $1/k_2$ from *Formula 8.9*
- flexibility of the shear connection.

It has been found (Johnson and Fan, 1991) that this last flexibility can be neglected in design to EN 1994-1-1. This leads to *Formula 8.7* for stiffness k_s.

This 'continuous U-frame' concept has long been used in the design of steel bridges. There is a similar 'discrete U-frame' concept, which would be relevant to composite beams if the steel sections had vertical web stiffeners. The shear connectors closest to those stiffeners would then have to transmit almost the whole of the bending moment Fh (Figure 8.9(a)), where F is now a force on a discrete U-frame. The last of the three flexibilities listed above might then not be negligible, nor is it certain that the shear connection and the adjacent slab would be sufficiently strong (Johnson and Molenstra, 1990). Where stiffeners are present, the resistance of the connection above each stiffener to repeated transverse bending needs to be established, as there is a risk of local shear failure within the slab. There is at present no simple method of verification. The conditions for the use of this method therefore say (*clause 8.4.3(1)(f)*) that the web is stiffened at a support, and 'Elsewhere, the web is unstiffened'. The problem can be avoided in bridge design by using transverse steel members, such as U- or H-frames (Figure 8.10(a)).

Figure 8.9. U-frame action and distortional lateral buckling

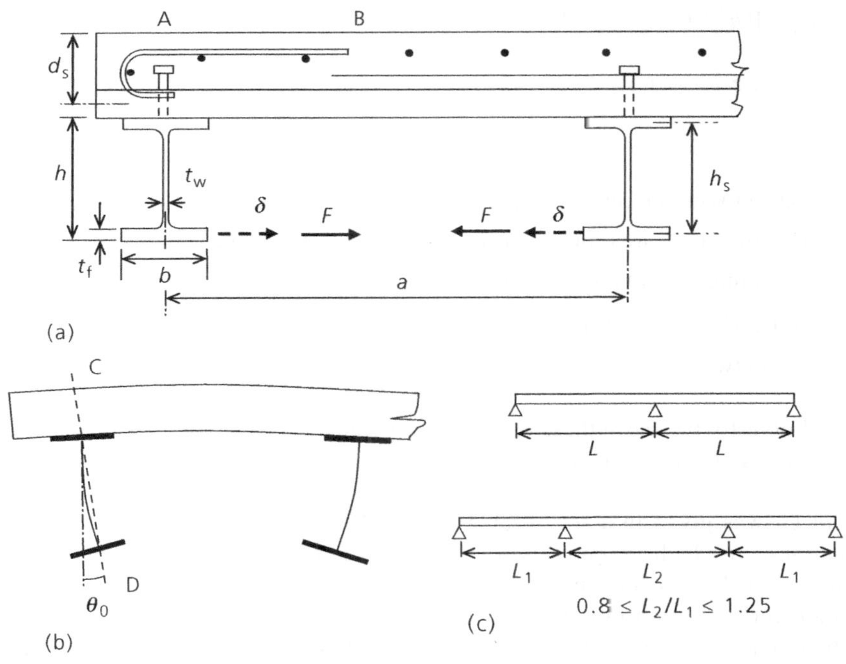

The conditions in *clause 8.4.3(1)* referred to from *clause 8.4.2(5)* are commonly satisfied in buildings by the beams that support composite slabs; but where these are secondary beams, the method is not applicable to the primary beams because condition (*e*) is not satisfied. Bottom flanges of primary beams can sometimes be stabilised by bracing from the secondary beams.

The calculation of k_s is straightforward, apart from finding $(EI)_2$, the cracked flexural stiffness of a composite slab. An approximate method, used in Example 8.6, is derived in Appendix A.

Concrete-encased web

Clause 8.4.2(7)
Clause 8.4.2(9)

Clauses 8.4.2(7) and ***8.4.2(9)*** allow for the additional stiffness provided by web encasement. This is significant for rolled steel sections – k_2 from *Formula 8.11* is from ten to 40 times the value from *Formula 8.10*, depending on the ratio of flange breadth to web thickness. Encasement will often remove any susceptibility to lateral-torsional buckling. The model used for *Formula 8.11* is explained in Appendix A.

Theory for the continuous inverted U-frame model

EN 1994-1-1 does not give a formula for the buckling moment M_{cr} for the U-frame model, as it is considered to be 'text-book material'. Subject to conditions discussed below, the elastic critical buckling moment at an internal support of a continuous beam is

$$M_{cr} = (k_c C_4 / L)[(G_a I_{at} + k_s L^2 / \pi^2)\, E_a I_{afz}]^{1/2} \quad \text{(X8.11)}$$

where

k_c is a property of the composite section, given below
C_4 is a property of the distribution of bending moment within length L
G_a is the shear modulus for steel: $G_a = E_a/[2(1 + \upsilon)] = 80.8$ kN/mm^2
I_{at} is the torsional moment of area of the steel section
k_s is the rotational stiffness defined in *clause 8.4.2(6)*
L is the length of the beam between points at which the bottom flange of the steel member is laterally restrained (typically, the span length)
I_{afz} is the minor-axis second moment of area of the steel bottom flange.

Where the cross-section of the steel member is symmetrical about both axes, the factor k_c is given by

$$k_c = (h_s I_y / I_{ay}) / [(h_s^2 / 4 + i_x^2)/e + h_s] \quad \text{(X8.12)}$$

$$\text{with } e = A I_{ay} / [A_a z_c (A - A_a)] \quad \text{(X8.13)}$$

where

h_s is the distance between the centres of the flanges of the steel section
I_y is the second moment of area for major-axis bending of the cracked composite section of area A
I_{ay} is the corresponding second moment of area of the steel section
i_x^2 $= (I_{ay} + I_{az})/A_a$, where I_{az} and A_a are properties of the steel section
z_c is the distance between the centroid of the steel beam and the mid-depth of the slab.

Four of the conditions for the use of these formulae are given in *paragraphs (c)* to *(f)* of *clause 8.4.3(1)*. Further explanation is given in Appendix A.

The coefficient C_4 was given in a set of tables within ENV 1994-1-1 (BSI, 1994), determined by numerical analyses, in which its range is 6.2 to 47.6. These values are given in Appendix A (see Figures A.3 and A.4). The coefficient accounts for the increased resistance to lateral buckling where the bending moment is not uniform along the member. When checking lateral stability, the distribution of bending moments corresponding to C_4 must be used as the action effects, and not an equivalent uniform value. The calculation method is shown in Example 8.6.

Where the resistance of a continuous composite beam would otherwise be governed by lateral buckling near an internal support, lateral bracing to the bottom flange may be provided. The bending moment M_{cr} should then be found by elastic critical analysis by computer, as the method using factors C_4 from Appendix A is not applicable.

Alternative theory for elastic critical moment

There is an analogy between the differential equations for distortional lateral buckling, taking account of restraint to warping, and those for a compressed member on an elastic foundation. This led to an alternative expression for the elastic critical moment (Hanswille, 2002). Like Equation X8.11, its use requires computed values that depend on the bending-moment distribution.

Predictions of M_{cr} by Equation X8.11 and by Hanswille's method were compared with results of FE analyses for beams with IPE 500 and HEA 1000 rolled sections. The latter method was found to agree with the FE results for both internal and external spans. Equation X8.11 was found to be satisfactory for internal spans, but to be less accurate generally for external spans, and unconservative by over 30% in some cases. This suggests that it needs further validation.

8.4.3 Simplified verification of lateral-torsional buckling of continuous composite beams with cross-sections in Class 1, 2 and 3, without direct calculation for beams in buildings

As calculations for the U-frame model are quite extensive, a simplified method has been developed from it. ***Clause 8.4.3(1)*** defines the structural steel rolled sections that may be used for continuous composite beams and cantilevers without lateral bracing to the bottom flange, except at supports. Its *Table 8.1* gives limits to the steel grade and overall depth of the steel member, provided that it is an IPE, HEA or HEB rolled section. The contribution from partial encasement is allowed for (*paragraph (h)*). *Clause 8.4.3(1)*

Clause 8.4.3(1)(i) gives a more general method for doubly symmetric I- and H-sections in terms of a section parameter, *Formula 8.11*, and limits (*Table 8.2*) for steel grades up to S460. These results are needed for the British UB sections. They are derived from Equation X8.11 for the elastic critical buckling moment, making assumptions that further reduce the scope of the method.

An account of its origin is available (Roik *et al.*, 1990). It is similar to that used in the treatment of lateral-torsional buckling of haunched beams (Lawson and Rackham, 1989). The basis is that there shall be no reduction, due to lateral buckling, in the resistance of the beam to hogging bending. It is assumed that this is achieved when $\bar{\lambda}_{LT} \leq 0.4$. This value is given in a Note to EN 1993-1-1, clause 8.3.2.3(1). It can be modified by a national annex. The UK's National Annex to EN 1993-1-1 is expected to confirm the value 0.4.

The slenderness $\bar{\lambda}_{LT}$ is a function of the variation of bending moment along the span. This was studied using various loadings on continuous beams of the types shown in Figure 8.9(c), and on beams with cantilevers. The limitations in *paragraphs (a)* and *(b)* of *clause 8.4.3(1)* on spans and loading result from this work.

Simplified expression for $\bar{\lambda}_{LT}$ and use of British UB rolled sections

The basis for the section parameter method is as follows. For uncased beams that satisfy the conditions that apply to Equation X8.11 for M_{cr}, have a double symmetrical steel section, and are not concrete encased, the slenderness ratio for a Class 1 or Class 2 cross-section may conservatively be taken as

$$\bar{\lambda}_{LT} = 5.0\left[1+\frac{t_w h_s}{4b_f t_f}\right]\left[\frac{h_s}{t_w}\right]^{0.75}\left[\frac{t_f}{b_f}\right]^{0.25}\left(\frac{f_y}{E_a C_4}\right)^{0.5} \quad \text{(X8.14)}$$

The derivation from Equation X8.11 is given in Appendix A of Johnson (2012). Most of the terms in Equation X8.14 define properties of the steel I-section, where b_f is the breadth of the bottom flange and C_4 is explained above. Other symbols are as in Figure 8.9(a).

To check if a particular section qualifies for 'simplified verification', a section parameter F is calculated. From Equation X8.14,

$$F = \left[1 + \frac{t_w h_s}{4 b_f t_f}\right]\left[\frac{h_s}{t_w}\right]^{0.75}\left[\frac{t_f}{b_f}\right]^{0.25} \quad \text{(X8.15)}$$

Limiting values of F, F_{lim} say, are given in *Table 8.2*. Some examples are given in Table 8.1, with the values of F_{lim} in the column headings. Many of the heavier wide-flange sections in S275 steel qualify for verification without direct calculation, but few UB sections in S355 steel do so. Welded cross-sections are not included. The method of *clause 8.4.2* should be used where the steel section does not qualify.

Use of UB rolled sections with encased webs to *clause 7.5.3(2)*

It has been shown (Johnson, 2012) that the effect of web encasement is to increase F_{lim} by at least 29%. All the sections shown in Table 8.1 then qualify for S275 steel, and all except one do so for S355 steel.

Table 8.1. Qualification of some UB rolled steel sections for verification of lateral-torsional stability, in a composite beam, without direct calculation

Section	Right-hand side of Expression X8.15	S275 steel, uncased	S355 steel, uncased	S275 steel, cased	S355 steel, cased
		(13.9)	(12.3)	(18.0)	(15.8)
457×152 UB52	16.4	No	No	Yes	No
457×152 UB67	14.9	No	No	Yes	Yes
457×191 UB67	13.6	Yes	No	Yes	Yes
457×191 UB98	11.8	Yes	Yes	Yes	Yes
533×210 UB82	14.4	No	No	Yes	Yes
533×210 UB122	12.5	Yes	No	Yes	Yes
610×229 UB125	14.1	No	No	Yes	Yes
610×229 UB140	13.5	Yes	No	Yes	Yes
610×305 UB149	12.2	Yes	Yes	Yes	Yes
610×305 UB238	9.83	Yes	Yes	Yes	Yes

Use of intermediate lateral bracing

Where the buckling resistance moment $M_{b,Rd}$ as found by one of the preceding methods is significantly less than the design resistance moment M_{Rd} of the cross-sections concerned, it may be cost-effective to provide discrete lateral restraint to the steel bottom flange. Where the slab is composite, a steel cross-member (Figure 8.10(a)) may be impracticable and a steel section with a wider bottom flange

Figure 8.10. Laterally restrained bottom flanges

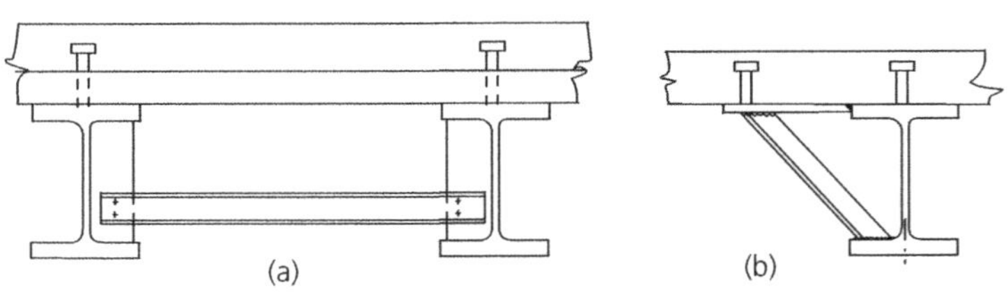

may be needed. For solid slabs, other solutions are possible (e.g. Figure 8.10(b)). Clause 8.3.2.1(2) of EN 1993-1-1 refers to 'beams with sufficient restraint to the compression flange', but 'sufficient' is defined in its Annex D only for continuous restraint.

There is no guidance in EN 1994-1-1 on the minimum strength or stiffness that a lateral restraint must have. There is guidance (Lawson and Rackham, 1989) which states that a discrete restraint should be designed for 2% of the *maximum* compressive force in the flange. The calculation of this compressive force involves an elastic analysis of the section, not otherwise needed, and the rule is unsafe near points of contraflexure. Advisory Note AD528 from the Steel Construction Institute recommends 2% for full lateral restraint and 1.25% for intermediate lateral restraints. An elastic analysis can be avoided by taking the stress in the flange as the yield stress.

The design methods based on Equation X8.11 for M_{cr} work well for complete spans but become unsatisfactory (over-conservative) for short lengths of beam between lateral bracings.

The design method of *clause 8.4.2,* with application of the theory in Appendix A, is illustrated in Example 8.6, which will be found after the comments on *clause 8.6*. The length of the method is such that the 'simplified verification' of *clause 8.4.3* is used wherever possible.

Flow charts for design of a span of a continuous beam

The flow charts in Figures 8.11 and 8.12 cover some aspects of design for ultimate limit states of an internal span of a composite T-beam for a building, assumed to be continuous over its supports. Their scope is limited to a beam with a uniform steel section of I or H shape in Class 1 or 2, an uncased web and a solid concrete unhaunched slab flange. Composite cross-sections in sagging bending are assumed to be in Class 1 or 2, and those at internal supports to be in Class 2 or 3. Shear connectors are in ductility category D1 or better. The loading is static, vertical and uniformly distributed, with the same action effects at supports A and C and no temperature effects. The flow chart refers only to cross-sections at support A and at mid-span, and to longitudinal shear in critical length AB (Figure 8.11).

It is assumed that the thickness of the slab and its longitudinal and transverse reinforcement have been chosen based on preliminary calculations. The following clauses and methods are not included.

- Construction with partial propping (e.g. of part of a continuous beam).
- Non-linear global analysis, *clause 7.4.3.*
- Redistribution of sagging bending moments, *clause 7.4.4(5)(b).*
- Structural steel of higher grade than S355, *clause 7.4.4(6).*
- Rigid-plastic global analysis, *clause 7.4.5.*
- Non-linear resistance to bending, *clause 8.2.1.5.*
- General method of shear connection using non-linear analysis, *clause 8.6.2.*

Figure 8.11 refers to the notes below.

Note 1. The class of the web at the supports depends on the area of longitudinal reinforcement in the slab and the hogging bending moment. Trial and error may be needed.

Note 2. Global analysis is needed at this stage if construction is unpropped, because elastic resistances to bending depend on action effects.

Note 3. It is assumed that an effective modular ratio is used, from *clause 7.4.2.2(11).*

Note 4. The elastic global analysis is simpler if the 'uncracked' model is used. The limits to redistribution of hogging moments include allowance for the effects of cracking, but redistribution is not permitted where allowance for lateral-torsional buckling is required (*clause 7.4.4(4)*). 'Cracked' analysis, to *clause 7.4.2.3(2)* or *(3)*, may then be preferred as it gives lower hogging moments at internal supports.

Note 5. Global analysis should include shrinkage effects if lateral buckling affects resistance or if the cross-section at A is Class 3. A second analysis may be needed.

Note 6. If fully propped construction is used, $M_{a,Ed}$ is zero at all cross-sections.

For Figure 8.12, the following assumptions have been made.

- A single shear span is considered, such as length AB of the beam in Figure 8.11.
- The design action effects at cross-sections A and B are known.
- The resistances to bending at cross-sections A and B have been determined, assuming full shear connection, so that initial values for the design compression in the slab at section B and the tension at section A are known.
- Plastic resistance moments of cross-sections in Class 2 do not exceed 2.5 $M_{pl,a,Rd}$, *clause 8.6.3.2(5)*.
- The connectors are not subjected to direct tensile force, *clause 8.6.8.2*.
- Global analysis does not take account of tension stiffening and overstrength of concrete, *clause 8.6.5(6)*.

Figure 8.11. Flow chart for design for ultimate limit state of an internal span of a continuous composite beam in a building, with a uniform steel H- or I-section and no cross-sections in Class 4

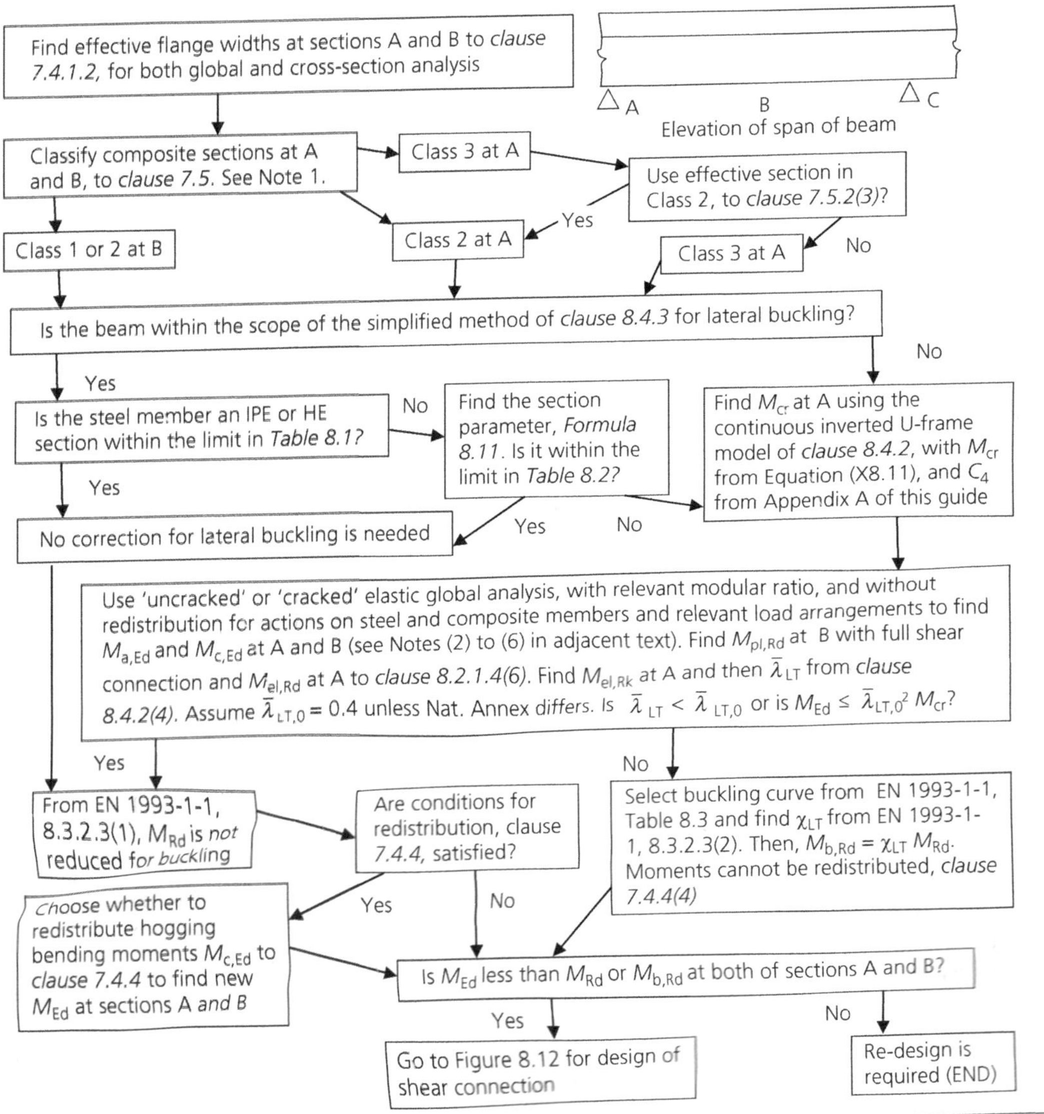

'Inelastic redistribution of shear' (***clause 8.6.1(3)***) is relied on in the many provisions that permit uniform spacing of connectors and partial shear connection. ***Clause 8.6.1(5)*** defines 'ductile' for connectors as being in category D2 or D3, meaning that they have deformation capacity sufficient to assume ideal plastic behaviour of the shear connection. In practice, designers may not be able to calculate required and available slip capacities. *Clause 8.6.3.3* enables this to be avoided in beams that are not subject to fatigue loading by linking the extent of partial shear connection to the ductility category of the shear connectors. *Clause 8.6.1(3)* *Clause 8.6.1(5)*

The need for compatibility of load/slip properties, ***clause 8.6.1(6)***, is one reason why neither chemical bond nor adhesives can be used to supplement the shear resistance of studs. The combined use of studs and block-and-hoop connectors has been discouraged for the same reason, though there is little doubt that projections into the concrete slab, such as bolt heads and ends of flange plates, being effectively rigid, contribute to shear connection. *Clause 8.6.1(6)*

'Separation', in ***clause 8.6.1(7)***, means separation sufficient for the curvatures of the two elements to be different at a cross-section, or for there to be a risk of local corrosion. None of the design methods in EN 1994-1-1 take account of differences of curvature, which can arise from a very small separation. *Clause 8.6.1(7)*

The tendency to separate arises from local variations in the flexural stiffnesses of the concrete and steel elements, and from the slab riding up on the weld collars. The standard heads of stud connectors have been found to be large enough to control separation, and the rule in ***clause 8.6.1(8)*** is intended to ensure that other types of connector, with anchoring devices if necessary, can do so. The authors are not aware of evidence from tests on connectors without it (unheaded studs, for example) that their performance is satisfactory after inelastic behaviour begins. *Clause 8.6.1(8)*

Resistance to uplift is much influenced by the reinforcement near the bottom of the slab, so if the resistance of an anchor is to be checked by testing, reinforcement in accordance with *clause 8.6.11* should be provided in the test specimens. Anchors are inevitably subjected also to shear.

Clause 8.6.1(9) refers to 'direct tension' – for example, a load from a travelling crane hanging from the steel member. In bridges, it can be caused by the differential deflections of adjacent beams under certain patterns of imposed load. Where it is present, its design magnitude should be determined. Further comment is given on the new *Annex H*, on tension resistance of headed studs. *Clause 8.6.1(9)*

Clause 8.6.1(10) on 'concentrated forces' is a principle that has led to many application rules. The shear forces on connectors are inevitably 'concentrated'. One research study (Johnson and Oehlers, 1981) found that 70% of the shear on a stud was resisted by its weld collar, and that the local (triaxial) stress in the concrete was several times its cube strength. Transverse reinforcement performs a dual role. It acts as horizontal shear reinforcement for the concrete flanges, and controls and limits splitting. Its detailing is critical where connectors are close to a free surface of the slab. *Clause 8.6.1(10)*

Larger concentrated forces occur where precast slabs are used, and connectors are placed in groups in holes in the slabs. This influences the detailing of the reinforcement near these holes. Further comment is given under *Annex J* on precast slabs.

Clause 8.6.1(12) refers to 'concentrated longitudinal shear forces' and refers to relevant design rules. Conventional calculation finds these forces at sudden changes of cross-section of a composite member. Their concentration is reduced by local slip. *Clause 8.6.1(12)*

Clause 8.6.1(13) says that events that 'significantly affect' longitudinal shear forces, such as the use of unpropped construction, should be 'taken into account'. Appropriate methods are given elsewhere in EN 1994. *Clause 8.6.1(13)*

Clause 8.6.1(14) is intended to permit the use of other types of connector, such as concrete dowels to CEN/TS 1994-1-102. ENV 1994-1-1 included provisions for many types of connector other than studs: block connectors, anchors, hoops and angles. They have all been omitted because of their limited use and to shorten the code. Some of these connectors, such as blocks, are non-ductile. The effects of this have been studied (Leskelä, 2017). This clause also enables research-based evidence to be used – for example, for demountable composite beams (Girão Coelho *et al.*, 2020). *Clause 8.6.1(14)*

8.6.2 General method using non-linear analysis

Clause 8.6.2 Shear connection for beams is normally designed using one of *clauses 8.6.3* to *8.6.6*. ***Clause 8.6.2*** provides general rules for use where neither plastic nor linear-elastic theory for longitudinal shear is applicable. This can occur where non-linear theory to *clause 8.2.1.5* is used for the bending resistance of cross-sections with partial shear connection.

Clause 8.6.2(1)
Clause 8.6.2(2)

These rules are not applicable to Class 3 sections with structural steel that yields in compression, presumably because they do not take account of buckling. This ***clause 8.6.2(1)*** also excludes sections in Class 4, for which elastic theory to *clause 8.6.5* should be used. ***Clause 8.6.2(2)*** shows that *clause 8.6.2* is about analyses for longitudinal shear. These are neither global nor section analyses, as are analyses for lateral buckling in beams.

Clause 8.6.2(4) The rule in ***clause 8.6.2(4)*** on taking the *ductility* of shear connectors into account is more precise than *clause 7.4.3* on global analysis, which refers to the *behaviour* of shear connectors. The phrase 'taken into account' means 'include in a calculation model'. It is sometimes (incorrectly) understood to mean 'consider'.

It can be difficult to prove, for a given shear connection, that the maximum slip required does not exceed the characteristic slip capacity. In practice, the application rules that follow are used. When using the methods of *clause 8.6.3* or *8.6.4*, there is no need to consider method of construction, as they are limited to beams with resistance to bending determined by plastic theory.

8.6.3 Beams in buildings where plastic theory is used for the resistance of the cross-section

Clause 8.6.3.1

This is the simplest and most widely used of the methods in *clause 8.6*, and the only one to provide detailed rules for partial shear connection. ***Clause 8.6.3.1*** states the scope of *clause 8.6.3*, as shown in Table 8.2. The reference to partial shear connection is relevant to the design for a beam with Class 3 sections at internal supports, where elastic analysis limits the curvature of those regions. The ultimate-load curvature at mid-span may then be too low for the full-interaction plastic bending resistance to be reached. The use of the hole-in-web method and partial shear connection is then appropriate, with M_{Rd} less than $M_{pl,Rd}$.

Clause 8.6.3.2 ***Clauses 8.6.3.2*** and *8.6.3.3* extend previous provisions for studs to other connectors in category D2 or D3. They do not enable connectors in category D1 to redistribute longitudinal shear force.

Clause 8.6.3.2(1) ***Clause 8.6.3.2(1)*** requires the design longitudinal shear to be 'consistent with the design bending resistance' – that is, with M_{Rd}, not M_{Ed}. This is relevant to the use of partial shear connection, designed to provide a resistance M_{Rd} not less than M_{Ed}.

Spacing of shear connectors in beams for buildings

Clause 8.6.3.2(3) In ***clause 8.6.3.2(3)*** the wording '*taking into account the deformation capacity of the shear connectors*' is a reference to the new ductility categories.

Clause 8.6.3.2(4)

Full shear connection is required for reinforcement in hogging moment regions within a continuous span (*clause 8.2.1.3(2)*). To enable uniform spacing to be used between a support and mid-span, ***clause 8.6.3.2(4)*** requires the lengths of bars of the tensile reinforcement to be related to the spacing of the shear connectors, taking account of the variation of bending moment along the beam. At flexural failure, the increased shear flow from longer bars could overload the connectors locally; hence the wording 'curtailed'.

Clause 8.6.3.2(5)
Clause 8.6.3.2(6)

The convenient use of uniform spacing requires the connectors to satisfy ***clause 8.6.3.2(5)***, which implies the use of plastic resistance moments. The connectors must be in ductility category D2 or D3 (*clause 8.6.3.1(1)*) and satisfy the conditions of ***clause 8.6.3.2(6)***. These, like the rules for factor β in *clause 8.2.1.2(2)*, depend on the depth ratio z_{pl}/h, but the limits are more severe. One of them, for $M_{pl,a,Rd}/M_{pl,Rd} > 0.65$, is uncommon in practice. The other limits vary with steel grade. The interaction with the rules for β is illustrated here for S355 steel.

- Reduction in $M_{pl,Rd}$ by factor β begins when z_{pl}/h reaches 0.2.

- for uniform spacing of connectors, subdivision of critical length begins at $z_{pl}/h = 0.4$
- elastic theory is required for bending resistance where $z_{pl}/h > 0.5$.

Where in composite slabs the conditions for uniform spacing are not satisfied, the regular pitch of the ribs within the profiled steel sheeting may make non-uniform spacing impracticable.

The reference in ***clause 8.6.3.2(7)*** to a 'number of connectors' means those within a length of span where uniform spacing has been used. *Clause 8.6.3.2(7)*

Clause 8.6.3.2(8), requiring further checks where $M_{pl,Rd} > 2.5\ M_{pl,a,Rd}$, could apply to a simply supported or a continuous beam with a large concrete slab and a relatively small steel top flange. It relates to situations where crushing of concrete could occur before either the bending resistance or the rotation capacity has been reached, on which uniform spacing of connectors relies. *Clause 8.6.3.2(8)*

Where composite dowel connectors are used, there may be no steel top flange. Connectors spaced uniformly along a critical length could then have insufficient slip capacity. Use of an additional critical section would lead to a more suitable distribution.

The method of ***clause 8.6.3.2(9)*** enables 'additional checks' to be omitted because it makes less demand on slip capacity than does uniform spacing. In calculating longitudinal shear, the tensile strength of concrete should be ignored over the length of span that is used in the determination of the bending-moment distribution. From *clause 8.6.5(3)*, it should not be ignored where elastic theory is used for resistance to bending. *Clause 8.6.3.2(9)*

Partial shear connection

Clause 8.6.3.3(1) refers to the 'calculated maximum slip', which is not easily calculated. The conditions in ***clause 8.6.3.3(2)*** for the use of partial shear connection are not always satisfied in beams for buildings. The condition of 'predominantly symmetric' loading would exclude non-central point loads from columns. It is not clear whether this exclusion applies where an additional critical cross-section is located at the column. *Clause 8.6.3.3(1)* *Clause 8.6.3.3(2)*

The minimum degree of shear connection for ductile connectors is given by

$$\eta \geq \eta_0 \rho_m^2 k_{up} \geq \eta_{min} \qquad (8.12)$$

where η_{min} is either 0.3 or 0.4, depending on the ductility category. The value η_0, defined in *clause 8.6.3.3(3)*, is usually higher than this, and can be difficult to exceed where profiled sheeting is used, so reduction factors, ρ_m and k_{up}, are provided. When Eurocode 4 was first drafted, η_{min} was set at 0.5 because of lack of evidence to support lower limits. This still appears as an alternative in *clause 9.3.1(4)(b)* on deflections. The reduced lower limits for ULS design counteract to some extent the more restrictive recent rules on detailing of stud connectors in troughs of profiled sheeting.

The new *moment utilisation factor* ρ_m ($0.8 \leq \rho_m \leq 1$) is defined as $M_{Ed}/(0.95\ M_{Rd}(\eta))$ where M_{Ed} is the maximum bending moment within the critical length considered, and $M_{Rd}(\eta)$ is the corresponding resistance allowing for partial shear connection. The factor allows for the non-linear decrease in maximum slip as M_{Ed} decreases below the design resistance moment $M_{Rd}(\eta)$ (Couchman, 2015; Vigneri *et al.*, 2021). Factor ρ_m provides reductions of up to 36%, reached where $M_{Ed} \leq 0.76\ M_{Rd}(\eta)$.

The factor k_{up} allows for the reduction in longitudinal shear where unpropped construction is used, and bending moment $M_{a,Ed}$ is resisted by the steel member alone. The condition $M_{Ed} \leq 0.95\ M_{Rd}(\eta)$ avoids the calculation of trivial reductions. The maximum reduction is 15%, available where $M_{a,Ed} \geq 0.15\ M_{pl,Rd}$, which is quite common for unpropped construction. Direct calculation of the lowest permitted degree of shear connection thus involves iteration. In practice, it is simpler to choose a convenient value for η and check against *Formula 8.12*.

The expressions for η_0 as a function of effective span L_e and yield strength f_y are given in ***clause 8.6.3.3(3)***, without the former limits to the dimensions of studs. For composite beams in sagging bending, the steel top flange can often be smaller than the bottom flange. A smaller flange lowers the *Clause 8.6.3.3(3)* plastic neutral axis of the composite section, and increases the slip required by the model for partial-interaction design. This is why *clause 8.6.3.3(3)* gives limits on degree of shear connection that are less liberal than those for beams with equal steel flanges. Interpolation can be used for flange-area ratios between 1 and 3.

Shear connectors in composite slabs

Where composite slabs that span transverse to the beam are used, there may be insufficient space in the trough for the studs required by the former rules, so there was pressure to relax them. In the UK, this led the Steel Construction Institute to sponsor tests and parametric studies on simply supported beams of this type and push tests on studs in trapezoidal sheeting (Hicks and Smith, 2014; Smith and Couchman, 2010).

Clause 8.6.3.3(5) This work led to the relaxation of limiting effective spans at which low degrees of shear connection can be used, given in ***clause 8.6.3.3(5).*** *Formula 8.17* can be more liberal than the general *Formula 8.12* in *clause 8.6.3.3(2)*. It leads to a lower minimum degree of shear connection for spans less than a value ranging from 16.5 m to 25 m, depending on the value for ρ_m. The upper limit to span L_e that may be used is that given in the rules for η_0. The use of this clause is subject to the conditions given for *Formula 8.12*. The numerical analyses that led to this rule assumed $\delta_{uk} \geq 8$ mm and propped construction (Aribert, 1997).

Clause 8.6.3.3(6) The new rule in ***clause 8.6.3.3(6)*** is intended to ensure that the proportions of the composite section are such that its resistance to flexure is not reduced by premature crushing of the concrete slab.

Examples of these rules for minimum degree of shear connection are shown in Figure 8.13 for S355 steel, equivalent spans L_e, connectors in ductility categories D2 and D3, equal-area steel flanges except where noted, and propped construction. Comparison of the lines labelled 0.8 and 1 show that where the detailing requires a low degree of shear connection, design with $\rho_m < 1$ enables a significant reduction in the minimum required for the longer spans. Example 8.5 on the use of partial shear connection is given after the text on *clause 8.6.11* and Example 8.4.

These rules may seem optimistic when compared with the results of push tests using solid slabs, but connectors behave better in beams reinforced and detailed as required by the Eurocode than in small push-test specimens, where splitting can cause premature failure. One might expect a lower available slip from studs in very strong concrete, but the 6-mm limit for category D2 has been confirmed in nominally identical push tests with 19, 22 and 25 mm diameter studs embedded in C70/85 concrete (Jähring, 2008).

Figure 8.13. Minimum degree of partial shear connection for beams in S355 steel

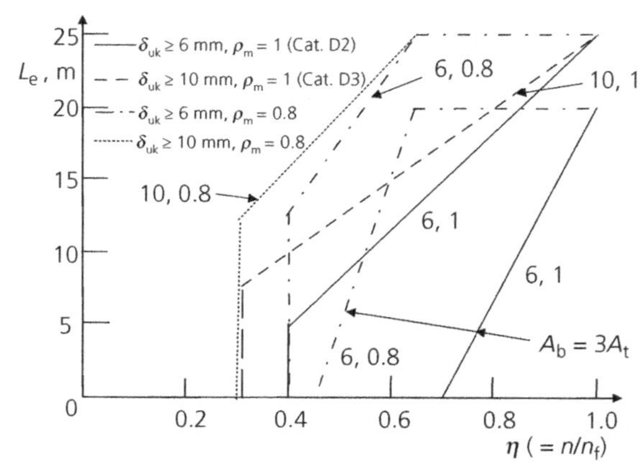

8.6.4 Other beams where plastic theory is used for the resistance of the cross-sections

Clause 8.6.4 is more complex than *clause 8.6.3*, and unlikely to be used where the connectors are in category D2 or D3. It provides a method for connectors in category D1, 'flexible'. *Clause 8.6.4*

Shear connection is based on the longitudinal force in the slab at critical cross-sections. Where $M_{Ed} > M_{el,Rd}$, it is unconservative to calculate this force by elastic theory, because the lever arm for bending reduces as plasticity within the section increases (***clause 8.6.4(3)***). In the previous EN 1994-1-1, the force was (conservatively) based on $M_{pl,Rd}$. This clause, previously in EN 1994-2, is based on M_{Ed}. *Clause 8.6.4(3)*

Clause 8.6.4(4) refers to the compressive force N_{cd} as given by *Figure 8.14*, which is a simpler alternative to *Figure 8.8* in *clause 8.2.1.5*. It is presented there as a method for finding the inelastic bending resistance, assuming the compressive force in the slab is known. That involves a quite long calculation, which in practice one tries to avoid. By assuming that this resistance equals the (known) action effect M_{Ed}, *Figure 8.14* can be used to find the compression in the slab (and hence the longitudinal shear) for given M_{Ed}. It is slightly more conservative than *Figure 8.8*. *Clause 8.6.4(4)*

For a slab in tension in a region of hogging bending, it has been usual to follow *clause 8.2.1.3(2)* and take the tensile force above a support as the yield force for the tensile reinforcement $A_s f_{yd}$. The more complex method of ***clause 8.6.4(5)*** will require more shear connection, and may be appropriate for beams subjected to fatigue loading. Its purpose is to ensure that the force attracted to the shear connection is not underestimated, possibly resulting in excessive slip. As an alternative, ***clause 8.6.4(6)*** permits the shear flow to be determined using elastic analysis based on the uncracked cross-section, which is assumed to offset the unconservatism that exists from ignoring inelasticity. *Clause 8.6.4(5)* *Clause 8.6.4(6)*

Flow chart for number and spacing of shear connectors within a critical length of a beam for a building

This flow chart (Figure 8.14) is for a region of sagging bending with known longitudinal shear force, and no sudden change of section or concentrated longitudinal shear forces. Its reference to cross-sections in Class 2 includes those in effective Class 2. It does not include

- 'advanced methods', referred to in clauses as alternatives
- prestressing by tendons
- biaxial bending of beams.

8.6.5 Beams in which elastic theory is used for resistances of cross-sections

Clause 8.6.5(1) refers to the 'required' longitudinal shear resistance. This means the resistance required by the design method being used, not that $v_{L,Rd}$ must at least equal $v_{L,Ed}$ at all cross-sections. That is required only for connectors in category D1, in *clause 8.6.5(4)*. *Clause 8.6.5(1)*

The use of the uncracked section, ***clause 8.6.5(3)***, overestimates longitudinal shear where the section is likely to be cracked. This allows for possible over-strength of concrete and avoids the need to estimate the extent of cracking. To allow for cracking involves calculations of a complexity best avoided (*clause 8.6.5(6)*). *Clause 8.6.5(3)*

Clause 8.6.5(7) says that 'more advanced calculations are required' for partly encased and shallow floor beams. This would apply to sections in Class 3 or 4, which are unlikely in practice. For sections in Class 1 and 2, plastic theory is normally used, from *clause 8.3.2* for partly encased sections and *Annex I* for shallow floor beams. Elastic theory would be needed for fatigue checks. *Clause 8.6.5(7)*

8.6.6 Beams in which non-linear theory is used for resistances of cross-sections

Clause 8.6.6 refers to *clause 8.2.1.5*, where comment is given. The use of this long method seems to offer no advantages in practice. *Clause 8.6.6*

Figure 8.14. Flow chart for number and spacing of shear connectors within a critical length of a beam for a building

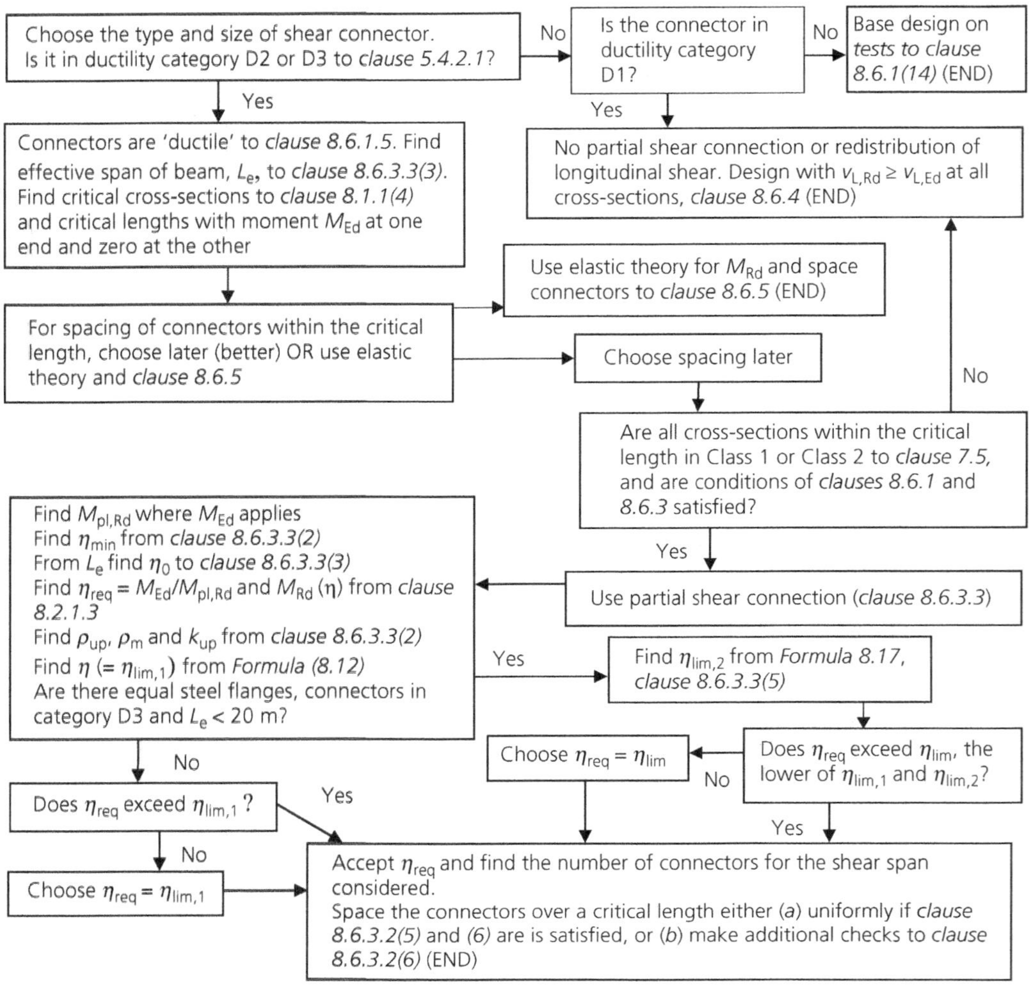

Notes: This chart does not include some further checks that are explained in the text on *clause 8.6.3.2(8)* and *clause 8.6.3.3(4)*. Cross-sections where $\beta < 1$ in *Figure 8.3* are not included.

Also not included are the restrictions on spacing of connectors caused by the troughs in profiled steel sheeting, discussed in the text that follows *clause 8.6.3.3(3)* and in Example 8.5.

Symbol η_{lim} appears in EN 1994-1-1 as η. Subscript req means 'required'.

8.6.7 Local effects of concentrated longitudinal shear force

Clause 8.6.7.1

Clauses 8.6.7.1 and *8.6.7.2* on this subject were previously in EN 1994-2 and have been revised to take account of the introduction of ductility categories. Where concentrated longitudinal forces are applied to a composite section, the stress state can easily be determined some distance away from the point of application from considerations of equilibrium and from the usual assumption that plane sections remain plane. Plane sections do not remain plane in the vicinity of the force and accurate determination of the local state of stress requires complex analysis. These simpler methods are based on parametric finite-element analyses (Johnson and Ivanov, 2001) and existing practice.

The subject of *clause 8.6.7.1* is a concentrated longitudinal force, F_{Ed}, applied to the concrete or steel part of a cross-section – for example, by a bonded or unbonded tendon' (*paragraph (1)*), so it rarely applies to structures for buildings. Part of this force, $V_{L,Ed}$, is the longitudinal shear transferred from one part to the other, with the distribution of shear force per unit length at the interface, $v_{L,Ed}$, given in this clause.

Clause 8.6.7.1(2)

Clause 8.6.7.1(2) assumes that the force is applied to a concrete flange and distinguishes between application within a length of flange or at an end. The length over which the force is assumed to be distributed depends on the ductility category of the connectors and is, as before, for connectors in category D2 or D3. It has been reduced for connectors in category D1. In *Figure 8.15(a)* and *(b)*, the y-axis is $v_{L,Ed}$, the longitudinal shear per unit length, and the shaded areas are the action effect $V_{L,Ed}$.

For forces applied at an end of a concrete flange, the distribution length is halved. The reference to effective width 'for global analysis' means that the simple provisions of *clause 7.4.1.2(4)* can be used, rather than the effective width appropriate to the cross-section where the force is applied.

Stud shear connectors are assumed to be sufficiently ductile to permit a uniform distribution of shear flow over the above lengths at the ultimate limit state. This leads to *Formulae 8.20* and *8.21* in ***clause 8.6.7.1(3)***. For fatigue limit states, the distributions of *Formulae 8.18* and *8.19* should be used. *Clause 8.6.7.1(3)*

Clause 8.6.7.2 provides simple rules for the determination of the design shear flow between steel and concrete at ends of slabs where *Clause 8.6.7.2*

- the primary effects of shrinkage or differential temperature are developed (***clause 8.6.7.2(1)***) *Clause 8.6.7.2(1)*
- there is an abrupt change of cross-section (***clause 8.6.7.2(2)***), such as that shown in *Figure 8.15d*. *Clause 8.6.7.2(2)*

The shear $V_{L,Ed}$ transferred across the concrete and steel interface due to shrinkage or temperature may be assumed to be distributed over the length defined in ***clause 8.6.7.2(3)***, with uniform distribution for stud connectors. *Clause 8.6.7.2(3)*

Figure 8.15. Stresses from primary shrinkage near an end of a composite beam

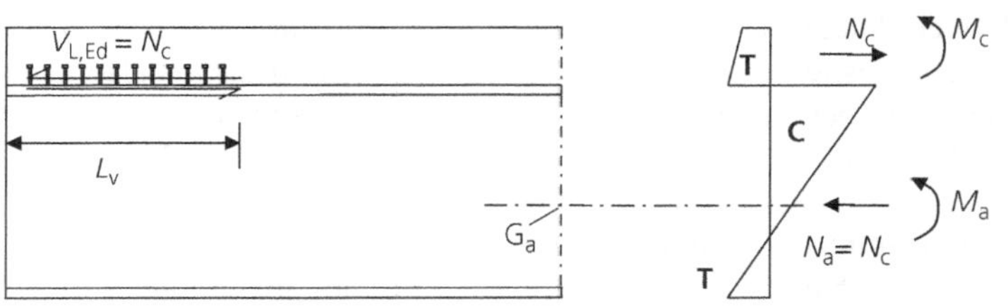

Figure 8.15 shows one end of a simply supported beam where shrinkage of the concrete caused equal and opposite longitudinal shear forces $V_{L,Ed}$ acting over lengths L_v at the two ends of the span. In the rest of the span the difference between the strains in the steel and concrete at their interface is the shrinkage strain of the concrete. There is no other loading. The longitudinal shear can be assumed to disperse into the concrete slab at the angle $2\beta_L$ given by *clause 8.6.7.2(6)*. The elastic distribution of longitudinal stress in the beam is then as shown in the figure, with tensile stress at both faces of the cross-section, and a bending moment $M_a + M_c$ ($= N_c z$), which causes sagging curvature.

Figure 8.16 shows an internal length AB of a beam that is composite over part of its length and subjected to uniform sagging bending moment M_{Ed}. The longitudinal compression in the slab at section B–B is transferred to the steel by the shear force N_c over a length L_v, as shown. The distribution of elastic strain at section B–B is linear, with stresses corresponding to moment M_{Ed}.

Figure 8.16. Longitudinal shear caused by bending moment at a sudden change of cross-section

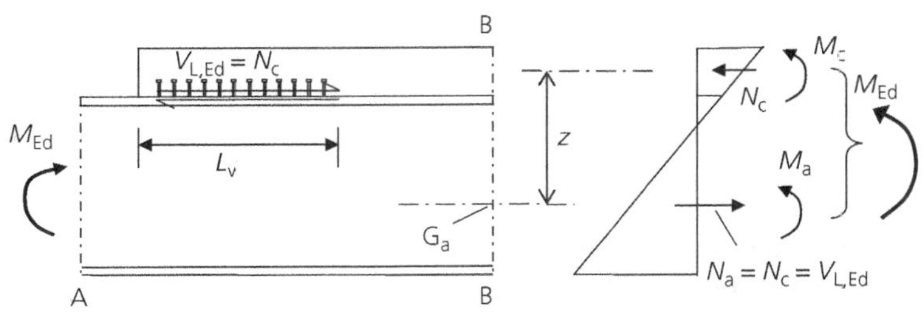

8.6.8 Headed stud connectors in solid slabs and concrete encasement

Resistance to longitudinal shear

The scope of EN 1994-1-1 includes all shear connectors that satisfy *clause* 5, but stud connectors are the only type for which detailed rules are given. There is no simple mechanical model for the shear resistance of a stud connector. The design equations given in ***clause 8.6.8.1(1)*** are based on a simplification: that a stud with a shank diameter d and an ultimate strength f_u, set in concrete with a characteristic strength f_{ck} and a mean secant modulus E_{cm}, fails either in the steel alone or in the concrete alone.

Clause 8.6.8.1(1)

The concrete failure is found in tests to be influenced by both the stiffness and the strength of the concrete. This led to *Formula 8.24*, in which the numerical constant and partial safety factor γ_V have been deduced from analyses of test data. It is as in the previous Eurocode 4, except that the reduction factor k_{cc} has been added. This arises from an unresolved difference of opinion on whether relaxation during compression tests on concrete should be allowed for, and is discussed in *Section 5.1*. The recommended value, 1.0, can be altered in a national annex. It is expected that the UK's Annex will give the value 1.0, as in the earlier Eurocode 4.

It was concluded from the initial study (Stark and van Hove, 1991) that the coefficient in *Formula 8.24* should be 0.26. In a more recent study, a larger push-test database with a wider range of stud diameters and stronger concrete was considered (Hicks, 2017). The results from that study showed that the existing Eurocode 4 model was justified up to C90/105 and stud diameters between 12.7 and 31.75 mm when the coefficient in *Formula 8.24* was 0.25, and the design resistance given by *Formula 8.23* could be increased by using a coefficient of 0.94 in place of 0.8.

However, both these studies were based on push tests, where the mean number of studs per specimen was only six, and where lateral restraint from the narrow test slabs was usually less stiff than in the concrete flange of a composite beam. Strength of studs in many beams is also increased by the presence of hogging transverse bending of the slab. For these reasons the coefficient in *Formula 8.23* was unchanged at 0.8 and that in *Formula 8.24* was increased from 0.26 to 0.29.

Design resistances of 19-mm stud connectors in solid slabs, given by *clause 8.6.8.1(1)*, are shown in Figure 8.17. It is assumed that $k_{cc} = 1$ and $\gamma_V = 1.25$. For any given values of f_u and f_{ck}, the figure shows which failure mode governs. It can be used for this purpose for studs of other diameters.

The previous EN 1994-1-1 reduced the shear resistance of studs with $h_{sc}/d < 4$, where h_{sc} is the 'overall nominal height' of the stud. This has been replaced by the rule $h_{sc}/d \geq 3.9$. There are other restrictions in *clauses 8.6.9.1(2)* and *F.1(1)*.

Figure 8.17. Design shear resistances of 19-mm studs in solid slabs; $h/d \geq 3.9$, $k_{cc} = 1$

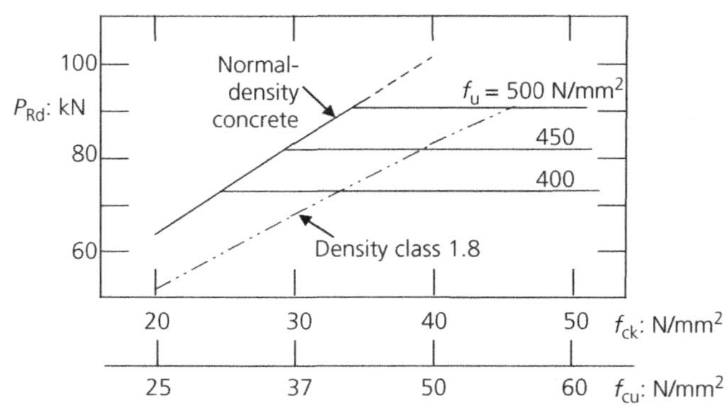

For studs subjected to fatigue loading, research has found (Hanswille *et al.*, 2007) that gradual reduction in static shear strength occurs during the fatigue life. In theory, it falls to the fatigue strength at the end of the fatigue life. No account is taken of this in current design methods, mainly because of the absence of failures that can be attributed to this effect.

The resistances in *clause 8.6.8.1(1)* are not applicable to studs of diameter exceeding 25 mm. Data on the static and fatigue strength of studs up to 30 mm in diameter are available (Hicks, 2017; Lee *et al.*, 2005). Studs of diameter exceeding 20 mm are rarely used in buildings, as welding through sheeting becomes more difficult due to the limited range of ceramic ferrules and more powerful welding plant is required.

Stud welding

Clause 8.6.8.1(2) refers to two ISO standards for arc stud welding and weld collars (BS EN ISO 14555:2017; BS EN ISO 13918:2018+A1:2021). The 'length after welding' (LAW) of 19-mm diameter studs welded through profiled steel sheeting can be up to 5 mm less than the values given in EN ISO 13918. For ease of use the formulae in *clause 8.6.9* and *Annex G* account for this reduction in stud height, so the designer only needs to consider a single dimension, irrespective of the welding detail. For a 19-mm diameter × 105-mm stud, LAW is assumed in examples here to be 100 mm. *Clause 8.6.8.1(2)*

It is known that for studs with normal weld collars, a high proportion of the shear is transmitted through the collar. It should not be assumed that the shear resistances of *clause 8.6.8.1* are applicable to studs without collars (e.g. where friction welding by high-speed spinning is used). A normal collar should be fused to the shank of the stud. Typical collars in the test specimens from which the design formulae were deduced had a diameter not less than 1.25d and a minimum height not less than 0.15d, where d is the diameter of the shank.

The collars of studs welded through profiled sheeting can be of different shape from those for studs welded direct to steel flanges, and the shear strength may also depend on the effective diameter of the weld between the sheeting and the flange, about which little is known. The resistances of *clause 8.6.8.1* are applicable where the welding is in accordance with EN ISO 14555, as is required by EN 1090-2 (BS EN 1090-2:2018+A1:2024).

Studs that cause splitting forces in the direction of the slab thickness

Clause 8.6.8.1(3) refers to these splitting forces. They occur where the axis of a stud lies in a plane parallel to that of the concrete slab – for example, if studs are welded to the web of a steel T-section that projects into a concrete flange. There has been research on the local reinforcement needed to prevent or control splitting (Kuhlmann and Breuninger, 2002). The subject, not in the previous EN 1994-1-1, was then in EN 1994-2, and is now in Normative *Annex F*. *Clause 8.6.8.1(3)*

This clause gives straightforward detailing rules of limited scope. The condition in *paragraph 3(b)* requires the use of unusually long studs. Further comment is given under *Annex F* and in Hendy and Johnson (2006). A similar problem occurs in composite L-beams with studs close to a free edge of the slab. This is addressed in *clause 8.6.10.3(2)*.

Influence of axial tension on the shear resistance of stud connectors

Pressure under the head of a stud connector and friction on the shank normally causes the stud weld to be subjected to vertical tension before shear failure is reached. This is why *clause 8.6.1(8)* requires shear connectors to have a resistance to tension that is at least 10% of the shear resistance. ***Clause 8.6.8.2(2)*** therefore permits tensile forces that are less than this to be neglected. Greater tension, outside the scope of the previous EN 1994, is now treated in Normative *Annex H*, where comment is given. *Clause 8.6.8.2(2)*

Biaxial loading of shear connectors

The interaction formula of *clause **8.6.8.3(1)*** is based on vectorial addition of two shear forces. Biaxial horizontal loading, referred to in *clause 8.6.9.2(5)*, occurs where stud connectors are used to provide end anchorage for composite slabs, as shown in *Figure 10.1(3)*. This subject is discussed under *Clause 10.7.4(3)*. *Clause 8.6.8.3(1)*

8.6.9 Design resistance of headed studs used with profiled steel sheeting in buildings

The load-slip behaviour of a stud connector in a trough of sheeting or rib of concrete (both terms are used) is more complex than in a solid slab. It is influenced by

- the direction of the ribs relative to the span of the beam
- their mean breadth b_0 and depth h_p
- the diameter d and height h_{sc} of the stud

- the number n_r of the studs in one trough, and their spacing
- whether or not a stud is central within a trough, and if not, by its eccentricity and the direction of the shear.

The interactions between these parameters have been explored by testing, with sheeting continuous across the beam. It is clear that the most significant are the ratios h_{sc}/h_p and b_0/h_p and, for ribs transverse to the supporting beams, n_r and the eccentricity, if any. In EN 1994-1-1, as in earlier codes, reduction factors k (≤ 1.0) are given for application to the design resistances of studs in solid slabs. They are based entirely on testing and experience.

Figure 8.18. Detail of a haunch with parallel sheeting (dimensions in mm)

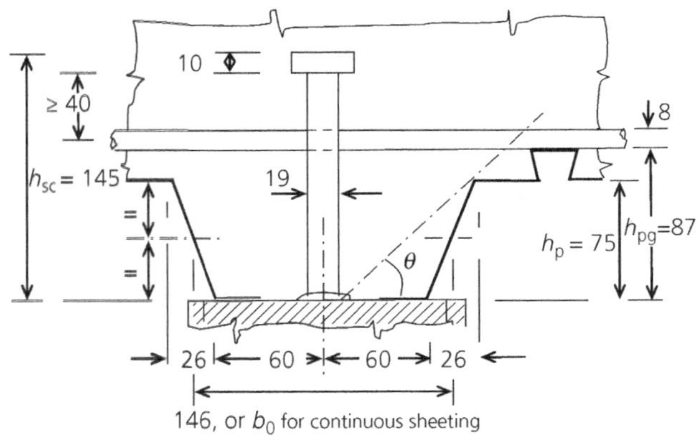

Sheeting with top ribs

Since EN 1994 was first written, sheetings with a small top rib, as shown in Figure 8.18, have come into use, so a profile has two depths. The one to be used depends on the verification required. Symbol h_p is defined in *clause 3.2.2* as 'overall depth excluding longitudinal stiffener', with symbol h_{pg} where the stiffener is included. This depth is used or referred to in many places within EN 1994-1-1. Guidance on whether to use h_p or h_{pg} is given in this guide wherever the question arises. The choice is often obvious.

It has been found that 'small' top ribs (defined in *clause 8.6.10.8(3)*) do not contribute to shear failures that occur within troughs.

Sheeting with ribs parallel to the supporting beams (*clause 8.6.9.1*)

There are two situations. The sheeting may be continuous across the beams – its side walls then provide lateral restraint to the concrete around the studs – or it may be discontinuous, as shown in Figure 8.18. The haunch may then be wider than the breadth b_0 of a trough. From *clause 8.6.9.1(3)* it can be treated as a haunch not formed by sheeting, to *clause 8.6.10.4*.

A haunch that just complies with *clause 8.6.10.4* is shown, to scale, in Figure 8.18.

The rules specify

- the angle θ ($\leq 45°$)
- the concrete side cover to the connector (≥ 50 mm)
- the depth of the transverse reinforcement below the under-side of the head (≥ 40 mm).

This last rule cannot be satisfied using studs of normal height (100 or 125 mm), and in the UK it has not been applied in this situation. Another rule on projection of studs above reinforcement is discussed under *clause 8.6.10.1(1)*.

Clause 8.6.9.1(2) The application of ***clause 8.6.9.1(2)*** to this haunch is now considered. Its rib is 'small', and the reduction factor is

$$k_\ell = 0.6\ (b_0\ /\ h_p)\ [(h_{sc}\ /\ h_p) - 1] \leq 1 \qquad (8.28)$$

where h_{sc} may not be taken as greater than $h_p + 75$ mm. This limit rarely governs. For b_0, defined in *Figure 8.17*, the top rib is irrelevant. *Formula 8.28* is from 1977 (Grant *et al.*, 1977).

There is little recent research on ribs parallel to the beam. Factor k_ℓ is strongly influenced by h_p. It gives for the haunch in Figure 8.18

$$k_\ell = 0.6(146/75)\ [(145/75) - 1] = 1.09 \leq 1$$

so there is no reduction. For a stud of more typical height, say 125 mm, the result is $k_\ell = 0.78$.

For discontinuous sheeting, edge fixings provided for erection may not provide much lateral restraint, and the trough may be too narrow for the rules of *clause 8.6.10.4* to be applied. The reference to '*appropriately anchored*' in ***clause 8.6.9.1(3)*** provides an alternative. The basis for advice from the Steel Construction Institute is that the anchorage to each edge of the sheeting should be as strong as the force, transverse to the beam, that would 'unfold' the profile by means of plastic hinges at the top and bottom edges of its troughs. This force is found not to exceed 4 kN/m. Anchorage can be provided by self-drill self-tap screws or shot-fired nails at about 250 mm spacing. *Clause 8.6.9.1(3)*

The rules of *clause 8.6.9.1* for continuous or anchored sheeting should normally ensure good detailing. There is no penalty for off-centre studs in troughs of parallel sheeting. Concentrations of studs should be avoided, as there is too little research to provide guidance on the use of more than one at a cross-section.

Where sheeting is continuous across a beam and through-deck stud welding is used, there may be shear transfer from the sheeting to the beam. No guidance is given on this complex situation, which may be ignored in design.

Sheeting with ribs transverse to the supporting beams (*clause 8.6.9.2*)

In the preceding paragraphs, the likely mode of failure was loss of restraint to the base of the stud due to bursting (lateral expansion) of the haunch. A rib transverse to the beam, shown in Figures 8.19 and 8.20, with specified limits given for some dimensions, is more highly stressed, as it has to transfer much of the shear from the base of the stud to the continuous slab above. Three modes of failure for trapezoidal troughs are shown in Figure 8.19(a) to Figure 8.19(c)

(*a*) failure surface above a stud that has insufficient embedment ('concrete pull-out')
(*b*) rib that is too slender, with plastic hinges at each end of the stud
(*c*) eccentricity on the 'weak' side of the centre, the 'unfavourable' location, with crushing of concrete in the region of width e_k.

Figure 8.19. Failure modes and placing of studs for troughs of transverse profiled sheeting. The alternate favourable and unfavourable placing of pairs of studs is shown in (d), and the diagonal ('staggered') placing of pairs of studs in (e)

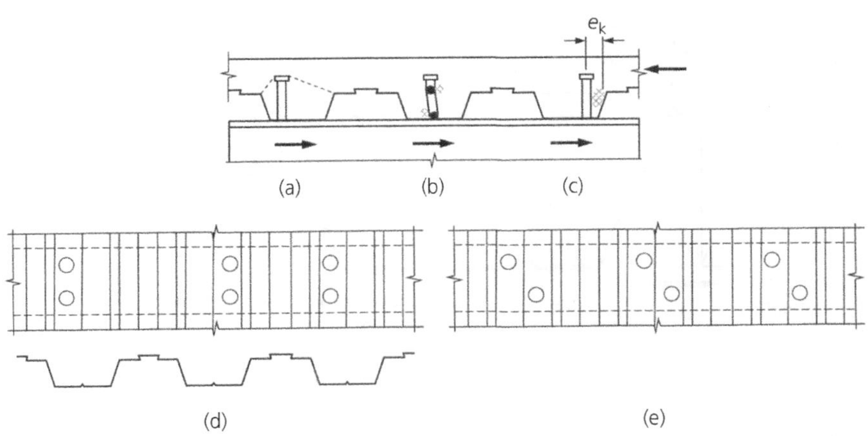

Eccentricity (*c*) reduces the effective breadth of the rib, and causes 'rib punching' failure (Degtyarev and Hicks, 2023; Konrad *et al.*, 2020; Vigneri *et al.*, 2022).

The shear resistance of a stud in a solid slab is multiplied by a reduction factor k_t given by *Formula 8.29* with format similar to that for k_ℓ. The formula is unchanged from the previous EN 1994-1-1. Tests and numerical modelling showed that its scope had to be reduced, and further rules added, for the following reasons.

- The variety and the size of troughs in profiled sheeting have increased.
- Many sheetings have troughs with a central stiffening rib, which forces stud connectors to be placed closer to one side of the trough than to the other, leading to the 'weak' failure mode shown in Figure 8.19(c).
- Tests have shown that the previous rule for embedment of the stud in the slab, dimension h_A in *Figure 8.19*, was insufficient to prevent pull-out failure (Figure 8.19(a)) for trapezoidal sheetings with certain profiles.
- Pull-out failure is resisted partly by transverse reinforcement below the head of each stud. This led to the '30-mm' rule in *clause 8.6.10.1(1)*, not to be confused with the 40-mm rule shown in Figure 8.18. For sheetings with depth greater than 50 mm, the widely used 100-mm studs were not long enough. Rather than using longer studs, it has sometimes been assumed that the sheeting acted as this reinforcement, with the bar or mesh reinforcement in the slab placed too high to be effective. However, recent research has found (Degtyarev and Hicks, 2023) that the resistance of studs in push tests is insensitive to position of the reinforcement.
- In *Formula 8.29*, n_r is the number of studs at one intersection of a rib with a supporting beam. Its definition is clearer than before, because it was being incorrectly assumed that the previous wording, $n_r \leq 2$ '*in calculations*' did not exclude use of $n_r > 2$ in practice, with n_r left as 2 when calculating k_t (Hicks and Smith, 2014).
- The previous Eurocode 4 had no rules limiting the possible layouts of studs in a trough, other than the minimum spacings now given in *clause 8.6.10.7(4)*. Two studs can be arranged to act in parallel, Figure 8.19(d), or in series, Figure 8.20, or can be 'staggered', Figure 8.19(e), at various spacings, all of which affect resistance to shear.
- When the *Formula 8.29* was originally developed, it assumed that all the variability was accounted for within *Formulae 8.23* and *8.24* because of the small number of tests (Stark and van Hove, 1991). As more push test data has become available, the scope was revised following analyses according to EN 1990 (Vigneri *et al.*, 2022).

Clause 8.6.9.2(3) The changes of scope are that more restrictive limits are now given in ***clause 8.6.9.2(3)*** on b_0, h_A, e_k and the thickness of the sheeting, t_p. There is a new condition, that 'a reinforcement layer is placed underneath the head of the stud', which implies that there must be bottom reinforcement. It is unclear whether sheeting may be deemed to be 'reinforcement'.

Figure 8.20. Details of trough transverse to the supporting beams with positions of stud connectors (dimensions in mm)

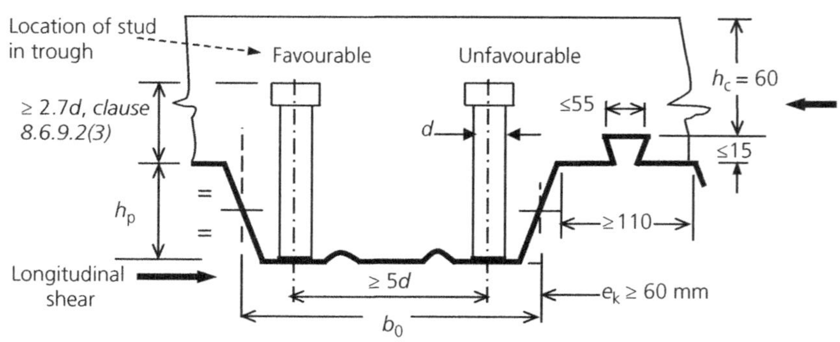

Figure 8.20 shows a pair of studs in series, for which the minimum spacing is 5*d*. The new rule for minimum e_k (60 mm) would require dimension b_0 to be at least 215 mm, which is unlikely for

trapezoidal sheeting. Use of the staggered layout would enable a narrower trough to be used. Eurocode 4 does not specify a minimum spacing in a diagonal direction, but $5d$ could be used, which reduces the minimum b_0 to 177 mm.

There is only one new rule in this *clause 8.6.9.2* that affects sheetings with re-entrant troughs: that the nominal thickness of the sheeting should be at least 0.70 mm. Most of the twentieth century test data is from sheeting exceeding 1.0 mm in thickness, with through-deck welding of studs up to 20 mm in diameter. Later work led to the reductions in $k_{t,max}$ for other situations given in *Table 8.3*. According to *clause 5.4.2.1(3)*, headed studs in decking may be deemed to be at least ductility category D2 when the design resistance is calculated using *Formulae 8.28* or *8.29*.

Clause 8.6.9.2(6) refers to the new Informative *Annex G*, which provides a different design method where the sheeting or connectors are outside the reduced scope of *clause 8.6.9*. It takes account of additional parameters: the characteristic tensile strength of concrete, factors n_y and k_u that depend on the layout of a pair of studs, the width of the top of the trough, and the transverse spacing of a pair of studs. To prevent its use as an alternative to the method of *clause 8.6.9.2*, it has the condition that at least one of the dimensions h_A and e_k should be below its lower limit in that clause. Both methods state that for situations outside the scope of either, tests to *Annex B* should be used. Comments on the *Annex G* method are given in Chapter 17 of this book. Background is available: Vigneri *et al.* (2022). *Clause 8.6.9.2(6)*

Example 8.2: comparison of two methods of prediction of P_{Rk} for a stud in sheeting

The methods of *clause 8.6.9.2* and *Annex G* are not alternatives. Their scopes make them mutually exclusive. This is done in the definitions of the limiting values of h_A and e_k in *clauses 8.6.9.2(3)* and *G.2(1)*.

Calculation of resistance P_{Rk} by both methods, for data at the boundary between their scopes, may not give the same result because they are based on different research. This is now shown for a single 19-mm stud in a central position in trapezoidal sheeting. In the usual notation, the data for *clause 8.6* are $d = 19$ mm, $h_p = 70$ mm, $h_A = 2.7d = 51$ mm, $f_u = 450$ MPa, $f_{ck} = 30$ MPa, $t = 1.0$ mm, $h_{sc} = 51 + 70 = 121$ mm.

The width b_0 is 120.4 mm, giving $e_k = 60.2$ mm for a central stud.

$E_{cm} = 32$ GPa (normal-density concrete with quartzite aggregate)

For *Annex G*, the only change is to reduce e_k to 60.0 mm, with negligible change in width b_0.

Using *clause 8.6.8.1* with $k_{cc} = 1$, the solid-slab resistance is 102 kN, from *Formula 8.24*. The limit in *clause 8.6.9.2* to the factor k_t (0.85) governs, and $\boldsymbol{P_{Rk}} = \mathbf{86.8\ kN}$.

Using *Annex G, Formula G.1* gives $P_{Rk} = 74.0$ kN for failure determined by f_u. For concrete failure, *Formula G.2* is far more complex than before. It gives $\boldsymbol{P_{Rk}} = \mathbf{71\ kN}$. Through-deck welding is assumed, and $b_{top} = 150$ mm. The agreement is not close. The two methods make different approximations and are here each at an edge of their range of validity.

Example 8.3: reduction factors for two studs in troughs for transverse sheeting

The method of *clause 8.6.9.2* is used here. The sheeting is as shown in Figure 8.20 with an assumed thickness $t = 0.9$ mm, including zinc coating. The shoulder depth h_p is 55 mm. For two 19-mm-diameter studs in parallel per trough, placed centrally and side by side (not shown in the Figure), at lateral spacing $\geq 4d$ from *clause 8.6.10.7(4)*, and $e_k = 60.2$ mm as before, assuming $b_0 = 120.4$ mm. The minimum height h_{sc} is $55 + 2.7d = 106$ mm; 125 mm is assumed. Then, from *Formula 8.29* and *Table 8.3*,

$$k_t = (0.7/1.41)(120.4/55)(125/55-1) = 1.38 \text{ (but } \leq \mathbf{0.7}, \text{ or } \leq 0.85 \text{ for one stud)}$$

For two studs in series, as in Figure 8.20, $b_0 \geq 5\,d + 120.4 = 215$ mm, which exceeds the width of many sheeting profiles. Then, the side-by-side (parallel) layout would be used. This is not possible for sheetings with a central stiffening rib. The rule $e_k > 60$ mm (*clause 8.6.9.2(3)*) is for the higher-moment side of the stud, which suggests that lower concrete cover, say 30 mm, could be used on the other side. However, in mid-span regions, the longitudinal shear can act in either direction, so this is impracticable.

The 60-mm rule, which is new, would require the use of *Annex G*, rather than *clause 8*, for most sheeting profiles with a central rib, even where only one stud per trough is needed. For these profiles, the use of a diagonal layout could be considered, with pairs offset so widely that the $5d$ rule for longitudinal spacing could be ignored. The code gives no rule for this.

8.6.10 Detailing of the shear connection and influence of execution

It is rarely possible to prove the general validity of application rules for detailing because they apply to so great a variety of situations. They are based partly on previous practice. An adverse experience causes the relevant rule to be made more restrictive. In research, existing rules are often violated when test specimens are designed, in the hope that good experience may enable existing rules to be relaxed.

Resistance to separation and 'bottom reinforcement'

Clause 8.6.10.1(1) The purpose of ***clause 8.6.10.1(1)*** on resistance to separation is to ensure that failure surfaces in the concrete from longitudinal shear cannot pass above the connectors and below the reinforcement, intersecting neither. Tests show that these surfaces are not plane. The local reinforcement is just one of the many factors that affect their shear resistance. The dimension 30 mm was intended to ensure that a failure surface that met this condition had a certain minimum slope, where both the bars and the studs were at their maximum permitted spacing.

Where a composite slab has only one layer of mesh reinforcement, it is best placed high in the slab for crack-width control and fire resistance rather than 30 mm below the studs, so the rule has not always been followed. It has now been made applicable only to solid slabs, because it has been found that for this purpose the sheeting is equivalent to reinforcement.

The required embedment of studs in the slab has been increased (the rule $h_A \geq 2.7d$, *clause 8.6.9.2(3)*). Research has found (Degtyarev and Hicks, 2023) that the resistance of studs in transversely oriented ribs is insensitive to position of the reinforcement and that decking has some effect as transverse reinforcement.

Cover to connectors

Stud connectors have occasionally been used with zero top cover. This is now prevented by
Clause 8.6.10.2(2) ***clause 8.6.10.2(2)***.

Loading of shear connection during execution

As almost all relevant provisions on execution appear in standards for either steel or concrete struc-
Clause 8.6.10.2(3) tures, there is no section 'Execution' in EN 1994-1-1. This is why ***clause 8.6.10.2(3)*** appears here. The method of construction of composite beams and slabs (i.e. whether propped or unpropped), and also the sequence of concreting, affects stresses calculated by elastic theory and the magnitude of deflections. It is significant, therefore, in verifications of resistance of cross-sections in Class 3 or 4 and in serviceability checks.

Where propped construction is used, it is usual to retain the props until the concrete has achieved a compressive strength of at least 75% of its characteristic value. If this is not done, verifications at removal of props should be based on a reduced compressive strength.

Clause 8.6.10.2(3) gives a lower limit for this minimum concrete strength. It also relates to the staged casting of a concrete flange for an unpropped composite beam, setting, in effect, a minimum time interval between successive stages of casting.

Figure 8.21. Longitudinal shear reinforcement in an L-beam

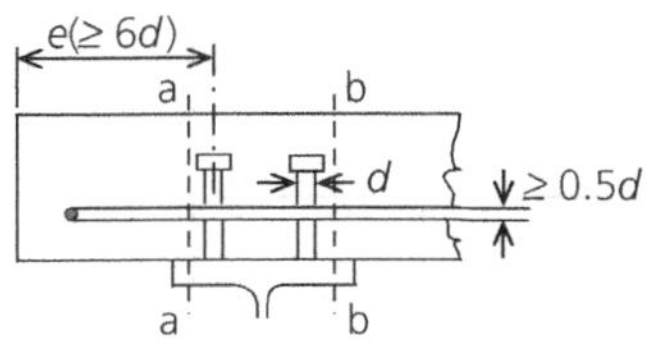

Local reinforcement in the slab

Where shear connectors are close to a longitudinal edge of a concrete flange, use of U-bars is almost the only way of providing the full anchorage required by ***clause 8.6.10.3(1)*** beyond surfaces such as a–a in Figure 8.21. The U-bars can be in a horizontal plane, as shown here, or, where top reinforcement is needed, in a vertical plane. *Clause 8.6.10.3(1)*

The splitting referred to in ***clause 8.6.10.3(2)*** is a common mode of failure in push-test specimens, which have narrow slabs. It was also found, in full-scale tests, to be the normal failure mode for composite L-beams constructed with precast slabs (Johnson and Oehlers, 1982). *Clause 8.6.10.3(2)*

Detailing rules are given in *clause 8.6.10.3(2)* for slabs where the edge distance *e* in Figure 8.21 is less than 300 mm. The required area of bottom transverse reinforcement should be found using *clause 8.6.11*.

In the unhaunched slab shown in Figure 8.21 failure surface b–b will be critical because the shear on surface a–a is low in an L-beam with an asymmetrical concrete flange.

No rules are given for the effectiveness as transverse reinforcement of profiled sheeting transverse to the supporting beams, with ribs that extend to a cantilever edge of the slab. The length needed to develop the full tensile resistance of the sheeting will be known from the design procedure for the composite slab. It is always greater than the minimum edge distance of 6*d*, and usually greater than 300 mm. Where it exceeds the length *e* available, bottom transverse reinforcement may be needed. The situation can be improved by placing all the connectors near the inner edge of the steel flange. Sheeting with ribs parallel to the free edge should not be assumed to resist longitudinal splitting of the slab.

Figure 8.22. Examples of details susceptible to longitudinal splitting

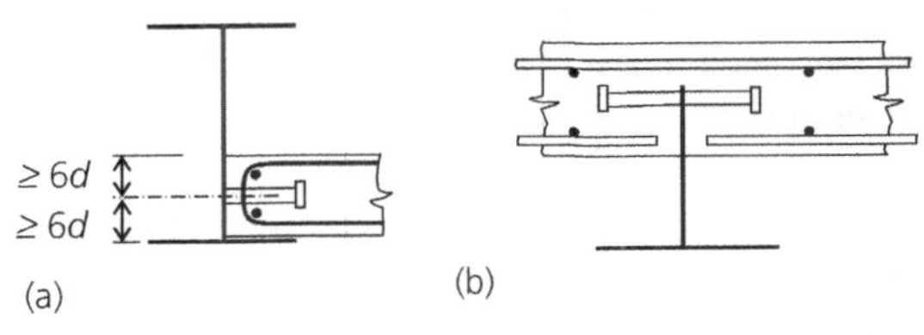

Horizontally lying studs

There is also a risk of splitting where the shank of a stud (a 'horizontally lying stud') is parallel and close to a free surface of the slab – for example, as shown in Figure 8.22. The U-bars should then be in a vertical plane. Research has found (Kuhlmann and Breuninger, 2002) that the '6*d*' rule for edge distance in *clause 8.6.10.3(2)* and the provision of U-bars may not be sufficient to ensure that the design shear resistance of the stud is reached. The height of the stud and the longitudinal bars shown are also important.

The slip capacity may be less than 6 mm, so the shear connection may not be 'ductile'. The stud shown in Figure 8.22(b) may be subjected to simultaneous axial tension and vertical and longitudinal shear, so details of this type should be avoided. Design rules are given in Normative *Annex F*, 'Headed studs that cause splitting forces in the direction of the slab thickness'. Further information is given in Chapter 16 of this book and in Kuhlmann and Kürschner (2006).

Figure 8.23. Reinforcement at the end of a cantilever

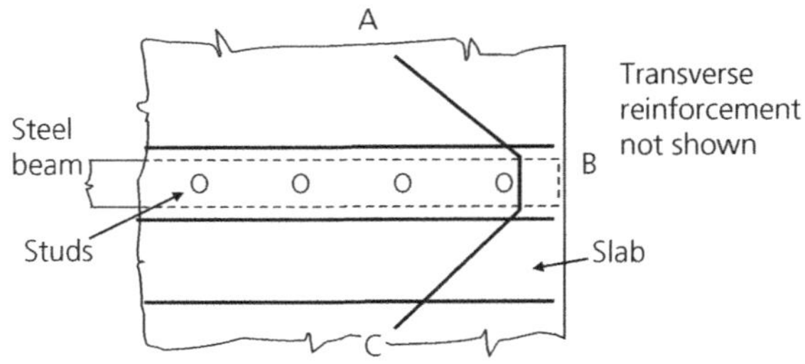

Reinforcement at the end of a cantilever

At the end of a composite cantilever, the force on the concrete from the connectors acts towards the nearest edge of the slab. The effects of shrinkage and temperature can add further stresses that tend to cause splitting in region B in Figure 8.23, so the local reinforcement needs careful detailing. *Clause 8.6.10.3(3)* ***Clause 8.6.10.3(3)*** can be satisfied by providing 'herringbone' bottom reinforcement (ABC in Figure 8.23) sufficient to anchor the force from the connectors into the slab, and by ensuring that the longitudinal bars provided to resist that force are anchored beyond their intersection with ABC.

The situation where a column is supported at point B is particularly critical. Incompatibility between the vertical stiffnesses of the steel beam and the slab could cause local shear failure of the slab, even where the steel beam alone is locally strong enough to carry the column.

Haunches

Clause 8.6.10.4 The detailing rules of ***clause 8.6.10.4*** are based on experience, rather than evidence from tests, and are long-established. In regions of high longitudinal shear, deep haunches should be used with caution because there may be little warning of failure. Haunches formed by profiled sheeting are considered under *clause 8.6.9.*

Maximum spacing of connectors

Situations where the stability of a concrete slab is ensured by its connection to a steel beam are unlikely to occur in buildings. The converse situation, stabilisation of the steel flange, is of interest only where the steel compression flange is not already in Class 1, so it rarely arises where rolled I-sections are used.

Where the steel beam is a plate girder, there is unlikely to be any need for a top flange in Class 1. Its proportions can usually be chosen such that it is in Class 2, unless a wide thin flange is needed to avoid lateral buckling during construction.

Clause 8.6.10.5(2) ***Clause 8.6.10.5(2)*** is not restrictive in practice. As an example, a plate girder is considered, in steel with $f_y = 355$ N/mm^2, where the top flange has $t_f = 20$ mm, an overall breadth of 350 mm, and an outstand c of 165 mm. The ratio ε is 0.81 and the slenderness is

$$c/(t_f\varepsilon) = 165/(20 \times 0.81) = 10.2$$

so from Table 7.4 of EN 1993-1-1, the flange is in Class 3. From *clause 8.6.10.5(2)*, it can be assumed to be in Class 1 if shear connectors are provided within 146 mm of each free edge, at longitudinal spacing not exceeding 356 mm, for a solid slab.

The factor 22 in this clause is based on the assumption that the steel flange cannot buckle towards the slab. Where there are transverse ribs (e.g. due to the use of profiled sheeting), the assumption may not be correct, so the factor is reduced to 15. The maximum spacing in this example is then 243 mm. These rules are long-established. The factors 22 and nine were in British codes over 50 years ago.

Clause 8.6.10.5(3) The maximum longitudinal spacing of connectors in buildings, given in ***clause 8.6.10.5(3)***, $6h_c \leq 800$ mm, is based on behaviour observed in tests, particularly those where composite slabs have studs in alternate ribs only. Some uplift then occurs at intermediate ribs. The execution standard EN 1090 'Execution of steel structures and aluminium structures' states, in clause 8.5.2 of Part 4, that ribs of transverse profiled sheeting with wavelength between 100 and 400 mm should be attached to the steel beam at every second rib, or more closely. Shear connection will always do this and is usually provided in every rib. Where studs are welded through holes in sheeting, their heads provide resistance to uplift.

Detailing for shear connectors

The rules of ***clause 8.6.10.6*** are intended to prevent local overstress of a steel flange near a shear connector and to avoid problems with stud welding. ***Clauses 8.6.10.7(1)*** and ***8.6.10.7(2)*** are concerned with resistance to uplift. The rules for resistance of studs, minimum cover and projection of studs above bottom reinforcement often lead to the use of studs of height greater than the minimum in *clause 8.6.8.1(1)*, $3.9d$. *Clause 8.6.10.6* *Clause 8.6.10.7(1)* *Clause 8.6.10.7(2)*

The limit 1.5 for the ratio d/t_f in ***clause 8.6.10.7(3)*** influences the design of shear connection for closed-top box girders in bridges. For buildings, the more liberal limit of *clause 8.6.10.7(5)* normally applies. *Clause 8.6.10.7(3)*

In ***clause 8.6.10.7(4)***, the low minimum lateral spacing of studs in 'solid slabs', $2.5d$, facilitates the use of large precast slabs supported on the edges of the steel flanges, with projecting U-bars that loop over pairs of studs. There is much experience, validated by tests (Lam *et al.*, 2000), of this form of construction, especially in multi-storey car parks. Closely spaced pairs of studs must be well confined laterally. The words 'solid slabs' should, therefore, be understood to exclude haunches. *Clause 8.6.10.7(4)*

Further detailing rules for studs placed within troughs of profiled sheeting are given in ***clause 8.6.10.8***. The first rule, which requires studs to project $2d$ above sheeting, applies to all sheeting, and to design to *Annex G*. For open-trough profiles, the use of resistances to *clause 8.6.9.2* requires a higher projection, $2.7d$. The rule in ***clause 8.6.10.8(2)*** is intended to ensure good compaction of concrete around studs. *Clause 8.6.10.8* *Clause 8.6.10.8(2)*

Clause 8.6.10.8(3) defines the largest top-flange stiffener for profiled sheeting that can be assumed to be 'small', which enables the depth of the sheeting to be taken as h_p, rather than h_{pg} (*Figure 8.21*) in some cases. There is further comment under *clause 8.6.9*. *Clause 8.6.10.8(3)*

8.6.11 Longitudinal shear in concrete slabs

The subject of ***clause 8.6.11*** is the avoidance of local failure of a concrete flange near the shear connection by the provision of appropriate reinforcement transverse to the span of the beam. These bars enhance the resistance of a thin concrete slab to in-plane shear in the same way that stirrups strengthen a concrete web in vertical shear. A cross-section such as a–a in both Figure 8.21 and *Figure 8.22* usually has to resist hogging bending as well as longitudinal shear. The same situation occurs in reinforced concrete T-beams. The rules for shear surfaces of type a–a are based on a truss analogy. It is an application of strut-and-tie modelling. *Clause 8.6.11*

Transverse reinforcement is also needed to prevent longitudinal splitting of the slab and local shear failure around a group of shear connectors. As the shear transfer from a web to a flange is less uniform for Eurocode 4 than for Eurocode 2, it provides additional rules. The shear surfaces to be checked and the basic design method of *clause 8.6.11* are unchanged from the previous Eurocode 4.

Shear surfaces

Clause 8.6.11.1(3) refers to 'shear surfaces'. Those of types b–b, c–c and d–d in *Figure 8.22* are different from a type a–a surface because they resist almost the whole longitudinal shear. The relevant reinforcement intersects them twice, as shown by the factors 2 in the table in *Figure 8.22*. The shear for a surface of type a–a close to the connectors is the proportion of the total longitudinal shear given by the relative effective widths of the slab on the two sides of the beam – for example, half, for a symmetrical T-beam. *Clause 8.6.11.1(3)*

Clause 8.6.11.1(4) requires the design longitudinal shear to be 'consistent with' that used for the design of the shear connectors. This means that the distribution along the beam of resistance to in-plane shear in the slab should be the same as that assumed for the design of the shear connection – for example, uniform resistance to longitudinal shear flow (v_L) should be provided where the connectors are uniformly spaced, even if the vertical shear over the length considered is not constant. It does not mean that if, for reasons concerning detailing, the shear resistance of the connectors provided exceeds the longitudinal shear assumed for their design, the transverse reinforcement must provide the same degree of over-strength. *Clause 8.6.11.1(4)*

Clause 8.6.11.1(5) In applying ***clause 8.6.11.1(5)***, it is sufficiently accurate to assume that the longitudinal bending stress in the concrete flange is constant across its effective width and zero outside it. The clause is relevant, for example, to finding the shear on plane a–a in the haunched beam shown in *Figure 8.22*, which, for a symmetrical flange, is less than half the shear resisted by the connectors. The use here of the Eurocode 2 symbol τ_{Ed} for longitudinal shear stress is a change from the previous Eurocode 4, which used v_{Ed}, though the use of $v_{L,Ed}$ for longitudinal shear per unit length is unchanged.

Figure 8.24. Truss analogy for in-plane shear in a concrete flange

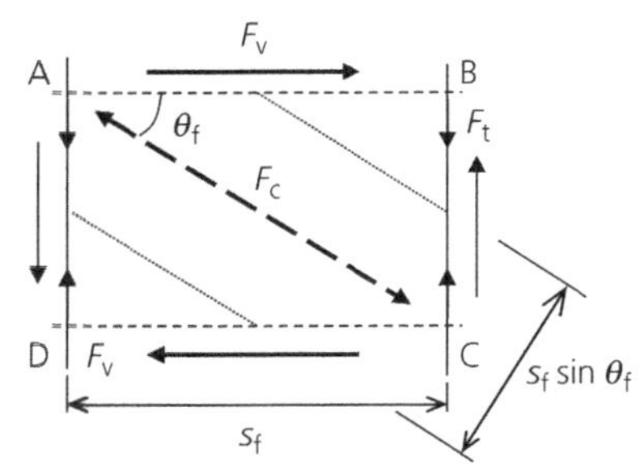

Resistance of a concrete flange to longitudinal shear

Clause 8.6.11.2 ***Clause 8.6.11.2*** refers to the method of EN 1992-1-1: 8.2.5, which is written for a design longitudinal shear stress τ_{Ed} acting on a cross-section of thickness h_f. The truss analogy for transfer of longitudinal shear from the shear connectors laterally across the effective width of the concrete flange is shown in Figure 8.24. This is a plan of a width AD and a spanwise length AB equal to the spacing s_f of the transverse reinforcement, with bars each of area A_{sf}. The effective thickness is h_f, with longitudinal shear stress τ_{Ed}.

The shear force per transverse bar is $F_v = \tau_{Ed}\, h_f\, s_f$, acting on side AB of the rectangle shown. It is transferred to side CD by a concrete strut AC at angle θ_f to AB, and with edges that pass through the mid-points of AB and so on, as shown, so that the width of the strut is $s_f \sin \theta_f$. Failure is assumed to occur by either crushing of the concrete strut or yielding of the reinforcement.

For equilibrium at C in Figure 8.24, $F_t = F_v \tan \theta_f$. To avoid yielding of the reinforcement,

$$F_t \leq A_{sf} f_{yd}, \text{ so this check is } A_{sf} f_{yd} \geq \tau_{Ed} h_f s_f \tan\theta_f \qquad \text{(8.69 in EN 1992-1-1)}$$

For equilibrium at A, $F_v = F_c \cos \theta_f$. To avoid crushing of the strut,

$$F_c = \tau_{Ed} h_f s_f \sec\theta_f \leq \nu f_{cd} h_f s_f \sin\theta_f$$

which is given in EN 1992-1-1, clause 8.2.5(4) as

$$\tau_{Ed} \leq \nu f_{cd} / (\cot\theta_f + \tan\theta_f) \qquad \text{(8.70 in EN 1992-1-1)}$$

where ν is a strength reduction factor, given in EN 1992-1-1 as 0.5.

If this check fails, θ_f, the angle between the diagonal strut and the axis of the beam, should be increased, which increases the area of transverse reinforcement required. The angle θ_f is chosen by the designer. The limits specified for it in clause 8.2.5(3) of EN 1992-1-1, 18° to 45°, are wider than before for compression flanges. The tighter limits for tension flanges, 39° to 45°, are unchanged. Earlier methods assumed $\theta_f = 45°$, which usually leads to more transverse reinforcement than necessary.

The symbol ν (Greek nu) is a strength reduction factor to allow for the difference in strength of concrete determined from a standard cylinder compared to the reduced strength of concrete within a structure, which has been affected by cracking and duration of loading. For angles within the specified limits, clause 8.2.5(4) of EN 1992-1-1 says that value 0.5 *may* be used. The method of clause 8.2.3(5) of EN 1992-1-1 for lower angles is unlikely to be needed.

Clause 8.6.11.2(3) gives the areas of reinforcement effective for the shear planes to be considered. The rules for shear surfaces of type b–b and c–c in *Figure 8.22* usually ensure that at least half of the reinforcement required to resist longitudinal shear will be placed quite low in the cross-section. *Clause 8.6.11.2(3)*

Planes of type a–a will usually also need reinforcement for hogging bending moment from the loading on the slab. Clause 8.2.5(6) of EN 1992-1-1 gives simple rules for its design and references to more complex methods. Its rules enable the truss reinforcement to be assumed effective, to the extent defined, for both longitudinal shear and transverse bending.

The reference in ***clause 8.6.11.2(4)*** to clause 8.2.6 of EN 1992-1-1, 'Shear at interfaces' relates to longitudinal shear between a precast slab and in situ concrete. Longitudinal shear in floors using precast slabs is the subject of *clause J.6*, which defines the shear surfaces to be considered in design using *clause 8.6.11*. *Clause 8.6.11.2(4)*

Minimum transverse reinforcement

The reference in ***clause 8.6.11.3(1)*** is to a rule in clause 12.2(4) of EN 1992-1-1 on minimum transverse reinforcement for 'slabs requiring shear reinforcement'. Its purpose is to ensure that failure does not occur immediately upon shear cracking and that truss action can develop. It is used in Example 8.6. It will usually be possible to assume that transverse axial tension or compression in the slab can be neglected. *Clause 8.6.11.3(1)*

Longitudinal shear in beams with composite slabs (*clause 8.6.11.4*)

These design rules are mostly worded for sheeting with troughs that run transverse to the span of the steel beam, though *Figure 8.23* also shows parallel sheeting. Where sheets intersect a beam at some other angle, one can either use the more adverse of the two sets of rules or combine them in an appropriate way. Dimension h_f in **clause *8.6.11.4(1)*** is the effective thickness for the truss analogy, so is found using the gross depth of the sheeting. Symbol h_f comes from EN 1992-1-1. It has the same value here as the Eurocode 4 symbol h_c, which is used in *Formula 8.30*. *Clause 8.6.11.4(1)*

The contribution made by sheeting to resistance to longitudinal shear depends not only on its direction, but also on whether the designer can determine the position of the ends of individual sheets, and whether these ends are attached to the steel beam by through-deck welding of stud connectors. If these decisions are to be made subsequently by the contractor, the designer may be unable to use transverse sheeting as reinforcement for shear in the plane of the slab. Its contribution to shear resistance is substantial where it is continuous across the beam (***clause 8.6.11.4(4)***). *Clause 8.6.11.4(4)*

Where discontinuous sheeting is attached to a supporting steel flange by through-deck welding, ***clause 8.6.11.4(5)***, the contribution to resistance to longitudinal shear is smaller, but useful. It is given as $P_{pb,Rd}/s_x$, which, from *clause 10.7.4*, is $k_\phi\ d_{do}\ t_p f_{yp,d}/s_x$, where s_x is the longitudinal spacing of studs with a weld collar of diameter d_{do} that anchor an edge of the sheet to the beam flange. *Clause 8.6.11.4(5)*

Factor k_ϕ in *clause 10.7.4* (where further comment is given) depends on the distance from the through-welded stud to the free end of the sheet. This must be at least 1.5 d_{do}. The resistance increases until it reaches $5d_{do}$.

Clause 8.6.11.4(5) is applied in Example 8.6, where k_ϕ is found to be 2.91, and this contribution provides 60% of the longitudinal shear resistance required. Sheets attached only by fixings used for erection and those with span parallel to that of the beam are ineffective as transverse reinforcement.

Through-deck welding of studs is also used to enhance the resistance of a composite slab to longitudinal shear. EN 1994-1-1 does not state whether both procedures can be used at once. The question is discussed with reference to Figure 8.25, which shows an exploded view of the base of a stud welded to a steel flange through a layer of sheeting, which spans towards the right-hand side of the diagram. The concrete in the mid-span region of the composite slab applies the force A to the upper part of the stud, which enables it to anchor a tension A in the sheeting. The force B is due to the action of the steel top flange and the sheeting as equivalent transverse reinforcement.

Figure 8.25. A through-welded stud acting as end anchorage for a composite slab, and also providing continuity for transverse reinforcement

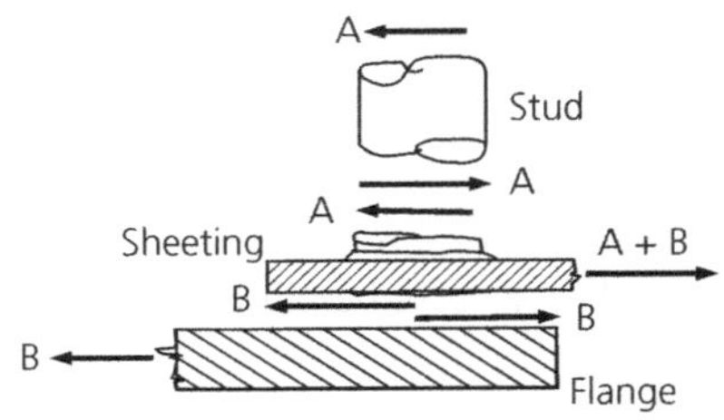

It is clear from Figure 8.25 that forces A and B are not additive for the stud, because A exists above the sheeting and B below it, but they are additive for the sheeting. The model in *clause 10.7.4(3)* for the calculation of resistance $P_{pb,Rd}$ is not shear failure of the stud, but bearing failure of the sheeting, which depends on the distance from the stud to the edge of the sheet. This analysis shows that if the two procedures are used for the same stud, the available resistance $P_{pb,Rd}$ should be split between them, in whatever ratio may be required.

Clause 8.6.11.4 does not refer to profiled sheeting with ribs parallel to the span of the beam. The designer is unlikely to have control of the location of the lap joints in the sheeting, and its stiffness in the transverse direction is very low. Its contribution to longitudinal shear resistance is, therefore, neglected.

Example 8.4: transverse reinforcement for longitudinal shear

For detailing, the reinforcement required for longitudinal shear will be combined with that required for transverse bending, assumed here to be hogging. The shear reinforcement is denoted A_s/s, where s is the spacing of bars of area A_s, and is assumed to vary little along a shear span. It is useful to find its value first. The greater of A_s/s and A_{sf}/s_f (which depends on the angle θ_f) is the reinforcement required for a layer near the top surface of the slab (clause 8.2.5(6) of EN 1992-1-1), so an initial value for θ_f is found from Formula 8.69 of Eurocode 2 by assuming A_{sf}/s_f equals the now-known value of A_s/s.

Figure 8.26. Cross-section for Example 8.4

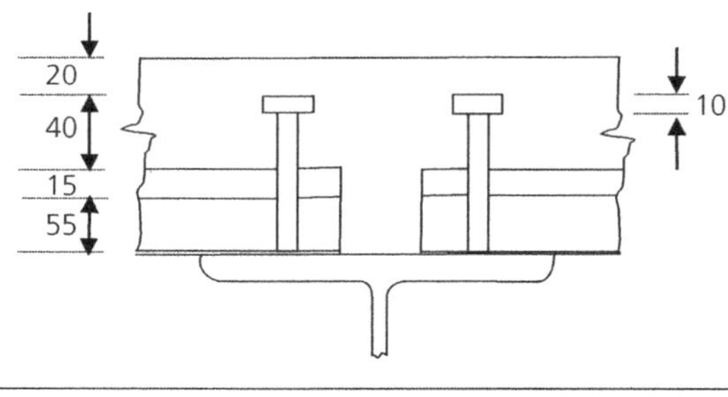

As an example, Figure 8.26 shows a part cross-section of a continuous composite beam supporting trapezoidal sheeting with the depths shown. The composite slab has been designed with continuous spans, and the transverse hogging bending moment requires top reinforcement with $A_s/s \geq 360$ mm^2/m for the whole span. It is assumed that from *clause 8.6.10.2(2b)*, the required top cover for the studs is 20 mm, and that C30/37 concrete is used, so $f_{cd} = 17$ MPa.

One condition for the use of the design method of *clause 8.6.9.2* for studs in troughs is that they should project at least 2.7d above the sheeting. This presumably excludes any small top rib, though this is not clear from *Figure 8.19* as none is shown. The chosen thickness of slab is 130 mm. Using studs with the non-standard length of 110 mm, their projection above the sheeting is 110 – 55 = 55 mm = 2.9d.

There is little space for two layers of reinforcement so one layer is assumed, giving an effective depth of about 105 mm for hogging bending, used in the calculation of A_s/s.

The design longitudinal shear flow, F_v in Figure 8.24 is $F_{v,Ed} = 217$ kN/m. For the truss analogy, the thickness of slab, h_f, is 60 mm, so $\tau_{Ed} = 217/60 = 3.62$ MPa. To estimate θ_f, we assume $A_{sf}/s_f = 360$ mm^2/m.

From Formula 8.69 of EN 1992-1-1 with $f_{yd} = 435$ MPa,

$$\tan\theta_f = 360 \times 435 / (3.62 \times 60\ 000) = 0.722, \text{ whence } \theta_f = 35.6°.$$

For a flange in tension, θ_f should not be less than 38.7°, so it is increased to this value. To check the resistance of the concrete strut, Formula 8.70 then gives

$$\tau_{Ed} \leq \nu f_{cd} / (\cot\theta_f + \tan\theta_f) = 4.15 \text{ MPa } (> 3.62 \text{ MPa, so satisfactory})$$

Formula 8.69 of EN 1992-1-1 then gives

$$A_{sf} / s_f \geq \tau_{Ed} h_f \tan\theta_f / f_{yd} = 3.62 \times 60 \times 0.801 / 0.435 = 400 \text{ mm}^2 / \text{m}$$

This exceeds 360 mm^2/m and is the reinforcement required for this single layer – for example, **10-mm bars at 180-mm spacing** (436 mm^2/m).

Example 8.5: partial shear connection and arrangement of shear connectors

This example is limited to simply-supported beams, because *clause 8.2.1.3(1)* applies only to the use of partial shear connection to regions of sagging bending, with the implied assumption that the slab is above a steel I-section. A beam with span 10.4 m has distributed loading, equal steel flanges, a uniform cross-section in Class 2, and S355 steel. A composite slab is assumed, with transverse trapezoidal sheeting, to show how profiled steel sheeting can restrict the layout of connectors. The design may be governed by the minimum degree of shear connection. Two types of shear connector are considered: 19-mm studs and an imagined new type, designed to be demountable.

19-mm studs, with proposed length after welding 95 mm, in category D2 or D3

It should be checked first that the design satisfies the conditions in *clause 8.6.9.2(3)* for the reduction factor k_t of *Formula 8.29* to be applicable. It may be found, for example, that the troughs in the sheeting require studs longer than 95 mm. The use of two studs per trough should be avoided, because the second stud increases the shear resistance from *Formula 8.29* by only 41%.

The method of *clause 8.6.3.3* is used. It has the problem that the degree of shear connection required affects the bending resistance of the beam, and the required minimum degree is a function of this resistance. A suggested route is to find first the shear resistance of a single stud welded through the sheeting, allowing for factor k_t; then calculate the number of studs needed for full shear connection, n_f. The ratio of number of troughs in a half span to n_f is the degree of shear connection with one stud per trough, η_1, say. This is then compared with the required minimum degree. If it is high enough, the bending resistance provided, $M_{Rd,1}$, say, is compared with M_{Ed}. Where the resistance more than sufficient, omission of studs from some troughs is not recommended. It may be prevented by rules on their maximum spacing in *clause 8.6.10.5*. Also, the rules for partial shear connection were derived assuming uniformly spaced connectors.

Application

The following data may be unrealistic, as they are chosen to illustrate methods, but for the beam

$$L = 10.4 \text{ m}, M_{a,pl,Rd} = 377 \text{ kN m}, M_{pl,Rd} = 989 \text{ kN m, and } M_{Ed} = 550 \text{ kN m at mid-span.}$$

The profiled sheeting is 1.1 mm thick, with three troughs per metre, each of depth 60 mm and mean width 161 mm. The studs have resistance $P_{Rd} = 60.2$ kN, and are assumed to be in category D2, so that the minimum degree of shear connection is given by *Formula 8.12*.

From *clause 8.6.9.2(3)*, they must extend 2.7d above the sheeting, so their minimum height after welding is 60 + 2.7 × 19 = 111 mm, and 120-mm studs are needed.

Formula 8.29 with $n_r = 1$ gives $k_t = 1.9$, and *Table 8.3* gives $k_{t,max} = 1.0$, so P_{Rd} is not reduced. (For sheeting 1.0 mm thick, $k_{t,max} = 0.85$, and this design would need two studs per trough.) For this beam, full shear connection needs 35 studs per half span, but only 15 troughs are available, and then the degree of shear connection, η = **15/35 = 0.43**.

The simplified method of *Figure 8.5*, *clause 8.2.1.3*, is used to find the resistance with one stud per trough and gives $M_{Rd,1}(\eta) = 640$ kN m. This is sufficient if 0.43 exceeds the minimum degree of shear connection, η, from *Formula 8.12*, which depends on the method of construction, and is highest for propped construction. Then, $k_{up} = 1.0$ and η is η_0 reduced, within limits, by the factor

$$\rho_m^2 = [M_{Ed} / (0.95\, M_{Rd}(\eta))]^2 = (550 / (0.95 \times 640))^2 = 0.818$$

From *Formula 8.13*, $\eta_0 = 0.25 + 0.03 \times 10.4 = 0.562$

Formula 8.12 with $\eta_{min} = 0.4$ gives $\eta \geq 0.562 \times 0.818 \times 1 = \mathbf{0.46} \geq \eta_{min}$

From *clause 8.6.3.3(1)*, the shear connection is not sufficient, because $0.46 > 0.43$.

Suppose now that unpropped construction is used, with $M_{a,Ed} = 140$ kN m, giving

$$k_{up} = (1 - 140 / 989) = 0.858 \text{ and } \eta \geq 0.46 \times 0.858 = 0.394 \geq 0.4$$

with 0.4 the limit for category D2 studs. As $0.40 < 0.43$, the shear connection is sufficient.

To conclude, one 19 × 125 mm (nominal) stud per trough provides sufficient shear connection only where unpropped construction is used.

Normally one would design using the method above. If that is not possible, changing the beam cross-section is likely to be simpler than using the alternative given in *clause 8.6.5* designing for longitudinal shear, calculated by elastic theory. This means using

$$v_L = V_{Ed} A\bar{z} / I$$

where v_L is the shear force per unit length of beam and V_{Ed} is the vertical shear on the composite section.

For uniform loading, this gives a triangular distribution of longitudinal shear, or separate distributions, to be superimposed, if creep is allowed for by using several modular ratios. The force in the slab at mid-span now depends on the proportion of M_{Ed} that is applied to the composite member, and on the proportions of the cross-section. Connectors corresponding to this force are then spaced accordingly, possibly with extra ones near mid-span to satisfy the rules of *clause 8.6.10.5* on maximum spacing of studs.

For continuous beams, where M_{Ed} at mid-span may be much less than $M_{pl,Rd}$, design is more complex, as partial shear connection is permitted only where the slab is in compression. A general method would use the envelope of design vertical shear from the global analysis. Example 8.6 shows that a simpler method can sometimes be used. From *clause 8.6.5(3)*, the properties of the uncracked cross-section are used throughout, because this gives an over-estimate in cracked regions.

A new shear connector that is not a standard stud

The connectors would obviously have to have a size and performance suitable for the sheeting troughs available. The main difference from the previous example is that evidence is needed from load-slip testing to show that the characteristic slip δ_{uk} of the connector is not less than 6 mm, so that it is 'ductile' to *clause 8.6.1(5)*, with reference to *Table 5.1*. Another condition, in *Table 5.1*, is that slip δ_{ek} should be less than 2.5 mm. This is the slip given by the method shown in Figure B.3.

Example 8.6: two-span beam with composite slab – ultimate limit state

A floor structure for an office building consists of composite beams of uniform cross-section at 2.5 m spacing, fully continuous over two equal spans of 12.0 m (Figure 8.27). The floor consists of a composite slab of overall thickness 130 mm, spanning between the beams. The three supports for each beam may be treated as point supports, providing lateral and vertical restraint. A design is required for an internal beam, subjected to vertical loading only. The design service life is 50 years.

Figure 8.27. Elevation and cross-sections of the two-span beam: (a) section D–D; (b) section E–E; (c) elevation of beam (dimensions in mm)

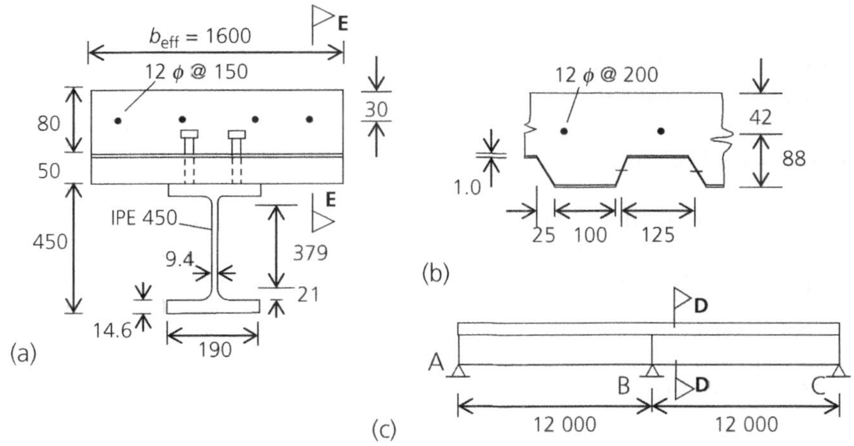

This particular concept is unlikely to be used in practice. The design will be found to be governed by bending resistance at the internal support and only a small part of the bending resistance at mid-span can be used. However, it illustrates the use of many of the provisions of EN 1994-1-1 and is chosen for that reason. The checking of the beam for serviceability limit states and the influence of semi-continuous beam-to-column connections are the subjects of worked examples in later chapters of this guide. An alternative design, treating each span as simply supported, is discussed after Example 9.1. The composite slab assumed here is not the one designed in Example 10.1. It will be assumed initially that unpropped construction is used, and that the whole 24-m length of concrete flange is cast before significant composite action is developed in any of this length.

Loadings and materials

The characteristic imposed load, including partitions, is $q_k = 7.0$ kN/m^2, with $\gamma_F = 1.50$ where γ_F is a partial safety factor. The floor finish adds: $g_{k1} = 1.20$ kN/m^2; $\gamma_F = 1.35$. These are applied to the composite structure (Table 8.3).

Lightweight-aggregate concrete is used in this example to illustrate changes in EN 1992-1-1, where this material is now treated in Annex M. This gives coefficients η for modifying the properties of normal-weight concrete in clause 5.1. Density class D1.8, used here, is taken from Table M.1. Its coefficients are $\eta_{lw,fc} = 0.82$ for compressive strength, $\eta_{lw,fct} = 0.89$ for tensile strength and $\eta_{lw,Ec} = 0.67$ for elastic modulus. Strength class C25/30 is chosen, from Table 5.1. From the data there, the mean compressive strength is $\boldsymbol{f_{cm} = 0.82 \times 33 = 27.1}$ **MPa**, and the mean tensile strength is $\boldsymbol{f_{ctm} = 0.89 \times 2.6 = 2.31}$ **MPa.**

The secant elastic modulus depends on the type of aggregate used, and Eurocode 2 gives an 'approximate indicative' value only for quartzite aggregate of $9500\,f_{cm}^{\ 1/3}$. The units are not stated and are N/mm^2. Other values can be given in national annexes, but *clause 5.1(6)* says that coefficients higher than 9500 should not be used. Assuming that this aggregate is used, and with Young's modulus for steel (Eurocode 3) as 210 GPa, the short-term modular ratio is $\boldsymbol{n_0 = 210/20.4 = 10.3}$. For creep, the simplified method of *clause 7.4.2.2(11)* is used, with the nominal modular ratio for all loadings taken as $2n_0$, so $\boldsymbol{n = 20.6}$.

For this concrete, Annex M gives the coefficient η_{cc} for design compressive strength as $(0.82 \times 40/f_{ck})^{1/3} = 1.09$, but ≤ 1, so $\eta_{cc} = 1$. *Formula 5.1* for f_{cd} includes a factor k_{tc} for sustained loading and time of loading. It is given in a note to clause 5.1.6(1) of EN 1992-1-1 as 1.0 for class CN concrete (normal Portland cement) with $t_{ref} = 28$ days.

However, Eurocode 4 uses a rectangular compressive stress block extending to the neutral axis, simpler than the one in Eurocode 2, and its *clause 5.1(3)* gives a reduced upper limit to $k_{tc}\ \eta_{cc}$ (0.85). The limit 1.0 is retained only for concrete-filled sections and elastic analyses. Hence, for plastic resistances,

$$f_{cd} = \eta_{cc} k_{tc} f_{ck} / \gamma_C = 0.85 \times 1 \times 25 / 1.5 = \mathbf{14.2\ N/mm^2}$$

BS EN 206 (2013) gives the design density for Class D1.8 concrete as between 1600 and 1800 kg/m^3. A value 1800 kg/m^3 is assumed here. The sheeting weighs 0.10 kN/m^2, and supports 0.105 m^3/m^2 of concrete, so its mass per unit volume of concrete is [0.10/(0.105 × 9.81)] × 1000 = 97 kg/m^3. A total of 1950 kg/m^3 is assumed below.

The chosen steel section is IPE 450, with dimensions shown in Figure 8.27. Its weight is g_{k3} = 0.76 kN/m, with γ_F = 1.35. Table 8.3 gives the loadings for a beam spacing of 2.5 m. The structural steel is grade S355, with γ_M = 1.0, so

$$f_y = f_{yd} = 355\ \text{N/mm}^2 \tag{X8.16}$$

and $\varepsilon = \sqrt{235/355} = 0.81$.

Table 8.3. Loadings per unit length of beam (dimensions in kN/m)

Type of load	Characteristic	Ultimate (minimum)	Ultimate (maximum)
Composite slab	5.02	6.78	6.78
Steel beam	0.76	1.02	1.02
Total, on steel beam	5.78	7.80	7.80
Floor finishes	1.20	1.62	1.62
Imposed load	17.50	0	26.25
Total, for composite beam	18.7	1.62	27.9
Total, for vertical shear	24.5	9.42	35.7

At this stage, the shape for the profiled sheeting is assumed to be as in Figure 8.27. The mean thickness of the floor is 0.105 m, giving a characteristic load

$$g_{k2} = 0.105 \times 1.95 \times 9.81 = 2.01\ \text{kN/m}^2$$

with a partial safety factor γ_F = 1.35. The transverse reinforcement is provisionally shown as 12-mm bars at 200 mm, for review when crack-width control is considered in Example 9.1.

From Informative *Annex C*, and a 'dry' environment, the 'nominal total final free shrinkage strain' is $\boldsymbol{\varepsilon_{cs} = -500 \times 10^{-6}}$

This value is used here to illustrate its consequences. It is so high that in practice it would be helpful to make a more accurate assessment, using EN 1992-1-1.

For the reinforcement, f_{sk} = 500 N/mm^2, γ_S = 1.15, and $\boldsymbol{f_{sd}}$ = **435 N/mm^2**

For ductility, the reinforcement is to be in Class B or C, from *clause 7.5.1(5)*. The shear connectors are assumed to be 19-mm studs, welded through the steel sheeting, with ultimate tensile strength $\boldsymbol{f_u}$ = **500 N/mm^2** and γ_V = 1.25.

Durability

Depending on floor finishes, the top of the slab may be exposed to 'low air humidity', for which the exposure class is XC1 for a service life of 50 years, from Table 6.1 of EN 1992-1-1. A more cautious assumption, 'moderate humidity' would require Class XC3. The minimum cover from clause 6.5.2 is governed by the bond requirement, and equals the diameter of the bars, at first assumed here to be 20 mm. The allowance for deviation for Class XC1 is given by clause 6.5.3 as 5 mm (an NDP). Hence **minimum cover** is assumed to be **25 mm generally** and can be reduced for smaller bars.

At the internal support, the 12 mm longitudinal bars are placed above the transverse bars, as shown in Figure 8.27.

Properties of the IPE 450 cross-section

From section tables

- area: $A_a = 9880$ mm^2, fillet radius: $r = 21$ mm
- second moments of area: $10^{-6} I_{ay} = 337.4$ mm^4, $10^{-6} I_{az} = 16.8$ mm^4
- torsional moment of area: $10^{-6} I_{at} = 0.659$ mm^4
- radii of gyration: $i_y = 185$ mm, $i_z = 41.2$ mm, $i_x = 190$ mm
- section moduli: $10^{-6} W_{ay} = 1.50$ mm^3, $10^{-6} W_{az} = 0.176$ mm^3
- plastic section modulus: $10^{-6} W_{pl,a,y} = 1.702$ mm^3.

It has been found convenient in calculations to use the factor 10^{-6} in this way. It enables moments in kN m to be related to stresses in N/mm^2 without further adjustment and gives numbers of convenient size. Another method is to use units of m^2 mm^2 for I.

Effective widths of the concrete flange

In *Figure 7.1* of *clause 7.4.1.2*, $L_1 = L_2 = 12.0$ m; for $b_{eff,1}$, $L_e = 10.2$ m

Assume $b_0 = 0.1$ m, then $b_1 = b_2 = 2.5/2 - 0.05 = 1.20$ m

At mid-span, $\mathbf{b_{eff}} = 0.1 + 2 \times 10.2/8 = 2.65$ m (but $\leq$ **2.5 m**), so $b_{ei} = 2.4/2 = 1.2$ m

At the internal support, for $b_{eff,2}$, $L_e = 0.25 \times 24 = 6$ m; $\mathbf{b_{eff}} = 0.1 + 2 \times 6/8 =$ **1.60 m**

At an end support, $b_{ei} = 1.20$ m; $\beta_i = (0.55 + 0.025 \times 10.2/1.20) = 0.762$

$$\mathbf{b_{eff}} = 0.1 + 2 \times 0.762 \times 1.2 = \mathbf{1.83\ m}$$

Classification of composite cross-section

For the compression flange, from Table 7.4 of EN 1993-1-1,

$$c = (190 - 9.4)/23 - 21 = 69.3 \text{ mm}$$

Hence, $c/t_w\,\varepsilon = 69.3/(14.6 \times 0.81) = 5.86$

The limit for a Class 1 flange is 9.0, so **the bottom flange is Class 1** for hogging bending.

The class of the web is sensitive to the area of longitudinal reinforcement in the slab at the internal support. It is (inconveniently) necessary to assume a value for this before the checks that govern it can be made. Design is more complex for Class 3 sections than for Class 2, so the longitudinal reinforcement at the boundary between Class 2 and Class 3 is found next.

From Table 7.3 of EN 1993-1-1, with $\alpha_c > 0.5$, this is when $c/(t_w\varepsilon) = 188/(6.53\alpha_c - 1)$

With dimensions from Figure 8.27 and $\varepsilon = 0.81$, $\alpha_c = 0.731$.

The depth of web to change stress by $2f_{yd}$ would be $(0.731 - 0.5) \times 379 = 87.6$ mm

Hence, for a Class 2 web

$$A_s < (87.6 \times 9.4 \times 355 \times 2)/435 = 1345 \text{ mm}^2$$

Large-diameter bars may not give sufficient control of crack widths, so the reinforcement is assumed to be 12-mm bars at 150 mm, or ten bars within the 1.6-m effective width. Hence,

$$\mathbf{A_s} = 36 \times 3.14 \times 10 = \mathbf{1131\ mm^2}$$

The effective area of concrete slab is 1.6 m by 80 mm, so $\mathbf{A_s/A_c = \rho_s = 0.0088}$

The requirement of *clause 7.5.1(5)* for minimum ρ_s will be checked during the design for crack-width control (see Example 9.1).

The force in these bars at yield is

$$F_{s,y} = 1131 \times 0.435 = \mathbf{492\ kN} \qquad \text{(X8.17)}$$

At mid-span, the cross-section is obviously in Class 1 or 2, so **the beam is in Class 2**.

It follows from *clause 7.4.2.4(2)* that for global analyses for ultimate limit states (provided that lateral-torsional buckling does not govern), the whole of the loading may be assumed to act on the composite member; and from *clause 8.2.1.1(1)* that rigid-plastic theory may be used for resistances to bending at all cross-sections of the beam. Partial shear connection (*clause 8.2.1.3(1)*) may be used in sagging regions. Design is thus much simpler when there are no spans with cross-sections in Class 3 or 4. This can usually be achieved in buildings, but rarely in multi-span bridges.

Plastic resistance to hogging bending (*clause 8.2.1.2*)

The plastic stress distribution for $M_{pl,Rd}$ is now found (Figure 8.28(a)). Starting from the stress distribution for $M_{pl,a,Rd}$, the depth of steel web that changes from tension to compression is

$$492/(9.4 \times 2 \times 0.355) = 74 \text{ mm}$$

The design plastic resistance to hogging bending is that of the steel section plus the effect of the reinforcing bars, shown in Figure 8.28(a)

$$M_{pl,a,Rd} = 1.702 \times 0.355 = \mathbf{604\ kN\ m} \text{ (hogging and sagging)}$$

$$M_{pl,Rd} = 604 + 492 \times 0.288 = \mathbf{746\ kN\ m} \text{ (hogging)}$$

The characteristic plastic resistance is also needed. With $\gamma_s = 1$ for the reinforcement, its force at yield increases to 492 × 1.15 = 566 kN, the depth of web to change stress is 85 mm and, by the method used for $M_{pl,Rd}$,

$$M_{pl,Rk} = \mathbf{764\ kN\ m} \text{ (hogging)}$$

These resistances may need to be reduced to allow for lateral-torsional buckling or vertical shear.

Figure 8.28. Plastic moments of resistance in (a) hogging and (b) sagging bending (dimensions in mm)

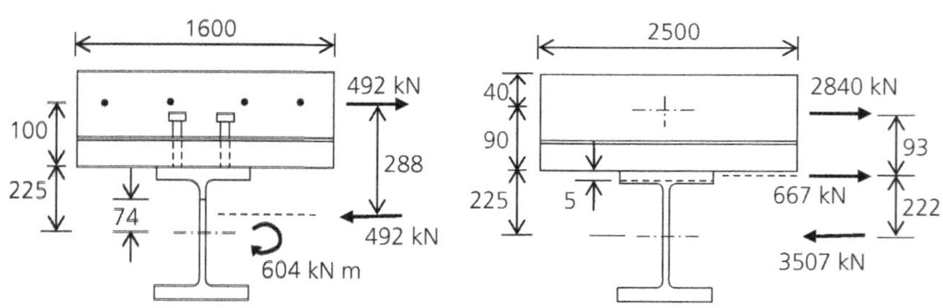

Resistance to sagging bending

For sagging bending at mid-span, reinforcement in compression is ignored and the available area of concrete is 2.5 m wide and 80 mm thick. If it is all stressed to $0.85 f_{cd}$, the compressive force is

$$F_c = 14.2 \times 2.5 \times 80 = 2840 \text{ kN} \qquad \text{(X8.18)}$$

If the whole steel section is at yield, the tensile force is

$$F_a = 9880 \times 0.355 = 3507 \text{ kN}$$

Assuming full shear connection, with the plastic neutral axis in the steel top flange, the thickness of flange to change from tension to compression is

$$t_{f,c} = (3507 - 2840) / (2 \times 0.355 \times 190) = 5.0 \text{ mm}$$

The longitudinal forces are then as shown in Figure 8.28(b), and

$$\boldsymbol{M}_{\mathbf{pl,Rd}} = 2840 \times 0.315 + 667 \times 0.222 = \mathbf{1043\ kN\ m} \text{ (sagging)} \qquad \text{(X8.19)}$$

This resistance may be reduced later by the use of partial shear connection.

Plastic resistance to vertical shear

Clause 8.2.2.2(1) refers to clause 8.2.6 of EN 1993-1-1. This defines the shear area of a rolled I-section as

$$A_v = A - 2bt_f + (t_w + 2r)t_f = 9880 - 380 \times 14.6 + 51.4 \times 14.6 = 5082 \text{ mm}^2$$

and gives the design plastic shear resistance as

$$V_{\mathbf{pl,Rd}} = A_v\left(f_y / \sqrt{3}\right) / \gamma_{M0} = 5082 \times 0.355 / \sqrt{3} = \mathbf{1042\ kN}$$

For shear buckling, *clause 8.2.2.4(1)* refers to EN 1993-1-5, which shows that no buckling check is required.

Flexural properties of elastic cross-sections

Several sets of elastic properties are needed, even where the steel beam is of uniform section, because of changes of modular ratio and effective width, and the use of 'cracked' and 'uncracked' sections. Here, slab reinforcement is ignored in the 'uncracked' analyses. The error is conservative (except for the shear connection) and usually very small. It is convenient to calculate all these properties at the outset (Table 8.4).

From *clause 7.4.2.2(11)*, creep may be allowed for by using a modular ratio $n = 2n_0 = 20.6$ for both short-term and long-term loadings. Results for $n = n_0$ are included in Table 8.4 for use in Chapter 9. Effects of shrinkage are unusually high in this example. For these, it helps to use the more accurate modular ratio from Eurocode 2.

Modular ratio for the effects of shrinkage

From *clause 7.4.2.2(4)*, the age of loading can be assumed to be one day. Creep coefficients for normal-weight concrete are given in Table 5.2 of EN 1992-1-1 in terms of h_n, the notional size of the cross-section. For a slab with both surfaces exposed, this equals the slab thickness; but the slab here has one sealed surface and h_n is twice its thickness. The mean thickness (Figure 8.27) is 105 mm, so $h_n = 210$ mm. The design life is 50 years.

For normally hardening cement and 'inside conditions', Table 5.3 gives the creep coefficient φ (50y, t_0) as 3.8, multiplied by a factor $(35/f_{ck})^A$ with A = 0.79, giving 4.96 for $f_{ck} = 25$ MPa. For lightweight concrete, Table M2 of EN 1992-1-1 gives a correction factor which in this case is $(18/22)^2 = 0.70$. The creep multiplier ψ_L in *clause 7.4.2.2(2)* takes account of the shape of the stress-time curve for the effect considered and is 0.55 for shrinkage. The modular ratio for shrinkage effects is

$$\boldsymbol{n}_{\mathbf{shr}} = n_0(1 + \psi_L\varphi) = 10.3(1 + 4.96 \times 0.7 \times 0.55) = \mathbf{29.2}$$

Calculations for cross-sections at internal support and mid-span

The calculation for the uncracked unreinforced properties at the internal support, with $n = 10.3$, is now given, as an example. In 'steel' units, the width of the slab is $1.6/10.3 = 0.155$ m, so the composite section is as shown in Figure 8.29. Its properties are

Area

$$A = 9880 + 155 \times 80 = 9880 + 12430 = \mathbf{22\,310\,mm^2}$$

Height of neutral axis above centre of steel section

$$z_{na} = 12\ 430 \times 315/22\ 310 = 175\ \text{mm}$$

Second moment of area

$$10^{-6}I_y = 337.4 + 9880 \times 0.175^2 + 12\ 430 \times 0.08^2/12 + 12\ 430\ (0.315 - 0.175)^2 = 891\ \text{mm}^4$$

Flexural stiffness

$$10^{-6}E_a I_y = 210 \times 891 \times 10^{-6} = 0.187\ \text{kN m}^2$$

Section modulus, top of slab, in concrete units, $n = 10.3$

$$10^{-6}W_{c,top} = 891 \times 10.3/180 = 51.0\ \text{mm}^3$$

Figure 8.29. Uncracked composite section at internal support, with $n_0 = 10.3$ (dimensions in mm)

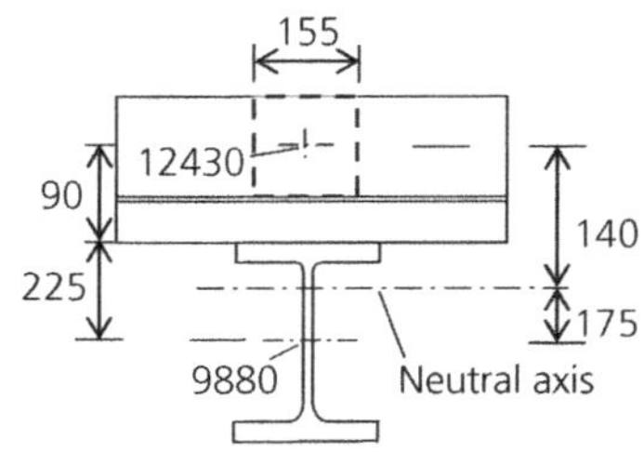

Similar calculations were done for the other elastic section properties required. The results are given in Table 8.4. The column 'Neutral axis' gives its distance above the centre of area of the steel beam, as shown in Figure 8.29.

Table 8.4. Elastic section properties of the composite cross-section

Cross-section	Modular ratio	b_{eff} (m)	Neutral axis (mm)	$10^{-6}I_y$ (mm^4)	$10^{-6}E_aI_y$ (kN m^2)	$10^{-6}W_{c,top}$ (mm^3)
(1) Support, cracked, reinforced	–	1.6	33	445	0.0934	–
(2) Support, uncracked	10.3	1.6	176	891	0.187	51.3
(3) Support, uncracked	20.6	1.6	122	720	0.151	63.7
(4) Mid-span, uncracked	10.3	2.5	209	998	0.210	70.4
(5) Mid-span, uncracked	20.6	2.5	156	829	0.174	85.8
(6) Mid-span, uncracked	29.2	2.5	129	743	0.156	96.0

Global analysis

It is found later that the resistance to lateral-torsional buckling of the beam near the internal support, $M_{b,Rd}$, exceeds $M_{pl,Rd}$ at that cross-section, so construction loads can be treated as loads on the composite beam. Wherever possible, the rules for global analysis in *clause 7* apply to both ultimate and serviceability limit states. Where alternatives are permitted, both types of limit state should be considered before making a choice.

For this beam, there is no need to take account of the flexibility of joints (*clause 7.1.2*). First-order elastic theory may be used (*clauses 7.2.2(1)* and *7.4.1.1(1)*). *Clause 7.2.2(4)* on imperfections is satisfied because lateral-torsional buckling is the only type of instability that need be considered and its resistance formulae take account of imperfections.

The simplest method of allowing for cracking, in *clause 7.4.2.3(3)*, is applicable here, and will be used. Cracking almost always occurs in continuous beams. In this example, calculations for uncracked sections, using the longer method of *clause 7.4.2.3(2)*, found that for the ultimate imposed loading on both spans, the tensile stress in the concrete at the internal support exceeded three times its tensile strength. This ignored shrinkage, which further increases the tension.

For this beam, *clause 7.4.2.3(3)* requires the use of cracked section properties over a length of 1.8 m each side of the internal support. For global analysis, *clause 7.4.1.2(4)* permits the use of the mid-span effective width for the whole span, but this does not give a uniform cross-section for the span, because of the use of 'cracked' properties, and so is not used. The longitudinal reinforcement for hogging bending is all within the effective width used here, 1.6 m, and reinforcement outside this width may be quite light. Each 12-m span thus has the non-uniform section shown in Figure 8.30. The ratios λ are given in Table 8.5.

From Table 8.3, 94% of the loading on the composite members is imposed load. Can this be assumed to cause negligible creep? For the quasi-permanent combination, the coefficient ψ_2 for variable load in office areas is given in Table A.1.7 of EN 1990 as 0.6, so some creep of the composite member is likely. In this example, $n = 2n_0 = 20.6$ will be used for all action effects except shrinkage, for which a more accurate value, 29.2, is calculated above.

Figure 8.30. Elastic propped cantilever with change of section at 0.15*L*

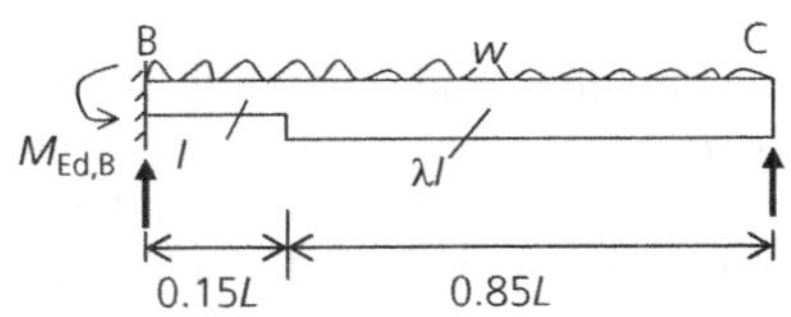

Bending moments

Although the cracked beam is of non-uniform section, calculation of bending moments algebraically is straightforward, because the two spans are equal. When both spans are fully loaded, only a propped cantilever need be considered (Figure 8.30). From the loads in Table 8.3, the permanent loading is $(5.78 + 1.20) \times 1.35 = 9.42$ kN/m and the variable loading is $17.5 \times 1.5 = 26.25$ kN/m.

For distributed loading w per unit length, and ratio of flexural stiffnesses λ, as shown, a formula for the elastic bending moment M_{Ed} at B is given by Johnson and Buckby (1986). It is

$$M_{Ed,B} = (wL^2/4)(0.110\lambda + 0.890)/(0.772\lambda + 1.228) \qquad \text{(X8.20)}$$

Results are given in Table 8.5 for $\lambda = 1.86$ and $w = 9.42 + 26.25 = 35.67$ kN/m.

Table 8.5. Design ultimate bending moments at the fixed end of the propped cantilever

Loading	w (kN/m)	n	λ	$M_{Ed,B}$ (kN m)
Permanent	9.42	10.3	2.24	130
Permanent	9.42	20.6	1.86	139
Variable	26.25	10.3	2.24	363
Variable	26.25	20.6	1.86	388

Hence, the design bending moment for the internal support, excluding shrinkage effects, with $n = 20.6$, is

$$\boldsymbol{M_{Ed,B}} = 388 + 139 = \mathbf{527\ kN\ m}$$

The vertical shear at the internal support is

$$\boldsymbol{V_{Ed,B}} = 35.67 \times 6 + 527/12 = \mathbf{258\ kN}$$

The plastic shear resistance of the IPE 450 section, $V_{pl,Rd}$, was found earlier to be 1042 kN. From a Note to *clause 8.2.2.5(1)*, bending resistance is not reduced by shear until $V_{Ed} > 0.5V_{pl,Rd}$ (0.5 is an NDP), so there is no reduction. Redistribution of moments is not used at this stage because *clause 7.4.4(4)* does not permit it if allowance for lateral-torsional buckling is required.

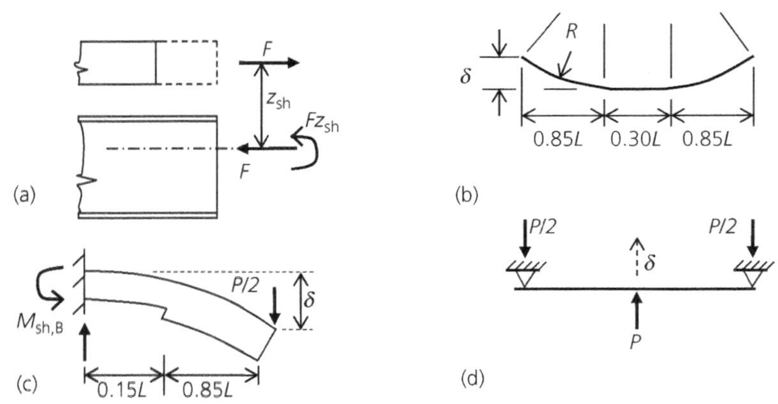

Figure 8.31. Secondary effects of shrinkage

Secondary effects of shrinkage

Shrinkage of the slab causes sagging curvature and shortening of the composite member, the 'primary effects'. In a continuous beam, the curvature causes bending moments and shear forces, the 'secondary effects'. In regions assumed to be cracked, both the curvature and the stresses from the primary effects may be neglected (*clause 7.4.2.2(8)*). This is done here. Including them can simplify calculations.

The important secondary effect in this beam, a hogging bending moment at the internal support, is now calculated. Shrinkage is a permanent action, and so is not reduced by a combination factor ψ_0. From clause 4.3.1 of EN 1992-1-1, the partial factor for shrinkage strain at ULS is 1.0 (an NDP).

The slab is imagined to be separated from the steel beam, Figure 8.31(a). Its area A_c is assumed to be that of the concrete above the sheeting. It shrinks. A force is applied to extend it to its original length. It is

$$F = A_c(E_a / n)|\varepsilon_{cs}| \tag{X8.21}$$

It acts at the centre of the slab, at distance z_{sh} above the centroid of the composite section. The parts of the beam are re-connected. To restore equilibrium, an opposite force F and a sagging moment Fz_{sh} are applied to the composite section. The radius of curvature of the uncracked part of the beam is

$$R = E_a I_{y,sag} / (Fz_{sh}) \tag{X8.22}$$

If the centre support is removed, the deflection δ at that point is

$$\delta = (0.85\ L)^2 / (2R) \tag{X8.23}$$

from the geometry of the circle (Figure 8.31(b)).

It remains to calculate the force P, applied at that point, to reduce the deflection to zero, so that the centre support can be replaced (Figure 8.31(d)). The secondary hogging bending moment at B is $M_{Ed,sh,B} = PL/2$ and the vertical shear is $P/2$.

Using slope and deflection coefficients for a cantilever, Figure 8.31(c), P can be found from the result

$$\delta = (P/2)\ L^3(0.129\lambda + 0.204) / (E_a I_{y,hog}\lambda) \tag{X8.24}$$

From Table 8.4, $\lambda = I_{mid}/I_{hog} = 743/445 = 1.67$

The calculation is as follows.

$$A_c = 2.5 \times 0.08 = 0.20\ m^2;\ E_a = 210\ kN/mm^2;\ n = 29.2;\ \varepsilon_{cs} = -\ 500 \times 10^{-6}$$

so, from Equation X8.21, $F = 719$ kN

From Equation X8.22 with $z_{sh} = 355 - 40 - 129 = 186$ mm,

$$R = 210 \times 743 / (719 \times 0.186) = 1167\ m$$

From Equation X8.23 with $L = 12$ m, $\delta = (0.85 \times 12)^2/(2 \times 1167) = 0.0446$ m

From Equation X8.24,

$$0.0446 = (P \times 864)(0.129 \times 1.67 + 0.204)/(210 \times 445 \times 1.67) = 0.00228\,P$$

whence $P = 19.5$ kN and $\boldsymbol{M_{\mathrm{Ed,sh,B}}} = 19.5 \times 6 = \mathbf{117\ kN\ m}$

If a more conservative result is acceptable, simpler calculations can be used – for example, one could assume that the hogging region has the same stiffness and shrinkage curvature as the sagging region.

Should shrinkage be neglected at ultimate load?

This significant bending moment, 117 kN m, results from the use of a material with high shrinkage and a continuous beam with two equal spans. It increases the design ultimate bending moment at B by 22%. *Clause 7.4.2.2(7)* permits it to be neglected for sections in Class 1 or 2 if resistance is not influenced by lateral-torsional buckling.

The reasoning is that as the bending resistances of the sections are determined by plastic theory, the ultimate condition approaches a collapse mechanism, in which elastic deformations (e.g. those from shrinkage) become negligible in comparison with total deformations.

However, if the resistance at the internal support is governed by lateral-torsional buckling, and if the buckling moment (to be calculated) is far below the plastic moment, the inelastic behaviour may not be sufficient for the shrinkage effects to become negligible before buckling occurs. Hence, the secondary shrinkage moment is not neglected at this stage, even though this beam happens to have a large reserve of strength at mid-span and would not fail until the support section was far into the post-buckling phase.

Resistance to lateral-torsional buckling

The top flange of the steel beam is restrained in both position and direction by the composite slab. Lateral buckling of the bottom flange near the internal support is accompanied by bending of the web, so the problem here is *distortional* lateral buckling.

The provisions in EN 1994-1-1 headed 'lateral-torsional buckling', *clause 8.4*, are in fact for distortional buckling. *Clause 8.4.1(3)* permits, as an alternative, use of the provisions in EN 1993-1-1 for steel beams. They are more conservative and based on an inappropriate model of behaviour, so the method of *clause 8.4.2* is used here. The detailed comments on it, both in the main text and in Appendix A, should be referred to as required.

The method requires the calculation of the elastic critical buckling moment at the internal support, M_{cr}, for which information was given in tables in ENV 1994-1-1. These are reproduced as graphs in Appendix A (Figures A.3 and A.4). The simple method of *clause 8.4.3* is not available, for two reasons. The loading does not conform to *paragraph (b)* of *clause 8.4.3(1)*, and the depth of the steel IPE section exceeds the limit given in *Table 8.1*.

Figure 8.32. Bending-moment distributions for ultimate limit state, excluding shrinkage

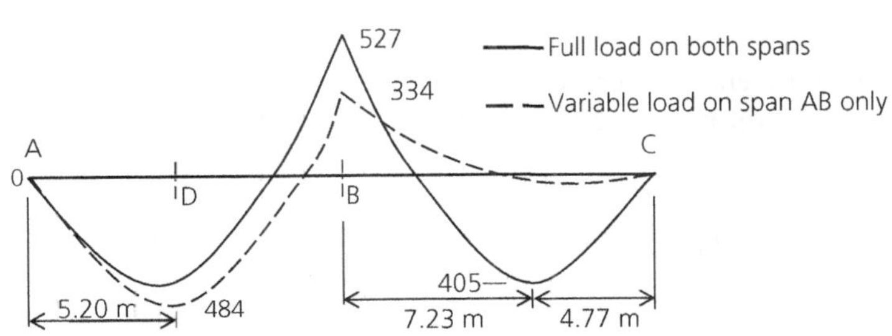

Buckling near an internal support is often most critical in a span with minimum load that is adjacent to a fully loaded span. In this beam, it was found that although the buckling moment $M_{b,Rd}$ is increased when both spans are loaded (because the length of bottom flange in compression is reduced), the increase in the applied moment M_{Ed} is greater, so the both-spans-loaded case is more critical. This case is now considered, with $n = 20.6$. All load is assumed to act on the composite member, because plastic resistances are based on this assumption.

From Table 8.5, the bending moments in the beam are as shown in Figure 8.32. Shrinkage is considered separately. The 'simply-supported' moment M_0 is

$$35.7 \times 12^2 / 8 = 643 \text{ kN m}$$

From Appendix A, Figure A.3 for C_4 with $\psi = M_B/M_0 = 527/643 = 0.834$, $C_4 = 28.3$.

The elastic critical buckling moment was given earlier (Equation X8.11) as

$$M_{cr} = (k_c C_4 / L)\,[(G_a I_{at} + k_s L^2 / \pi^2) E_a I_{afz}]^{1/2}$$

where I_{at} is the torsional moment of area of the steel section. Where a steel I-section is used, the term $G_a I_{at}$ above is negligible, because I-sections have little torsional stiffness. Including it, in this example, increases M_{cr} by only 2%. Equation X8.11 then becomes

$$M_{cr} = (k_c C_4 / \pi)(k_s E_a I_{afz})^{1/2} \qquad \text{(X8.25)}$$

where k_c is a property of the composite section, given by Equation X8.12 in section 8.4.2; k_s is defined in *clause 8.4.2(6)*, and I_{afz} is the minor-axis second moment of area of the steel bottom flange,

$$10^{-6} I_{afz} = 1.90^3 \times 14.6 / 12 = 8.345 \text{ mm}^4$$

The stiffness k_s is now found. It depends on the lesser of the 'cracked' flexural stiffnesses of the composite slab at a support and at mid-span, $(EI)_2$. The value at the support governs. An approximation for this, from Equation XA.1, derived in Appendix A, is

$$(EI)_2 = E_a[A_s A_e z^2 / (A_s + A_e) + A_e h_p^2 / 12]$$

where A_e is the equivalent transformed area per unit width of concrete in compression: $A_e = b_0 h_p / n b_s$, where b_0 is the mean width of the troughs, b_s is the spacing of the troughs, h_p is the depth of the sheeting (the net depth h_p, here), A_s is the area of top reinforcement per unit width of slab and z is the 'lever arm', as shown in Figure A.1 of Appendix A.

Figure 8.33. Cross-section of the composite slab

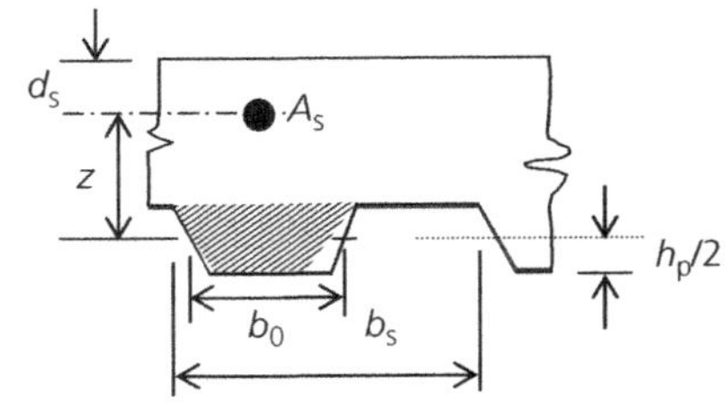

Calculation of $(EI)_2$ for the composite slab, and k_s

It is assumed that the transverse reinforcement above the steel beam will be below the longitudinal bars and not less than **12-mm bars at 200 mm**, giving $A_s = 565$ mm²/m and $d_s = 42$ mm, whence $z = 63$ mm (Figure 8.33).

From Figure 8.27, $b_0/b_s = 0.5$; $h_p = 50$ mm; and with $n = 20.6$, $A_e = 1214$ mm²/m. Hence,

$$(EI)_2 = 210\{[0.565 \times 1.214 \times 63^2 / (1214 + 565)] + 1.214 \times 50^2 / 12\,000\} = 374 \text{ kN mm}^2 / \text{m}$$

From *clause 8.4.2(6)*, for unit width of a slab continuous across the steel beams at spacing a, and assuming that the beam is an inner beam with at least four similar beams, so that $\alpha = 4$,

$$k_1 = 4(EI)_2 / a = 4 \times 374 / 2.5 = 598 \text{ kN m/rad per metre width}$$

$$\text{and } k_2 = E_a t_w^3 / [4\, h_s (1 - \upsilon_a^2)]$$

where h_s is the distance between the centres of the flanges of the IPE 450 section, 435 mm. Thus,

$$k_2 = 210 \times 9.4^3 / (4 \times 435 \times 0.91) = 110 \text{ kN/rad}$$

$$\text{and } k_s = k_1 k_2 / (k_1 + k_2) = 598 \times 110 / 708 = 93 \text{ kN/rad}$$

Calculation of k_c

Equations for k_c for a doubly symmetrical steel section are repeated from section 8.4.2:

$$k_c = h_s I_y / I_{ay}) / [(h_s^2 / 4 + i_x^2) / e + h_s] \quad \text{(X8.12)}$$

$$\text{with } e = A I_{ay} / [A_a z_c (A - A_a)] \quad \text{(X8.13)}$$

In these expressions, the symbols are properties of the steel section, given earlier, except that A is the area of the cracked composite section,

$$A = A_a + A_s = 11010 \text{ mm}^2$$

and z_c is the distance between the centroid of the steel beam and mid-depth of the slab. Here, 'slab' means the 130-mm deep composite slab, not the 80 mm depth of concrete that contributes to the composite section. It is the stiffness of the composite slab in the transverse direction that prevents rotation of the steel top flange; so

$$z_c = 225 + 130 / 2 = 290 \text{ mm}$$

Hence, $e = 11\,010 \times 337 \times 10^6/(9880 \times 290 \times 1131) = 1145$ mm

$$\text{and } k_c = (435 \times 445 / 337) / [(218^2 + 190^2) / 1145 + 435] = 1.13$$

Calculation of M_{cr} and $M_{b,Rd}$

From Equation X8.25 for M_{cr},

$$\boldsymbol{M_{cr}} = (1.13 \times 28.3 / \pi)\, [(93.0 \times 210 \times 8.345]^{1/2} = \mathbf{4109 \text{ kN m}}$$

Clause 8.3.2.3(1) of EN 1993-1-1 includes a rule that if $M_{Ed} \leq \bar{\lambda}_{LT,0}{}^2 M_{cr}$, lateral-torsional buckling effects may be ignored, with a note that for the end-fork support conditions assumed here, it may be assumed that $\bar{\lambda}_{LT,0} = 0.4$. This is expected to be confirmed in the UK's National Annex. Here, including shrinkage,

$$M_{Ed} = 644 \text{ kN m, and}$$

$$\bar{\lambda}_{LT,0}{}^2 M_{cr} = 0.16 \times 4109 = 657 \text{ kN m}$$

so lateral-torsional buckling need not be considered. From *clause 7.4.2.2(7)*, shrinkage effects may now be ignored, which reduces M_{Ed} at support B to 527 kN m.

It follows that the resistance moment at B, allowing for lateral-torsional buckling, is not required. However, it is now calculated, to illustrate the method. From *clause 8.4.2(4)*, the relative slenderness is

$$\bar{\lambda}_{LT} = \sqrt{M_{Rk} / M_{cr}} = (764 / 4109)^{1/2} = 0.431$$

For the reduction factor χ_{LT} for a rolled section, *clause 8.4.2(1)* refers to clause 8.3.2.3 of EN 1993-1-1, where buckling curve b in Figure 8.5 is specified for this IPE section. For $\bar{\lambda}_{LT} = 0.431$, it gives $\chi_{LT} = 0.91$, whence

$$\boldsymbol{M_{b,Rd}} = \chi_{LT} M_{pl,Rd} = 0.91 \times 746 = \mathbf{679 \text{ kN m}}$$

This exceeds M_{Ed} including shrinkage, 644 kN m, so is consistent with the simplification above.

Redistribution of hogging bending moment

Now that no allowance for lateral-torsional buckling is necessary, this beam meets all the conditions of *clause 7.4.4(4)* for redistribution of moments. 'Cracked' elastic analysis is used here, so from *Table 7.1* in *clause 7.4.4(5)*, M_{Ed} at support B could be reduced by up to 15%. This would enable the longitudinal reinforcement in this region to be reduced, which would reduce $M_{pl,Rd}$. The beam could then fail the condition, used above, that enabled lateral buckling to be ignored. A further reason for making no change is that deflections may be excessive, as shown later, and weakening the cross-section at B would increase them.

Design for sagging bending

The maximum sagging bending occurs in a span when the other span carries minimum load. The beam satisfies the condition of *clause 7.4.2.2(11)* for the use of $2n_0$ for all loading ($n = 20.6$), and this value is used. Removal of variable load from one span halves the bending moment it causes at the internal support, so from Table 8.5, the bending moment at the internal support is

$$M_{Ed,B} = 139 + 388/2 = 333 \text{ kN m}$$

For the fully loaded span AB, the end reaction is

$$V_{Ed,A} = 35.67 \times 6 - 333/12 = 186 \text{ kN}$$

so the point of maximum moment is distance 186/35.67 = 5.2 m from the support, and the maximum sagging moment is

$$\boldsymbol{M_{Ed}} = 186 \times 5.2/2 = \mathbf{484\ kN\ m}$$

This is so far below $M_{pl,Rd}$, 1043 kN m, that the least permitted degree of shear connection would be sufficient. M_{Ed} is even below the plastic resistance of the steel section, $M_{pl,a,Rd}$, which is 604 kN m. It does not follow that the design of this beam as non-composite would be satisfactory. Example 9.1 shows that its deflection would probably be excessive.

Design of the shear connection

First, the minimum permitted degree of shear connection is found. From *clause 8.6.3.3(3)*, the span length in sagging bending may be taken as 0.85 L, or 10.2 m here. From *Formula (8.13)* with $f_y = 355$ N/mm^2,

$$\eta_0 = n/n_f = 1 - (0.75 - 0.03 \times 10.2) = 0.56$$

The minimum practicable shear connection is one stud per trough, to be checked later against the minimum given by the rules. It is assumed that 19-mm studs in category D2 are used. Studs 100 mm high do not satisfy the condition $h_A \geq 2.7d$ for the use of *clause 8.6.9.2*, so 120-mm studs are chosen.

From *clause 8.6.9.2*, the shear resistance of a stud is governed in this case by concrete failure. From *Formula 8.24*, with k_{cc} (an NDP) taken as 1.0 in the UK, the design shear resistance per stud in a solid slab is

$$P_{Rd} = 0.29\, d^2 (f_{ck} E_{cm})^{1/2} / \gamma_V = 0.29 \times 19^2 (25 \times 20\ 400)^{1/2} / (1000 \times 1.25) = 59.8 \text{ kN}$$

This result is modified by a factor k_t. It depends on the height of the stud, h_{sc} (115 mm), the width b_0 and depth h_p of the trough in the sheeting (125 mm and 50 mm, respectively), the thickness of the sheeting (assumed to be 1.0 mm) and the number of studs per trough n_r, assumed to be one, centrally placed. From *Formula 8.29*,

$$\boldsymbol{k_t} = 0.7(b_0/h_p)[(h_{sc}/h_p) - 1] = 0.7 \times (125/50)\ (115/50 - 1) = 2.3, \text{ but } \leq \mathbf{0.85}.$$

Provided that the studs are not also required to anchor the profiled sheeting, their resistance is

$$\boldsymbol{P_{Rd}} = 0.85 \times 59.8 = \mathbf{50.8\ kN\ per\ stud} \qquad \text{(X8.26)}$$

The number of troughs in a shear span is 10.2/(2 × 0.25) = 20, so for a full-interaction force 2840 kN (Figure 8.28(b)) the degree of shear connection with one stud per trough is 20 × 50.8/2840 = 0.36.

From *clause 8.6.3.3(2)*, the rule for minimum degree of shear connection is

$$\eta \geq \eta_0 \rho_m^2 k_{up} \geq \eta_{min} \quad (8.12)$$

with $\rho_m = M_{Ed}/(0.95\, M_{Rd}(\eta))$ but $0.8 \leq \rho_m \leq 1$

Plastic analysis is used to find M_{Rd} for $\eta = 0.36$. The calculation is similar to that shown in Figure 8.28(b), except that the plastic neutral axis is within the steel web. It gives

$M_{Rd} = 875$ kN m and $\rho_m = 0.58$, so its minimum value (0.80) applies.

Factor $k_{up} = 1 - M_{a,Ed}/M_{pl,Rd}$, and is 0.91 here. Hence, $\eta \geq 0.56 \times 0.8^2 \times 0.91 =$ **0.33**

The length of the shear span depends on the load arrangement. For maximum hogging moment, it is 7.23 m (Figure 8.32). The tension in the reinforcement at yield is 492 kN. For studs spaced at 250 mm, the length of span needed for shear connection with this resistance is $492 \times 0.25/50.8 = 2.42$ m.

For maximum sagging moment, the compression in the slab for full shear connection is 2840 kN, so with $\eta = 0.33$, the length of span needed is

$$2840 \times 0.33 \times 0.25 / 50.8 = 4.61 \text{ m}$$

The total length, 7.03 m, is less than the 7.23 m available, so **one stud per trough** is sufficient. The shear span for maximum sagging moment (6.80 m) is lower than 7.03 m, but this is for a loading where the hogging bending moment is much lower, at 334 kN.

To conclude, calculation finds that one stud per trough throughout both spans is just sufficient. If it were not, a second stud per trough could be added where required.

Design of the transverse reinforcement

Clause 8.6.11.1(4) says that the design longitudinal shear for the concrete slab should be '*consistent with the design and spacing of the shear connectors*'. This is taken to mean that the resistance of the shear connection, rather than the design loading, determines the longitudinal shear. For one stud per trough,

$$v_{L,Ed} = 5 \times 50.8 = 254 \text{ kN/m}$$

Figure 8.34. Detail of a stud welded through discontinuous profiled sheeting (dimensions in mm)

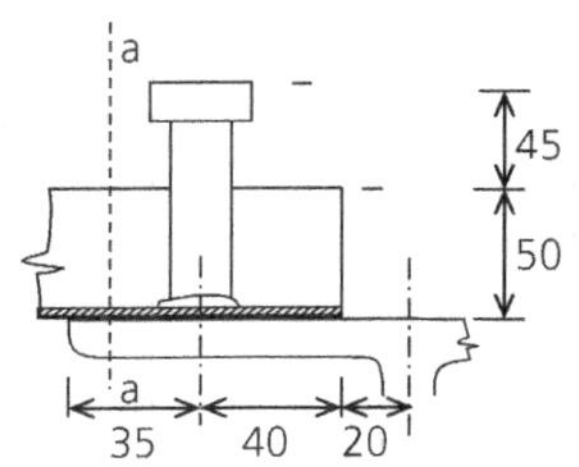

From *clause 8.6.11.4(2)* on transverse sheeting, shear surfaces that pass closely around a stud need not be considered. The critical situation is thus where sheeting is not continuous across a beam. It is at first assumed (as an example) to be anchored by a stud, as shown in Figure 8.34. For an internal beam, from symmetry, the critical shear plane, labelled a–a, should be designed to resist 127 kN/m. For an edge beam, this plane transfers almost all the shear, which would normally be about half the total for an internal beam.

The shear resisted by the sheeting is given in *clause 8.6.11.4(4)* and *(5)*. For the design bearing resistance of the sheeting, it refers to *clause 10.7.4*. For the detailing shown in Figure 8.34, the end distance a_{sc} is 40 mm. The diameter of the weld collar is taken as 1.1 $\times$ 19 = 20.9 mm; whence k_{φ} in *clause* 10.7.4(3) is $k_{\varphi} = 1 + 40 / 20.9 = 2.91$.

The thickness of the profiled sheeting is shown on Figure 8.27 as 1.0 mm; but the composite slab has not yet been designed. Here, it is assumed to be at least 0.9 mm thick, with yield strength 350 N/mm^2. The value recommended for γ_{M0} for the sheeting (in a note to clause 4.2(3) of EN 1993-1-3) is 1.0. It is expected to be confirmed in the UK's National Annex and is used here. From *Formula 10.16*,

$$P_{pb,Rd} = k_{\varphi} d_{d0} t f_{yp,d} = 2.91 \times 20.9 \times 0.9 \times 0.35 = 19.1 \text{ kN/stud}$$

For a stud spacing of 250 mm, the shear resistance provided by the sheeting is $v_{L,pd,Rd} = 19.1/0.25 = 76.4$ kN/m. This must not exceed the yield strength of the sheeting, $A_p f_{yp,d}$, which for this sheeting is over 400 kN/m.

Use of this method is optional (*clause 8.6.11.4(4)*). It is found above that use of one stud per trough is just sufficient shear connection. If this method were used, two studs per tough would be needed in some places, so it is decided to rely on the transverse reinforcement for the whole of the longitudinal shear in the slab, 127 kN/m.

The design of the transverse reinforcement is affected by rules for the control of cracking caused by the continuity of the floor slab across the beams (Example 9.1), so its area and spacing are not chosen at this stage.

Summary of Example 8.6

All important aspects of the design of this beam for persistent situations for ultimate limit states have now been considered, except the transverse reinforcement (see Example 8.4). Serviceability checks and further comments on the design are given in Example 9.1. The same beam with semi-continuous joints at support B is studied in Examples 11.1 and 12.1.

The calculations would have been more complicated if lateral buckling had governed resistance to hogging bending, or shear connection had required two studs per trough. For buildings, these problems are best avoided and, in practice, simply-supported spans are chosen where practicable.

Experience would enable many of the checks to be omitted. With normal-density concrete, shrinkage effects are often negligible. The composite slab is the main permanent load and would probably be designed before the beam. Here, the sequence of clauses in EN 1994 has been followed.

The surplus bending resistance at mid-span suggests that these 12-m spans could have been designed as simply supported. It is found that with the same loading, materials and method of construction, the same composite cross-section satisfies all the verifications for ultimate limit states. It is shown at the end of Example 9.1 that its problems are crack-width control and excessive deflection.

8.7 Fatigue

Clause 8.7.1(1)

As in the previous Eurocode 4, the detailed rules for fatigue verification are in EN 1994-2, because for many buildings they are not needed. ***Clause 8.7.1(1)*** refers to situations where this exemption applies. For structural steel, EN 1993-1-1 provides exemption where the design actions are static or quasi-static only, with a note that for buildings wind action 'does not normally lead to fatigue'. Its clause 10(3) gives the number of repeated stress cycles during the design life above which fatigue verification is required. It could apply to composite members that support fork-lift trucks or travelling cranes.

For concrete and reinforcement, clause 10.1(1) of EN 1992-1-1 lists three conditions for exemption. One of them is loading that causes less than 20 000 'significant load cycles'. The two types of loading quoted above could need fatigue checks to EN 1992-1-1.

From *paragraph (2)*, if fatigue assessment is required by either EN 1992-1-1 or EN 1993-1-1, then it is also required for shear connectors to EN 1994-2. Clause 8.7.1(3) of EN 1994-2 requires for studs that the characteristic shear force per connector should not exceed 0.75 P_{Rd}, with factor 0.75 an NDP. As $P_{Ed} \approx 1.4\ P_{Ek}$, this rule is $P_{Ed} \leq 1.05\ P_{Rd}$ approximately. The method of analysis is not specified. The clause was written for bridges and should be applied with P_{Ed} found using the methods of EN 1994-2, which do not include use of partial shear connection. No rules are given for connectors other than studs.

8.8 Composite columns and composite compression members

8.8.1 General

Scope

The only significant change in ***clause 8.8*** from the previous EN 1994-1-1 is the addition of the rules for non-linear analysis, used for the 'General method of design'. *Clause 8.8*

A composite column is defined in *clause 3.1.6* as '*a composite member subjected mainly to compression or to compression and bending*'. The title of *clause 8.8* includes 'compression members', to make clear that its scope is not limited to vertical members, but includes, for example, composite members in triangulated or Vierendeel girders. These girders may also have composite tension members, for which provisions are given in EN 1994-2.

In this guide, references to 'columns' includes other composite compression members, unless noted otherwise, and for buildings, 'column' means a length of column between adjacent lateral restraints; typically, a storey height.

Clause 8.8.1(1) refers to *Figure 8.24*, in which all the sections shown have double symmetry; but *clause 8.8.1(2)* makes clear that the scope of the general method of *clause 8.8.2* includes members of non-symmetrical section (Roik and Bergmann, 1990). *Clause 8.8.1(1)*

The bending moments in a column depend on the assumed location of the line of action of the axial force, *N*. Where the cross-section has double symmetry, this is the intersection of the axes of symmetry. In other cases, the choice made in the modelling for global analysis should be retained for the analysis of the cross-sections. A small degree of asymmetry can be allowed for by ignoring in calculations small areas of concrete such that symmetry is restored.

No shear connectors are shown in the cross-sections in *Figure 8.24* because, within a column length, the longitudinal shear is normally much lower than in a beam, and sufficient interaction may be provided by bond or friction. Shear connectors may be needed for load introduction, following *clause 8.8.4*.

The minimum compression for a member to be regarded as a column, rather than a beam, is not stated. As shown in Example 8.8, the use of the cross-sections of the types shown in *Figure 8.24* as beams without shear connectors is usually prevented by the low design shear strengths given in *clause 8.8.4.3*.

The strengths of materials in ***clause 8.8.1(3)*** are as for beams, except that class C60/75 and lightweight-aggregate concretes are excluded. For these, additional provisions (e.g. for creep, shrinkage and strain capacity) would be required (AS/NZS, 2017; Kilpatrick and Rangan, 1999; Wheeler and Bridge, 2002). The use of high-strength steel has been studied and design guidance is available, including complementary information from Singapore (AS/NZS, 2017; Liew *et al.*, 2021; Thai *et al.*, 2021). *Clause 8.8.1(3)*

Clause 8.8.1(3) concerns the scope of EN 1994-1-1. Its *Figure 8.24* does not include cross-sections with extreme ratios of areas of steel and reinforced concrete, which have been used in high-rise buildings. For these 'mixed' structures, additional consideration in global analysis of the effects of shrinkage, creep and column shortening may be needed. Many structures of this type have been built in China, with concrete-filled steel tube columns up to 1.6 m in diameter. In one comparison, designs to the 1999 Chinese code and to EN 1994-1-1 gave similar results (Zhong and Goode, 2001). *Clause 8.8.1(3)*

Recent practice includes sections with a massive steel core. Study of a 300-mm diameter solid steel section within a 400 × 6.3 mm circular hollow section found that residual compressive stress at the surface of the core reached the yield stress, and that column buckling curve d of EN 1993-1-1 was applicable (Hanswille *et al.*, 2017).

The steel contribution ratio (*clause 8.8.1(5)*) is, essentially, the proportion of the squash load of the section that is provided by the structural steel member. If it is outside the limits given, the member should be treated as reinforced concrete or as structural steel, as appropriate.

Figure 8.35. Independent bending moment and normal force (not to scale)

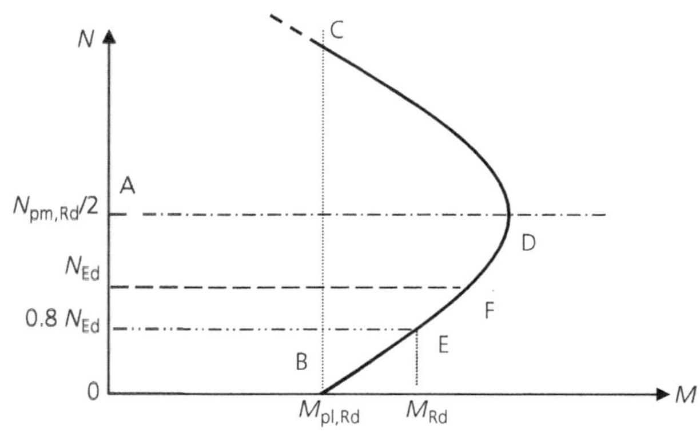

Independent action effects

Clause 8.8.1(7)

The interaction curve for the resistance of a column cross-section to combined axial force N and uni-axial bending M is shown in *Figure 8.27* and, as a polygon, in Figure 8.39 of this guide. It has a region BD where increase in N_{Ed} increases M_{Rd}. ***Clause 8.8.1(7)*** refers to a situation where at ultimate load, the factored bending moment $\gamma_F M_{Ek}$ could coexist with an 'independent' axial force that was less than its design value $\gamma_F N_{Ek}$. Its rule is illustrated in Figure 8.35, which shows region BDC of the interaction curve in *Figure 8.28* that is symmetrical about line AD. If

$$N_{Ed} < N_{pm,Rd} / 2 \tag{X8.27}$$

then M_{Rd} should be found for an axial force 0.8 N_{Ed}, as shown by point E. The reduction in M_{Rd} is usually small. Further discussion is given in Example 8.7.

Local buckling

Clause 8.8.1(8)

The principle of ***clause 8.8.1(8)*** is followed by its application rules. They ensure that the concrete (which will be reinforced in accordance with *clause 8.8.5.2*) restrains the steel and prevents it from buckling even when yielded.

For partially encased sections, the encasement prevents local buckling of the steel web and prevents rotation of the steel flange at its junction with the web, so that a higher b/t_f ratio may be used than for a bare steel section. *Table 8.4* gives the limit as 44ε, compared with about 22ε (from EN 1993-1-1) for a Class 2 flange. (In EN 1994, as in EN 1993, $\varepsilon = \sqrt{235/f_y}$, in units of newtons per square millimetre.)

For concrete-filled rectangular hollow steel sections (RHS), the limit of 52ε compares with about 41ε for a steel RHS. For a concrete-filled circular hollow section, the limiting d/t of $90\varepsilon^2$ compares with $70\varepsilon^2$ for Class 2 in EN 1993-1-1.

8.8.2 General method of design

Clause 8.8.2

Designers of composite columns will normally ensure that they fall within the scope of the simplified method of *clause 8.8.3*, but occasionally the need arises for a non-uniform or asymmetric member. The 'general method' of ***clause 8.8.2*** is provided both for this reason, and to enable advanced software-based methods to be used.

Clause 8.8.2 is more a set of principles than a design method. Development of software that satisfies these principles is a complex task. ***Clause 8.8.2(3)*** refers to 'internal forces'. These are the action effects within the column length, found from those acting on its ends, determined by global analysis to *clause 7.* It is essential to determine whether such an analysis includes or excludes the effects of member imperfections and second-order deformations, because this affects the use of its results. *Clause 8.8.2(3)* also refers to 'geometrical and material non-linear analysis with imperfections', which is defined in clause 6.1.2(8) of EN 1993-1-14. *Clause 8.8.2(3)*

As the three materials in a composite section follow different non-linear relationships, direct analysis of cross-sections is not possible. One has first to assume the dimensions and materials of the member, and then determine the axial force N and bending moment M at a cross-section from assumed values of axial strain and curvature ϕ, using the relevant material properties. The M-N-ϕ relationship for each section can be found from many such calculations. This becomes complex where biaxial bending is present.

Integration along the length of the column then leads to a non-linear member stiffness matrix that relates axial force and end moments to the axial change of length and end rotations.

8.8.3 Simplified method of design

Scope of the simplified method

The method has been calibrated by comparison with test results (Roik and Bergmann, 1992).

Its background has been published (Uy, 2003). Its scope (***clause 8.8.3.1***) is limited mainly by the range of results available, which leads to the restriction $\bar{\lambda} \leq 2.0$ in ***clause 8.8.3.1(1)***. For most columns, the method requires second-order analysis in which explicit account is taken of imperfections. The use of the strut curves of EN 1993-1-1 is limited to '*members in axial compression*', as explained in comments on *clause 8.8.3.5*. *Clause 8.8.3.1* *Clause 8.8.3.1(1)*

The restriction on unconnected steel sections in *clause 8.8.3.1(1)* reflects concern about loss of stiffness due to slip between the two sections, which would invalidate the formulae for EI of the column cross-section. This would apply to the use of a solid steel core within a concrete-filled circular hollow tube.

The limits to concrete cover in ***clause 8.8.3.1(2)*** arise from concern over strain softening of concrete invalidating the interaction diagram (*Figure 8.27*), and from the limited test data for columns with thicker covers. These provisions normally ensure that for each axis of bending, the flexural stiffness of the steel section makes a significant contribution to the total stiffness. Greater cover can be used by ignoring in calculation the concrete that exceeds the stated limits. *Clause 8.8.3.1(2)*

The limit of 6% in ***clause 8.8.3.1(3)*** on the reinforcement 'used in calculation' is more liberal than the 4% (except at laps) recommended in EN 1992-1-1. This limit and that on maximum slenderness are unlikely to be restrictive in practice. ***Clause 8.8.3.1(4)*** is intended to prevent the use of sections susceptible to lateral-torsional buckling. *Clause 8.8.3.1(3)* *Clause 8.8.3.1(4)*

Resistance of cross-sections

Calculations for composite sections, with three materials, are potentially more complex than for reinforced concrete, so simplifications to some provisions of EN 1992-1-1 are made in EN 1994-1-1. Reference to the partial safety factors for the materials is avoided by specifying resistances in terms of design values for strength, rather than characteristic values – for example, in *Formula 8.37* for plastic resistance to compression in ***clause 8.8.3.2(1)***. This resistance, $N_{pl,Rd}$, the 'squash load', is the design ultimate axial load for a short column assuming that the structural steel and reinforcement are yielding and the concrete is crushing. *Clause 8.8.3.2(1)*

For concrete-encased sections, the crushing stress is taken as 85% of the design cylinder strength, as explained in the comments on *clause 5.1.* For concrete-filled sections, the concrete component develops a higher strength because of the confinement from the steel section, and factor $\eta_{cc,min}$ is increased from 0.85 to 1.0. See also the comments on *clause 8.8.3.2(6)*.

Resistance to combined compression and bending

Clause 8.8.3.2(2)

The bending resistance of a column cross-section, $M_{pl,Rd}$, is calculated as for a composite beam in Class 1 or 2 (***clause 8.8.3.2(2)***). Points on the interaction curve shown in *Figures 8.27* and *8.28* represent limiting combinations of compressive axial load N and moment M which correspond to the plastic resistance of the cross-section.

The resistance is found using rectangular stress blocks. For simplicity, that for the concrete extends to the neutral axis, as shown in Figure 8.36 for resistance to bending (point B in *Figure 8.28* and Figure 8.39). As explained in the comments on *clause 5.1(1)*, this simplification is unconservative in comparison with stress/strain curves for concrete and the rules of EN 1992-1-1. To compensate for this, the plastic resistance moment for the column section is reduced by a factor α_M in *clause 8.8.3.6(1)*.

As axial compression increases, the neutral axis moves – for example, towards the lower edge of the section shown in Figure 8.36 and then outside the section. The interaction curve is, therefore, determined by moving the neutral axis in increments across the section and finding pairs of values of M and N from the corresponding stress blocks. This requires computer software, unless the simplification given in *clause 8.8.3.2(5)* is used. Simplified expressions for the coordinates of points B, C and D on the interaction curve are given in Appendix B. Further comment is given in Examples 8.7 and B.1.

Figure 8.36. Stress distributions for resistance in bending (tension positive)

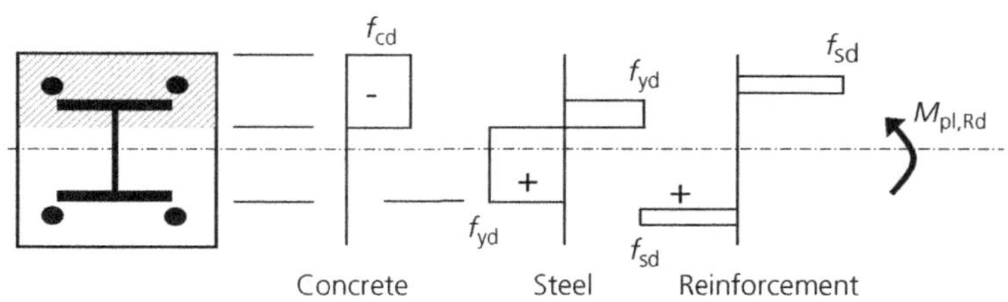

Influence of transverse shear

Clause 8.8.3.2(3)
Clause 8.8.3.2(4)

Clauses 8.8.3.2(3) and ***8.8.3.2(4)***, on the influence of transverse shear on the interaction curve, are generally the same as *clause 8.2.2.5* on moment-shear interaction in beams. One assumes first that the shear V_{Ed} acts on the structural steel section alone. If it is less than 0.5 $V_{pl,a,Rd}$, it has no effect. If it is greater, there is an option of sharing it between the steel and reinforced concrete sections, which may reduce that acting on the steel to below 0.5 $V_{pl,a,Rd}$. If it does not, then a reduced design yield strength is used for the shear area, as for the web of a beam. In a column, however, the shear area depends on the plane of bending considered and may consist of the flanges of the steel section. It is assumed that shear buckling does not occur.

Simplified interaction curve

Clause 8.8.3.2(5)

Clause 8.8.3.2(5) explains the use of the polygonal diagram BDCA in *Figure 8.28* as an approximation to the interaction curve, suitable for hand calculation. The method applies to any cross-section with biaxial symmetry, not just to encased I-sections.

First, the location of the neutral axis for pure bending is found, by equating the longitudinal forces from the stress blocks on either side of it. Let this be at distance h_n from the centroid of the uncracked section, as shown in *Figure 8.28(B)* and Figure B.2 in Appendix B. It is shown in Appendix B that the neutral axis for point C on the interaction diagram is at distance h_n on the other side of the centroid, and the neutral axis for point D passes through the centroid. The values of M and N at each point are easily found from the stress blocks shown in *Figure 8.28*.

Concrete-filled tubes of circular or rectangular cross-section

Clause 8.8.3.2(6) is based on the lateral expansion that occurs in concrete under axial compression. This causes circumferential tension in the steel tube and triaxial compression in the concrete. This increases the crushing strength of the concrete (Roik and Bergmann, 1992; Umamaheswari *et al.*, 2007) to an extent that outweighs the reduction in the effective yield strength of the steel in vertical compression. The coefficients η_a and η_c given in this clause allow for these effects. *Clause 8.8.3.2(6)*

This containment effect is not present to the same extent in concrete-filled rectangular tubes because less circumferential tension can be developed. In all tubes the effects of containment reduce as bending moments are applied. This is because the mean compressive strain in the concrete and the associated lateral expansion are reduced. With increasing slenderness, bowing of the member under load increases the bending moment and, therefore, the effectiveness of containment is further reduced.

For these reasons, η_a and η_c depend on the eccentricity of loading and on the slenderness of the member. The contribution from the concrete nowhere falls below 1.0 $A_c f_{cd}$, which is consistent with the version of *Formula 8.39* for concrete-filled tubes.

The 'steel' term has its lowest value, 0.75 $A_a f_{yd}$, where $\bar{\lambda} = 0$ and the eccentricity of loading $e_N = 0$. This is lower than in *Formula 8.39*, which can be used where it gives a higher total resistance because the use of *clause 8.8.3.2(6)* is optional.

Properties of the column

For columns in a frame, these properties of each column length are needed before or during global analysis of the frame

- the steel contribution ratio (***clause 8.8.3.3(1)***) *Clause 8.8.3.3(1)*
- the relative slenderness $\bar{\lambda}$ (***clause 8.8.3.3(2)***) *Clause 8.8.3.3(2)*
- the effective flexural stiffnesses (***clauses 8.8.3.3(3)*** and *8.8.3.4(2)*) *Clause 8.8.3.3(3)*
- the creep coefficient and effective modulus for concrete (***clause 8.8.3.3(4)***). *Clause 8.8.3.3(4)*

The steel contribution ratio is explained in the comments on *clause 8.8.1(4)*.

The relative slenderness $\bar{\lambda}$ is needed to check that the column is within the scope of the simplified method, *clause 8.8.3.1(1)*. It is defined, in *clause 8.8.3.3(2)*, in terms of N_{cr}, '*the elastic critical normal force for the relevant buckling mode*'. It has replaced the concept of 'effective length', used for columns in former codes, which now appears only in *clause 8.8.3.4(5)*, in relation to the definition of $N_{cr,eff}$.

As $\bar{\lambda}$ depends on the elastic critical normal force for the relevant buckling mode, the behaviour of the surrounding members needs to be taken into account. This could require a calculation of the load factor α_{cr} for elastic instability. It will often be possible though to make simplifying assumptions to show that a proposed column is within the scope for the method – for example, in a frame with a high stiffness against sway it would be reasonable to calculate N_{cr} assuming the member to be pin-ended. In any case, the upper limit on $\bar{\lambda}$ in *clause 8.8.3.1(1)* is somewhat arbitrary and does not justify great precision in N_{cr}. As calculating an elastic critical force does not involve the factors γ_M, the plastic resistance in *Formula 8.48* and the stiffness $(EI)_{eff}$ are found using characteristic values. The factor K_e (= 0.6) in *Formula 8.48* was previously presented in ENV 1994-1-1 as $0.8/\gamma_{CE}$, with $\gamma_{CE} = 1.35$.

The creep coefficient φ_t influences the effective modulus $E_{c,eff}$ (*clause 8.8.3.3(4)*), and hence the flexural stiffness of each column. It depends on the age at which concrete is stressed and the duration of the load. These will not be the same for all the columns in a frame. The effective modulus depends also on the proportion of the design axial load that is permanent. The design of a column is rarely sensitive to the influence of the creep coefficient on $E_{c,eff}$, so conservative assumptions can be made about uncertainties. Normally, a single value of effective modulus can be used for all the columns in a frame.

Verification of a column length

Clause 7.2.2(6) lists three methods for checking a column. The first one includes member imperfections in the global analysis, with the problem that at that stage it may be unclear whether the imperfection should be in the *xy* plane or the *xz* plane. The second method, (b), and ***clause 8.8.3.4*** are followed here. The third method has very limited scope.

Clause 8.8.3.4

The flow chart of Figure 8.37 shows a possible calculation route, intended to minimise iteration. It is used in Example 8.7. It is assumed that the column is in a frame not subject to sway instability and has details that can satisfy *clauses 8.8.1(8)* on local buckling, *8.8.3.1* on slenderness, *8.8.5.1* on concrete cover and *8.8.5.2* on reinforcement, so that it is within the scope of the simplified method of *clause 8.8.3*.

The relationship between the analysis of a frame and the stability of individual members is discussed in the comments on *clauses 7.2.2(3)* to *7.2.2(7)*. Conventionally, the stability of members is checked by analysis of individual members, using end moments and forces from a global analysis of the frame. Figure 8.37 follows this procedure, giving in more detail the procedures outlined in the lower part of the flow chart for global analysis (see Figure 7.1(a)).

Some of the comments now given are in the sequence of Figure 8.37 rather than in clause sequence. If bending is biaxial, the chart is followed for each axis in turn, as noted. It is assumed that lateral loading applied within the column length is not included in the global analysis.

The starting point is the output from the global analysis, listed at the top of Figure 8.37. The design axial compression is the sum of the forces from the two orthogonal frames of which the column is assumed to be a member. If, at an end of the column (e.g. the top) the joints to beams in these frames are at a different level, they could conservatively be assumed both to be at the upper level, if the difference is small. The flow chart does not cover situations where the difference is large (e.g. a storey height). The axial force N_{Ed} is normally almost constant along the column length. Where it varies, its maximum value can conservatively be assumed to be applied at the upper end.

However, for a member subject *only* to end compression, *clause 8.8.3.5(2)* enables the buckling curves of EN 1993-1-1 to be used. The word '*only*' is not in the clause, but that is the meaning, so this simplification, shown in Figure 8.37, is of little use in practice. The buckling curves can be useful as a preliminary check for columns with end moment; if the resistance to the normal force N_{Ed} is not sufficient, the proposed column is clearly inadequate.

Columns with transverse shear exceeding half the shear resistance of the steel element are rare. This is not included in Figure 8.37. If the shear is high, this should be checked first, because the interaction curve for the cross-section may be affected. Comment has been given earlier on *clauses 8.8.3.2(3)* and *8.8.3.2(4)* on shear, and on the interaction curve or polygon.

Initial imperfection

The initial imperfection, represented by a bow of amplitude e_0 is determined (*Table 8.7*). The imperfection is finally taken to lie in the most adverse plane and assumed initially to lie in the plane of the frame being analysed. The imperfection is the lateral departure at mid-height of the column of its axis of symmetry from the line joining the centres of symmetry at the ends of the column. Its value accounts, principally, for truly geometric imperfections and for the effects of residual stresses. It does not depend on the distribution of bending moment along the column. The curved shape is usually assumed to be sinusoidal, but use of a circular arc would be acceptable.

Clause 8.8.3.4(4)

For most columns, the member verification requires a second-order analysis in which explicit account is taken of imperfections (***clause 8.8.3.4(4)***). *Clause 7.2.1(3)* says that first-order analysis may be used if the deformations that it finds are not 'significant'. For judging this, it refers to conditions in EN 1993-1-1 which are quite severe. For a non-sway frame its clause 7.2.1(4) requires that the elastic critical buckling load is at least 25 times the design load. The ratio 25 can be changed in a national annex.

Figure 8.37. Flow chart for verification of a column length

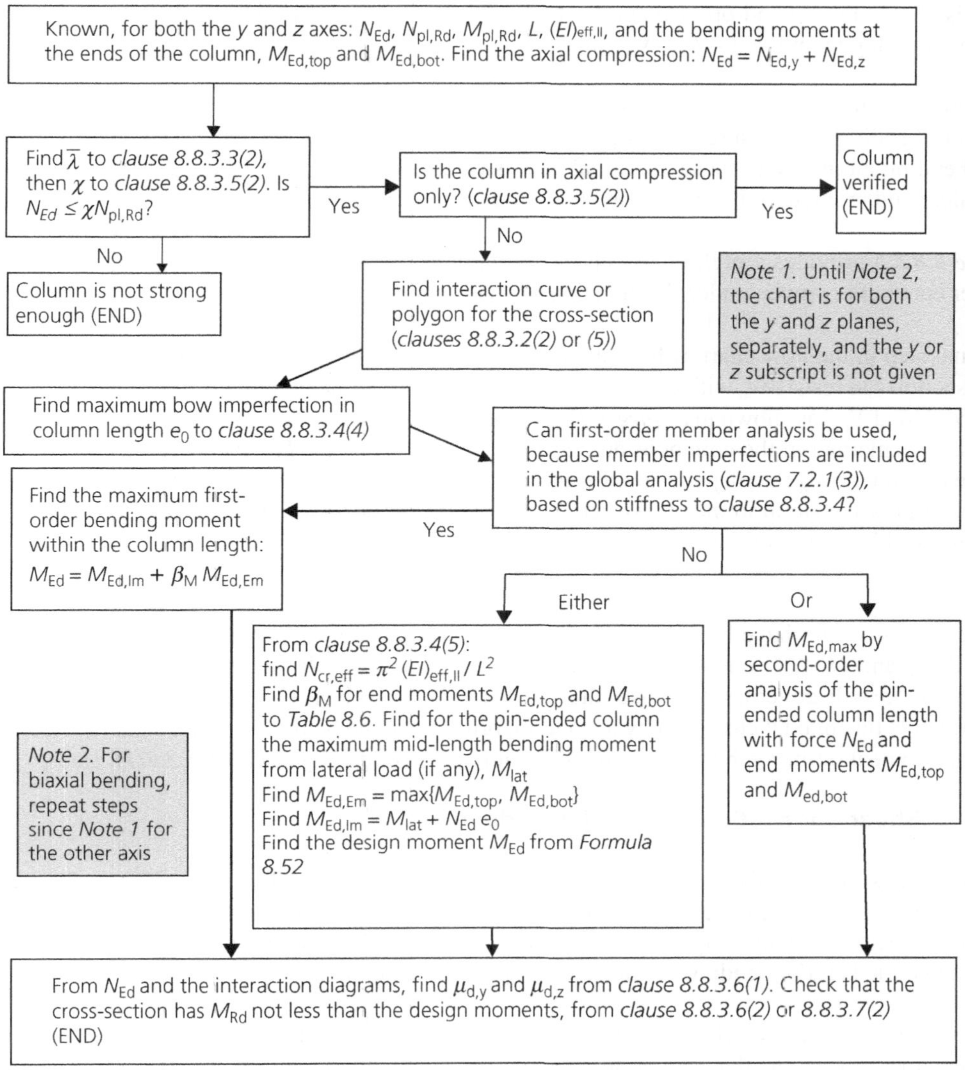

Where global analyses do not include member imperfections, they can be allowed for by the method of ***clause 8.8.3.4(5)***. Its use is shown in Example 8.7. Its *Formula 8.52* includes the elastic critical buckling load, which may be taken as $N_{cr,eff} = \pi^2 EI/L^2$, with L taken as the 'system length'. This term is not defined in EN 1994-1-1 and is understood to mean the length between effective lateral restraints, assuming the ends of the column to be pinned. *Clause 8.8.3.4(5)*

The flexural stiffness used for EI is $(EI)_{eff,II}$ (*clause 8.8.3.4(2)*), with the modulus of elasticity for concrete modified to take account of long-term effects. This flexural stiffness is lower than that defined in *clause 8.8.3.3(3)* because it is essentially a design value for ultimate limit states. The factor $K_{e,II}$ allows for cracking. The factor K_0 is from research-based calibration studies.

Design bending moment

The maximum first-order bending moment within the column length, $M_{Ed,Im}$, is the sum of the maximum moment from lateral loading on the column length, M_{lat}, and the moment from the imperfection, $N_{Ed}\ e_0$. These should be calculated assuming the column is pin-ended (though this is not stated in Eurocode 4) because the first-order contribution of end moments to the mid-length moment is included as the equivalent single-curvature end moment, $\beta_M\ M_{Ed,Em}$.

If first-order analysis is possible, the design moment is simply $M_{Ed,Im} + \beta_M M_{Ed,Em}$. It is now assumed that second-order analysis is needed, and that member imperfections have been neglected in the global analysis. Then, the second-order magnification of these moments is allowed for (*Formula 8.52* for M_{Ed}) by dividing them by $(1 - N_{Ed}/N_{cr,eff})$.

The condition in *Table 8.6*, $\beta_M \geq 0.44$, provides protection against snap-through buckling in double-curvature bending, and ensures that M_{Ed} will always exceed $M_{Ed,Im}$. No sign convention is given for these bending moments, because the initial bow is chosen to be in the most adverse direction and the equivalent end moment to be in the same direction as that caused by any lateral load.

Clause 8.8.3.5 **_Clause 8.8.3.5_** provides a preliminary check for columns with end moment, of little use in practice. Further comment is given under 'Verification of a column length', above.

The final step in the flow chart of Figure 8.37 for a column with uniaxial bending is to check that the cross-section can resist M_{Ed} with compression N_{Ed}. The interaction diagram gives a resistance $\mu_d M_{pl,Rd}$ with axial load N_{Ed}, as shown in *Figure 8.27*. This is unconservative, being based on rectangular stress-
Clause 8.8.3.6(1) blocks, as explained in the comment on *clause 8.2.1.2(1)(d)*, so in **_clause 8.8.3.6(1)_** it is reduced by the use of a factor α_M that depends on the grade of structural steel. This factor allows for the increase in the compressive strain in the cross-section at yield of the steel (which is adverse for the concrete), when the yield strength of the steel is increased.

Biaxial bending

Clause 8.8.3.7 Where values of M_{Ed} have been found for both axes, **_clause 8.8.3.7_** applies, in which they are written as $M_{y,Ed}$ and $M_{z,Ed}$. If one is much greater than the other, the relevant check for uniaxial bending, *Formula 8.55* will govern. Otherwise, the linear interaction given by *Formula 8.56* applies. If the member fails this biaxial condition by a small margin, it may be helpful to recalculate the less critical bending moment omitting the member imperfection, as permitted by *clause 8.8.3.7(1)*.

8.8.4 Shear connection and load introduction

Load introduction

The provisions for the resistance of cross-sections of columns assume that no significant slip occurs at
Clause 8.8.4.1(1) the interface between the concrete and structural steel components. **_Clauses 8.8.4.1(1)_** and **_8.8.4.1(2)_**
Clause 8.8.4.1(2) give the principles for limiting slip to an 'insignificant' level in the critical regions – that is, those where axial load and/or bending moments are applied to the column.

For any assumed 'clearly defined load path' it is possible to estimate stresses, including shear at the interface. In regions of load introduction, shear stress is likely to exceed the design shear strength from *clause 8.8.4.3*, and shear connection is then required (*clause 8.8.4.2(1)*). It is unlikely to be needed elsewhere, unless the shear strength τ_{Rd} from *Table 8.8* is very low, or the member is also acting as a beam (Abdullah and Sumei, 2009) or has a high degree of double-curvature bending.

Few shear connectors reach their design shear strength until the slip is at least 1 mm, but this is not 'significant' slip for a resistance model based on plastic behaviour and rectangular stress blocks. However, a long load path implies greater slip, so the assumed path should not extend beyond the introduction length given in *clause 8.8.4.2(2)*.

Where axial load is applied through a joint attached only to the steel component of a concrete-encased section, the force to be transferred to the concrete can be estimated from the relative axial loads in the two materials given by the resistance model. Accurate calculation is rarely practicable where the cross-section concerned does not govern the design of the column. In this partly plastic situation, the
Clause 8.8.4.2(1) more adverse of the elastic and fully-plastic models gives a safe result (**_clause 8.8.4.2(1)_**). In practice, it may be simpler to provide shear connection based on a conservative (high) estimate of the force to be transferred.

Load introduction can occur through the exterior of a concrete-filled steel tube, as shown in *Figure 8.31(b)*. The shrinkage of the concrete core tends to reduce the bond or friction at the interface between steel and concrete, and any truss action would cause lateral tension in the concrete. The details covered by *clause 8.8.4.2(6)* provide possible solutions.

Where axial force is applied by a plate bearing on both materials the proportion of the force resisted by the concrete gradually decreases, due to creep and shrinkage. It could be inferred from *clause 8.8.4.2(1)* that shear connection should be provided for a high proportion of the force applied. However, models based on elastic theory are over-conservative in this inherently stable situation, where large strains are acceptable. The application rules that follow are based mainly on tests.

In a concrete-filled tube, shrinkage effects are low, for only the basic shrinkage strain occurs, with a long-term value below 10^{-4}, from clause B.6 of EN 1992-1-1. Radial shrinkage is outweighed by the lateral expansion of concrete in compression, for its inelastic Poisson's ratio increases at high compressive stress. Friction then provides significant shear connection (***clause 8.8.4.2(3)***). Friction is also the basis for the enhanced resistance of stud connectors given in ***clause 8.8.4.2(4)***. *Clause 8.8.4.2(3)* *Clause 8.8.4.2(4)*

Detailing at points of load introduction or change of cross-section is assisted by the high bearing stresses given in ***clauses 8.8.4.2(5)*** and ***8.8.4.2(6)***. These rules are based mainly on tests (Porsch and Hanswille, 2006). As an example, the following data are assumed for the detail shown in *Figure 8.31(b)*, with axial loading (i.e. $e = 0$) *Clause 8.8.4.2(5)* *Clause 8.8.4.2(6)*

- steel tube with external diameter 300 mm and wall thickness 10 mm
- bearing plate 15 mm thick, with strength $f_{yd} = 355$ N/mm^2
- concrete with $f_{ck} = 45$ N/mm^2 and $\gamma_C = 1.5$.

Then, $A_c = \pi \times 140^2 = 61\,600$ mm^2; $A_1 = 15 \times 280 = 4200$ mm^2

From *Formula 8.57*,

$$\sigma_{c,Rd} / f_{cd} = [1 + (4.9 \times 10 / 300)(355 / 45)]\ (14.7)^{0.5} = 8.8$$

$$\sigma_{c,Rd} = 8.8 \times 30 = 260 \text{ N/mm}^2$$

This bearing stress is so high that if fully used, the bearing plate would need to be at least 180 mm deep to have sufficient resistance to vertical shear.

Figure 8.32 illustrates the requirement of ***clause 8.8.4.2(9)*** for transverse reinforcement for load introduction through a fully encased steel section. The transverse bars must be able to resist a force N_{c1}. If longitudinal reinforcement is ignored, this is given by *Clause 8.8.4.2(9)*

$$N_{c1} = [A_{c2} / (2\,nA)]\, N_{Ed}$$

where A is the transformed area of the cross-section 1–1 of the column in *Figure 8.32*, given by $A = A_s + (A_{c1} + A_{c2})/n$ and A_{c1} and A_{c2} are the unshaded and shaded areas of concrete, respectively, in cross-section 1–1.

Longitudinal shear

Clause 8.8.4.3 gives application rules (used in Example 8.8) relevant to the principle of *clause 8.8.4.1(2)*, for columns with the longitudinal shear that arises from transverse shear. The design shear strengths τ_{Rd} in *Table 8.8* are far lower than the tensile strength of concrete. They rely on friction, not bond, and are related to the extent to which separation at the interface is prevented – for example, in partially encased I-sections, lateral expansion of the concrete creates pressure on the flanges, but not on the web, for which $\tau_{Rd} = 0$; and the highest shear strengths are for concrete within steel tubes. *Clause 8.8.4.3*

Where small steel I-sections are provided, mainly for erection, and the column is mainly concrete, ***clause 8.8.4.3(4)*** provides a useful increase to τ_{Rd}, for covers up to 115 mm, more simply presented as *Clause 8.8.4.3(4)*

$$\beta_c = 0.2 + c_z / 50 \leq 2.5$$

Concern about the attachment of concrete to steel in partially encased I-sections appears again in ***clause 8.8.4.3(5)***, because under weak-axis bending, separation tends to develop between the encasement and the web. *Clause 8.8.4.3(5)*

8.8.5 Detailing provisions

If a steel I-section in an environment in class XC1 to EN 1992-1-1 has links in contact with its flange (permitted by *clause 8.8.5.2(3)*), the cover to the steel section could be as low as 25 mm. For a wide steel flange, this thin layer of concrete would have little resistance to buckling outwards, so the mini-
Clause 8.8.5.1(2) mum thickness is increased to at least 40 mm in ***clause 8.8.5.1(2)***.

Clause 8.8.5.2(1) Minimum longitudinal reinforcement (***clause 8.8.5.2(1)***) is needed to control the width of cracks, which can be caused by shrinkage even in columns with concrete nominally in compression.

Clause 8.8.5.2(4) ***Clause 8.8.5.2(4)*** refers to exposure class X0 of EN 1992-1-1. This is 'For concrete without reinforcement or embedded metal', with 'no risk of corrosion on attack'. Even buildings with very low air humidity, or some spaces within them, seem excluded by the 'embedded metal condition'. The minimum reinforcement provides robustness during construction.

Figure 8.38. Cross-section and properties of a composite column (dimensions in mm)

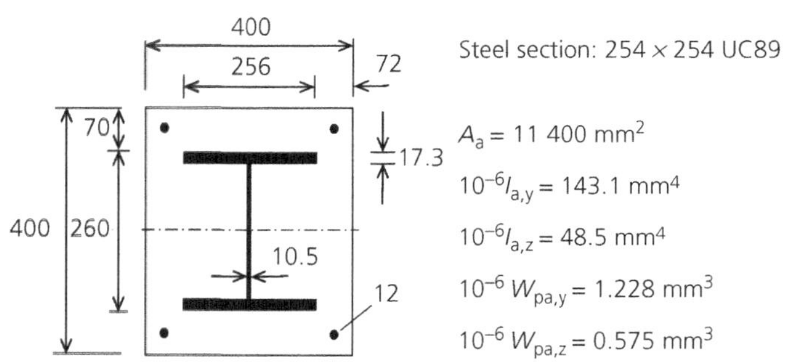

Example 8.7: composite column with bending about one or both axes

A composite column of length 7.0 m has the uniform cross-section shown in Figure 8.38. Its resistance to given action effects will be found. It is checked that the column is within the scope of the simplified method of *clause 8.8.3*. The flow chart in Figure 8.37 is less detailed than these calculations. The properties of the steel member, 254 × 254 UC89, are taken from section tables, and given here in Eurocode notation. The properties of the materials are taken from Eurocodes 2 and 3. Quartzite aggregate is assumed. For plastic resistance of this concrete-encased section, $\eta_{cc}k_{tc} = 0.85$. In the usual notation, the properties are

- structural steel: grade S355, $f_y = f_{yd} = 355$ N/mm², $E_a = 210$ kN/mm²
- concrete: strength class C25/30; $f_{ck,28} = 25$ N/mm², $f_{cd} = 0.85 \times 25/1.5 = 14.2$ N/mm²
- $E_{cm} = 30.5$ kN/mm², $n_0 = 210/30.5 = 6.8$
- reinforcement: ribbed bars, $f_{sk} = 500$ N/mm², $f_{sd} = 500/1.15 = 435$ N/mm²

Geometrical properties of the cross-section

In the notation of *Figure 8.24(a)*,

$$b_c = h_c = 400 \text{ mm} \qquad b = 256 \text{ mm} \qquad h = 260 \text{ mm}$$

$$c_y = 200 - 128 = 72 \text{ mm} \qquad c_z = 200 - 130 = 70 \text{ mm}$$

These satisfy the conditions of *clauses 8.8.3.1(2)*, *8.8.3.1(4)* and *8.8.5.1(2)*, so all the concrete casing is included in the calculations.

$$\text{Area of reinforcement} = 4 \times 36\pi = 446 \text{ mm}^2$$

$$\text{Area of concrete} = 400^2 - 11\,400 - 446 = 148\,150 \text{ mm}^2$$

The reinforcement has area 0.301% of the concrete area, so *clause 8.8.5.2(1)* permits it to be included in calculations. For simplicity, its small contribution will be ignored, so the values are

$$A_a = \mathbf{11\ 400 mm^2} \quad A_c = 400^2 - 11\ 400 = \mathbf{148\ 600\ mm^2} \quad A_s = 0$$

For the steel section: $10^{-6}I_{a,y} = 143.1\ \mathrm{mm}^4 \quad 10^{-6}I_{a,z} = 48.5\ \mathrm{mm}^4$

Design action effects, ultimate limit state

For the most critical load arrangement, global analysis including any global second-order effects (e.g. from sidesway of the frame) gives these values

$$N_{Ed} = \mathbf{1800\ kN}, \text{ including, from permanent loading, } N_{G,Ed} = 1200\ \mathrm{kN}$$

$$M_{y,Ed,top} = \mathbf{380\ kN\ m};\ M_{z,Ed,top} = 0$$

The bending moments at the lower end of the column and the lateral loading are zero. Later, the effect of adding $M_{z,Ed,top} = 50$ kN m is determined.

Properties of the column length

From *clause 8.8.3.2(1)*,

$$N_{pl,Rd} = A_a f_{yd} + A_c f_{cd} = 11.4 \times 355 + 148.6 \times 14.2 = 4047 + 2109 = \mathbf{6156\ kN}$$

From *clause 8.8.3.3(1)*, the steel contribution ratio is

$$\delta_c = 4047/6156 = 0.657$$

which is within the limits of *clause 8.8.1(5)*. Multiplying $N_{pl,Ed}$ by the partial factors gives the characteristic plastic resistance, with $\gamma_C = 1.5$,

$$N_{pl,Rk} = 4047 + 1.5 \times 2109 = \mathbf{7210\ kN}$$

Creep coefficient

From *clause 8.8.3.3(4)*

$$E_{c,eff} = E_{cm} / [1 + (N_{G,Ed} / N_{Ed})\varphi_t] \qquad (8.50)$$

The creep coefficient φ_t is $\varphi(t, t_0)$ to *clause 7.4.2.2*. The time t_0 is taken as 28 days and t is taken as 50 years, as creep reduces the stiffness, and hence the stability, of a column. From Table 5.2 of EN 1992-1-1, the 'perimeter exposed to drying' is

$$u = 2(b_c + h_c) = 1600\ \mathrm{mm} \quad \text{and} \quad h_n = 2A_c / u = 297\ 200 / 1600 = 186\ \mathrm{mm}$$

The correction factor for concrete strength, A, is 0.794 from Table 5.2 of EN 1992-1-1. Assuming 'dry atmosphere conditions' and the use of type CN ('normal') cement, EN 1992-1-1 gives

$$\varphi(50\mathrm{y}, 28\mathrm{d}) = (35/25)^{0.794} \times 2.2 = 2.9$$

The assumed 'age at first loading' has little influence on the result if it exceeds about 20 days. If, however, significant load was applied at age 7 days, φ_t would be increased to about 3.6.

From *Formula 8.50*,

$$E_{c,eff} = 30.5 / [1 + 2.9 \times (1200/1800)] = \mathbf{10.4\ kN/mm^2}$$

Elastic critical load, with characteristic stiffness

Although the loading applies major-axis bending, buckling about the minor axis could be the more critical, so both relative slendernesses $\bar{\lambda}_y$ and $\bar{\lambda}_z$ are needed. From *clause 8.8.3.3(3)*,

$$(EI)_{eff} = E_a I_a + 0.6\, E_{c,eff} I_c \qquad (8.49)$$

For the concrete,

$$10^{-6} I_{c,z} = 0.4^2 \times 400^2 / 12 - 48.5 = 2085 \text{ mm}^4$$

From *Formula 8.49*,

$$10^{-6} (EI)_{eff,z} = 210 \times 48.5 + 0.6 \times 10.4 \times 2085 = 23\ 200 \text{ kN mm}^2$$

In this example, the end conditions for the column are assumed to be such that lateral restraint is provided, but no elastic rotational restraint, so the effective length is the actual length, 7.0 m, and

$$N_{cr,z} = \pi^2 (EI)_{eff,z} / L^2 = 23\ 200\ \pi^2 / 49 = 4672 \text{ kN}$$

From *Formula 8.48*,

$$\bar{\lambda}_z = (N_{pl,Rk} / N_{cr,z})^{0.5} = (7210 / 4672)^{0.5} = \mathbf{1.24}$$

Similar calculations for the y-axis give: $10^{-6} I_{c,y} = 1990 \text{ mm}^4$

$$10^{-6} (EI)_{eff,y} = 42\ 470 \text{ kN mm}^2 \quad N_{cr,y} = 8554 \text{ kN} \quad \bar{\lambda}_y = \mathbf{0.92}$$

These slendernesses do not exceed 2.0, so *clause 8.8.3.1(1)* is satisfied.

Resistance to axial load, z-axis

From *clause 8.8.3.5(2)*, buckling curve c is applicable. From Figure 8.5 of EN 1993-1-1, for $\bar{\lambda}_z = 1.22$, $\chi_z = 0.41$

From *Formula 8.53*, $\mathbf{N_{Ed}} \leq \chi_z N_{pl,Rd} = 0.41 \times 6156 = \mathbf{2524\ kN}$

This condition is satisfied.

Transverse shear

For $M_{y,Ed,top} = 380$ kN m, the transverse shear is $V_{z,Ed} = 380/7 = 54$ kN

This is obviously less than 0.5 $V_{pl,a,Rd}$, so *clause 8.8.3.2(3)* does not apply.

Interaction curves and polygons

The interaction polygons corresponding to *Figure 8.28* are determined in Appendix B (see Example B.1), and reproduced in Figure 8.39. *Clause 8.8.3.2(5)* permits them to be used as approximations to the *N-M* interaction curves for the cross-section.

Member imperfections

From *Table 8.7* with $L = 7$ m, $e_{0,y} = L/200 = 35$ mm and $e_{0,z} = L/150 = 47$ mm

There is no lateral loading, so multiplying by N_{Ed}, 1800 kN,

$$\mathbf{M_{y,Ed,Im} = 63\ kN\ m} \text{ or } \mathbf{M_{z,Ed,Im} = 85\ kN\ m}$$

Figure 8.39. Interaction polygons for bending about the major and minor axes

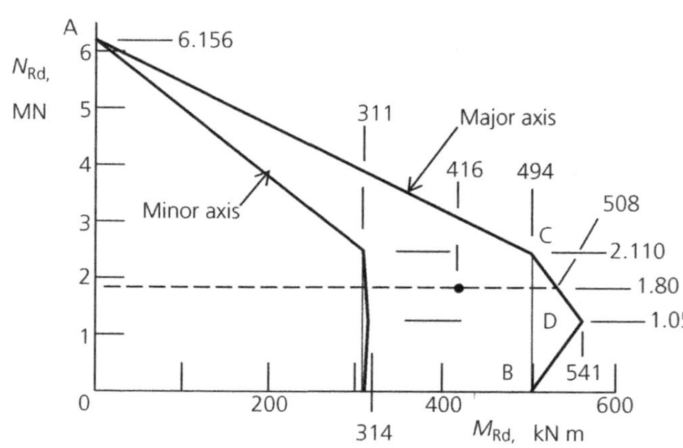

To check whether second-order moments can be neglected, a reduced value of N_{cr} is required, to *clause 8.8.3.4(3)*. From *Formula 8.51*,

$$(EI)_{y,eff,II} = 0.9(E_a I_a + 0.5\, E_{c,eff} I_c) = 0.9 \times 10^6 (210 \times 143.1 + 0.5 \times 10.4 \times 1990) = 3.64 \times 10^{10} \text{ kN mm}^2$$

Hence, $\mathbf{N_{cr,y,eff}} = 36\,400\,\pi^2/7^2 = \mathbf{7332\ kN}$

This is less than 25 N_{Ed}, so second-order effects should be considered (*clause 7.2.1(3)* and Formula 7.1 in EN 1993-1-1).

Figure 8.40. Design bending moments for a column: (a) global and (b) local

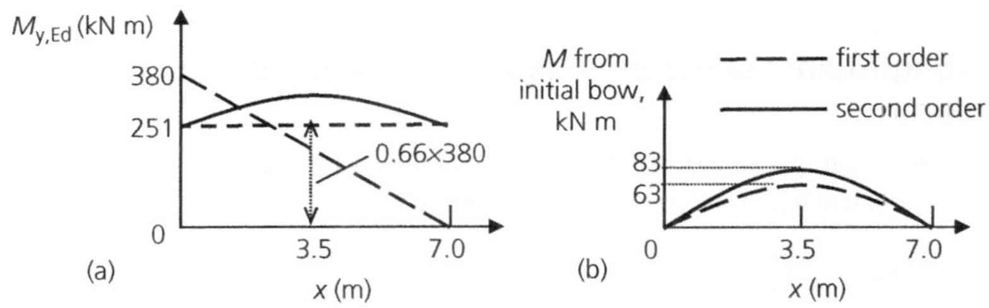

Second-order bending moments, *y*-axis

From *clause 8.8.3.4(5), Table 8.6,* for the end moments, $r = 0$, $\beta_M = 0.66$, so the equivalent single-curvature end moment is $M_{Ed,Em} = \beta_M\, M_{y,Ed,top} = 0.66 \times 380 = 251$ kN m

Figure 8.40 shows the first-order *y*-axis design moments by dashed lines. To allow for second-order effects, *Formula 8.52*, they are increased by the factor

$$1/(1 - N_{Ed}/N_{cr,y,eff}) = 1/(1 - 1800/7332) = 1.325$$

Hence, $M_{y,Ed} = 1.325 \times (251 + 63) = 416$ kN m

The point (N_{Ed}, $M_{y,Ed}$) is (1800, 416) on Figure 8.39. From the values shown in the figure,

$\mathbf{M_{pl,N,Rd} = 515\ kN\ m}$

From *Formula 8.55*, $M_{Ed}/M_{pl,N,Rd} = 416/515 = 0.81$

This ratio satisfies the condition of *clause 8.8.3.6(1)* that for S355 steel it should not exceed 0.9. There is no *z*-axis bending, unless the initial imperfection is applied in that plane, and then biaxial bending does not govern. The column is strong enough.

Clause 8.8.3.6(2) refers to values of μ_{dy} greater than 1.0. Here, from Figure 8.39,

$$\mu_{dy} = M_{pl,N,y,Rd}/M_{pl,Rd} = 515/494 = 1.04$$

but increase in axial compression reduces bending resistance (line CD), so no additional verification is needed. There is comment on this subject under *clause 8.8.1(7)*.

Biaxial bending

The method is illustrated by adding a minor-axis bending moment $M_{z,Ed,top}$ of 50 kN m. It is much smaller than $M_{y,Ed}$, so major-axis failure is assumed. From *clause 8.8.3.7(1)*, there is assumed to be no bow imperfection in the *x-z* plane, so $M_{z,Ed,Im} = 0$; but second-order effects have to be considered.

From *Formula 8.51*,

$$(EI)_{z,eff,II} = 0.9 \times 10^6 \times (210 \times 48.5 + 0.5 \times 10.4 \times 2085) = 1.89 \times 10^{10} \text{ kN mm}^2$$

Hence, $\mathbf{N_{cr,z,eff}} = 18\,900\,\pi^2/49 = \mathbf{3812\ kN\ m}$

As before, $\beta_M = 0.66$, and from *Formula 8.52* the second-order factor is

$$1/(1-1800/3812) = 1.89$$

so $\boldsymbol{M_{z,Ed}} = 1.89 \times 0.66 \times 50 = \mathbf{62.4\ kN\ m}$

From Figure 8.39 for $N_{Ed} = 1800$ kN, $\boldsymbol{M_{z,Rd}} = \mathbf{312\ kN\ m}$

Biaxial bending is checked, from *Formula* 8.56:

$$\boldsymbol{M_{y,Ed} / M_{pl,N,y,Rd}} + \boldsymbol{M_{z,Ed} / M_{pl,N,z,Rd}} = 416/515 + 62.4/312 = \mathbf{1.02}$$

This exceeds 1.0, so the check is just not satisfied, and the column cannot resist the additional bending moment.

Example 8.8: longitudinal shear outside areas of load introduction for a composite column

All of the data for this example are given in Example 8.7 and Figure 8.38 and are not repeated here. The design transverse shear was found to be $\boldsymbol{V_{z,Ed}} = \mathbf{54\ kN}$

The maximum shear that can be resisted without provision of shear connection is now calculated. The critical cross-section is B–B in Figure 8.41. *Clause 8.8.4.3(2)* permits the use of elastic analysis, 'considering long term effects and cracking of concrete'. Creep reduces the shear stress on plane B–B, so the short-term modular ratio $n_0 = 6.8$ is used. To allow for cracking would involve a difficult inelastic analysis of the cross-section with maximum bending moment, which the design model assumes is at mid-span. The resulting decrease in I_y would usually be small for a column. Uncracked section properties are used, and cracking is considered later.

Figure 8.41. Longitudinal shear on plane B–B in a column cross-section (dimensions in mm)

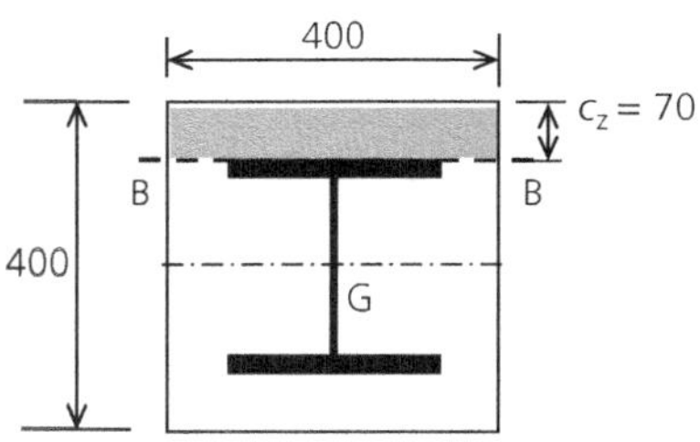

The cover $c_z = 70$ mm, so from *Formula 8.58* in *clause 8.8.4.3(4)*, $\beta_c = 1.60$. Assuming that the surface of the steel section satisfies the conditions of *clause 8.8.4.3(3)*, *Table 8.8* gives

$$\tau_{Rd} = 0.30 \times 1.60 = 0.48\ \text{N/mm}^2$$

From Example 8.7, $10^{-6}I_{a,y} = 143.1$ mm^4 $\quad 10^{-6}I_{c,y} = 1990$ mm^4

so for the uncracked section in 'concrete' units,

$$10^{-6} I_y = 1990 + 143.1 \times 6.8 = 2963\ \text{mm}^4$$

The 'excluded area' is $A_{ex} = 400 \times 70 = 28\,000$ mm^2

and its centre of area is $\bar{z} = 200 - 35 = 165$ mm from G in Figure 8.41. Hence,

$$\boldsymbol{V_{z,Rd}} = \tau_{Rd} I_y b_c / (A_{ex}\bar{z}) = 0.48 \times 2963 \times 400 / (28\ 000 \times 0.165) = \mathbf{123\ kN}$$

This exceeds $V_{z,Ed}$, so no shear connection is needed. The margin is so great that there is no need to consider the effect of cracking.

Column section used as a beam

The shear strengths given in *Table 8.8* are unlikely to be high enough to permit omission of shear connection from a column member with significant transverse loading. As an example, it is assumed that the column section of Figure 8.41 is used as a simply supported beam of span 8.0 m, with uniformly distributed loading. Using elastic analysis of the cracked cross-section with $n_0 = 6.8$ as before, but no axial load, the results are

$$10^{-6} I_{y,cracked} = 1441 \text{ mm}^4 \qquad V_{z,Rd} = 79 \text{ kN}$$

This corresponds to a design ultimate load of 19.7 kN/m, which is only 4.4 times the unfactored weight of the member, so it is likely that shear connection would be needed.

Designers' Guide to EN 1994-1-1: 2026: Eurocode 4 Design of composite steel and concrete structures

Roger P Johnson and Stephen J Hicks
ISBN 978-1-83662-921-4
https://doi.org/10.1108/978-1-83662-920-720253011

CHAPTER 9
Serviceability limit states

This chapter corresponds to *Section 9* of EN 1994-1-1, which has the same main clauses as the previous Eurocode 4, where it was Section 7. These are as follows.

- General *Clause 9.1*
- Stresses *Clause 9.2*
- Deformations in buildings *Clause 9.3*
- Cracking of concrete *Clause 9.4*

9.1 General

Clause 9 of EN 1994-1-1 is limited to provisions on serviceability that are specific to composite structures and are not in *clauses 5* (for structural analysis) or *10* (for composite slabs), or in Eurocodes 1990, 1991, 1992 or 1993. Some of these other more general provisions are briefly referred to here.

The initial design of a structure is usually based on the requirements for ultimate limit states. Serviceability is then checked, by methods that give designers and clients greater freedom to take account of factors such as the intended use of a building and the nature of its finishes. The consequences of unserviceability are less serious than those of reaching an ultimate limit state, and its occurrence is less easily defined – for example, a beam with an imposed-load deflection of span/300 may be acceptable in some situations but not in others.

The content of *clause 9* was influenced by the need to minimise calculations. Results already obtained for ultimate limit states are scaled or re-used wherever possible. Experienced designers know that many structural elements satisfy serviceability criteria by wide margins. For these, design checks must be simple, and it does not matter if they are conservative. For other elements, a longer but more accurate calculation may be justified. Some application rules, therefore, include alternative methods.

Serviceability verification and criteria

Clauses 9.1(1) and ***9.1(2)*** refer to clause 5.4 of EN 1990. This gives criteria for placing a limit state within the 'serviceability' group, with reference to deformations (including vibration), damage affecting durability, and the functioning of the structure. *Clause 9.1(1)* *Clause 9.1(2)*

The requirement for a serviceability verification is given in clause 8.4.1(1) of EN 1990 as $E_d \leq C_{d,SLS}$ where E_d is the design value of the effects of the specified actions and the 'relevant' combination, and $C_{d,SLS}$ is the limiting design value of the 'relevant' criterion. From clause 8.4.3 of EN 1990, serviceability is assessed using the characteristic, frequent or quasi-permanent combination, for serviceability limit states that are respectively, irreversible, reversible or a consequence of long-term effects. The quasi-permanent combination is also relevant for the appearance of the structure.

For buildings, the combinations of actions in Table A.1.6 of EN 1990 do not include a partial safety factor. Its value for actions for serviceability verifications is given as 1.0 under Formula 8.28 in EN 1990 in the definition of F_d.

Partial factors for properties of materials are not given. Unfactored characteristic or nominal mechanical properties of materials are used unless stated otherwise.

Clause 5.4(4) of EN 1990 says that serviceability criteria should be specified for each project; presumably by agreement between the client and the designer. Some criteria for buildings are given in clause A.1.8 of EN 1990 – for example, on satisfactory vibration behaviour. Further guidance may be found in specialised literature (Smith *et al.*, 2009).

No serviceability limit state of 'excessive slip of shear connection' has been defined. The effects of slip are considered in *clause 9.3.1* on deflection of beams. Generally, it is assumed that design of shear connection for ultimate limit states ensures satisfactory performance in service, but composite slabs can be an exception: ***clause 9.1(3)*** refers to relevant rules given in *clause 10.8.2*.

Clause 9.1(3)

No serviceability criteria are specified for composite columns, so, from here on, this chapter is referring to composite beams or, in some places, to composite frames.

9.2 Stresses

Clause 9.2.1
Clause 9.2.1(1)

Excessive stress is not itself a serviceability limit state, although stress calculations to ***clause 9.2.1*** are required for some verifications for deformation and cracking. ***Clause 9.2.1(1)*** on calculation of serviceability stresses includes inelastic behaviour of steel in its list of effects to be taken into account 'where relevant'. In the uncommon situations where stress limitation is required (fatigue, prestressing, jacking of supports) and elastic analysis for serviceability loads finds yielding, the design should be reviewed.

Clause 9.2.1(3)

For creep and shrinkage, ***clause 9.2.1(3)*** says that the modular ratios given in *clause 7.4.2.2 may* be used. For most situations for buildings, the simple method of *clause 7.4.2.2(11)* is applicable. For crack-width calculations, clause 9.1(5) of EN 1992-1-1 gives the even simpler method: $E_a/E_{c,eff} = 15$.

Clause 9.2.2

No checks on stresses are required for most buildings, ***clause 9.2.2***. No stress limits for buildings are given in the Eurocodes for concrete and steel structures, other than warnings in clause 9.2.1 of EN 1992-1-1, with Notes on recommended stress limits. The 'bridge' Parts of these Eurocodes include stress limits, which may be applicable for buildings that have prestressing or fatigue loading.

9.3 Deformations in buildings

9.3.1 Deflections

Global analysis

Deflections are influenced by the method of construction and may govern design, especially where beams designed as simply-supported are built unpropped. For propped construction, props to beams should not be removed until all of the concrete that would then be stressed has reached a strength equivalent to grade C20/25, from *clause 8.6.10.2(3)*. Then, elastic global analysis to *clause 7* is sufficient (*clause 9.3.1(2)*).

Clause 9.3.1(1)

Where unpropped construction is used and beams are not designed as simply-supported, the analysis may be more complex than is revealed by the reference to EN 1993 in ***clause 9.3.1(1)***. In a continuous beam or a frame, the deflection of a beam depends on how much of the structure is already composite when the slab for each span is cast. A simple and usually conservative method is to assume that the whole of the steel frame is erected first. Then, all of the concrete for the composite members is cast at once, its whole weight being carried by the steelwork; but more realistic multi-stage analyses may be needed for a high-rise structure and for long-span continuous beams.

Where falsework or re-usable formwork is supported from the steel beam, it will be removed after the member becomes composite. The small locked-in stresses that result can usually be ignored in buildings, but not always in bridges.

Clause 9.3.1(2)

Where elastic global analysis is used for ultimate limit states, it may be possible to obtain results needed for serviceability limit states by simple scaling using the ratio of the relevant loads. This ratio will depend on the method of construction, and also on which of the three serviceability load combinations is being used for the limit state considered. This scaling may be conservative if the ULS loading required allowance for second-order effects. Analysis for the lower SLS loading may not need this. ***Clause 9.3.1(2)*** on the use of elastic global analysis for deflections refers to *clause 7*. The method of *clause 7.4.2* 'linear elastic analysis' is available generally.

Redistribution of moments is permitted for most framed structures at SLS by *clause 7.4.4(8)*, which is new. The 10% permitted is essentially an allowance for the effects of cracking. Redistributions given in *clauses 9.3.1(7)* and *9.3.1(8)* are discussed below.

Redistribution of elastic moments, additional to that given in *clause 7.4.4*, is not forbidden. It would be necessary to show that any proposed novel method satisfied *clause 7.4.4(2)* on effects to be taken into account. The relevance of the restrictions in *clauses 7.4.4(4)* to *7.4.4(7)* should be considered.

Limits to deflection of beams

The specification of a deflection limit for a long-span beam needs care, especially where construction is unpropped and/or the steel beam is pre-cambered. Clause A.1.8 of EN 1990 gives guidance on defining serviceability criteria and enables national annexes to give limiting values. 'Suggested' values for maximum vertical deflections are given in Table A.1.10 (NDP). The UK's National Annex may give further guidance. The previous National Annex stated that limits should be specified for each project, and recommended the use of the characteristic combination where damage to non-structural elements (e.g. partition walls) could occur. No limits were given.

Clause A.1.8.2.2 of EN 1990 defines three components of deflection. Depending on circumstances, it may be necessary to set limits to any one of them, or to more than one, related to a defined load level. Prediction of long-term values should take account of creep of concrete, based on the quasi-permanent combination of actions, and may need to allow for shrinkage. Where precast floor units are used, it must be decided whether they should be cambered to compensate for creep. ***Clause 9.3.1(3)*** relates to the use of false ceilings, which conceal the sagging of a beam due to dead loading. Further comment is given in Example 9.1. *Clause 9.3.1(3)*

Longitudinal slip

Clause 9.3.1(4) refers to the additional deflection caused by slip at the interface between steel and concrete. This may be ignored if all of its three conditions are satisfied. Condition (b) relates to the minimum value of the degree of shear connection and gives a higher limit, 0.5, than the limit 0.4 in *clause 8.6.3.3(2)*. *Clause 9.3.1(4)*

It is prudent to conform to these conditions because the alternative, direct calculation to ***clause 9.3.1(5)***, is complex, and limited to uniform loading and spacing of shear connectors. For the stiffness of shear connectors, its reference to *Figure B.3* seeks information that users may not have. *Clause 9.3.1(5)*

Other methods for estimating increase of deflection due to longitudinal slip

There is an algebraic solution, with worked examples, for the additional deflection at mid-span of simply-supported beams, from slip (Aribert, 2010). It confirms a rule in ENV 1994-1-1 (BSI, 1994) that it can be neglected where $\eta \geq 0.5$. This work was based on the limiting spans given in the previous EN 1994-1-1, so the conclusion may be unreliable where a low degree of shear connection is used for the longer spans now permitted.

Its conclusion on degrees of shear connection below 0.5 is that the extra deflection can also be ignored where $0.4 \leq \eta < 0.5$, except as follows. Where the span is less than about 10 m, $\eta < 0.5$, a composite slab is present and propped construction is used for the beam, the extra deflection should be calculated. The total deflection δ is given by

$$\delta = \delta_c + (\delta_a - \delta_c)\,f(u) \qquad (X9.1)$$

where $f(u) = [(384/(5u^4)]\,(u^2/8 - 1)$

$$\text{and } u^2 = (kL/p)[1/E_aA_a + 1/E_cA_c + d^2/(E_aI_a + E_cI_c)] \qquad (X9.2)$$

Here, E, A and I have their usual meanings, with subscripts a and c referring to the steel and uncracked concrete cross-sections, respectively, which have centroids a distance d apart. Shear connectors of stiffness k are present at uniform spacing p along the span L. Symbol δ_c is the deflection for the composite beam with complete interaction, and δ_a is the common deflection of both components of the member (steel + concrete) with no shear connection, both calculated for the design loading (for SLS) for the composite member. Based on a load level of 0.7 P_{Rk}, Aribert (2010) gives the mean stiffness of 19 mm studs in C30/37 concrete in a solid slab as about 100 kN/mm, with other data; but k may be lower in a composite slab, and higher at the load levels expected in service. The UK's National Annex may give guidance. There is further comment in Chapter 12.

Equation X9.1 was found to be valid also for centre-point loading, and is conservative if creep is allowed for in its calculations. It is a useful approximation, for use wherever it appears likely that deflection may be excessive. There is further discussion in Example 9.1.

Cracking in global analysis

Apart from the different loading, global analysis for serviceability differs little from that for an ultimate limit state. *Clause 7.4.1.1(2)* requires 'appropriate corrections' for cracking of concrete. ***Clause 9.3.1(6)*** says that the general and simplified methods of *clause 7.4.2.3* apply to the effects of cracking.

Clause 9.3.1(6)

The use of *clause 7.4.2.3* involves analysis of beams with spans of non-uniform cross-section. The alternative method of ***clause 9.3.1(7)*** enables this to be avoided. It is based on Stark and van Hove (1990).

Clause 9.3.1(7)

Cracking causes transfer of hogging bending moment at internal supports to adjacent regions of sagging bending. The clause gives approximate values for this transfer in *Figure 9.1*, applicable to the results of 'uncracked' global analysis. They are useful where excessive deflection is not a problem. The clause 'may be used', and the words 'should' within it mean that it is for use as a package.

Clause 9.3.1(7) limits the scope of the greater reductions given by curve A, which depend on the ratio between the cracked and uncracked stiffnesses of the cross-section of the beam. Its condition, equal loadings on all spans, means maximum design loadings, because maximum hogging bending occurs with full load on both adjacent spans. Curve B is a uniform reduction of 40%, which could be excessive where the effect of cracking is small. Using the new end moments, M_{h1} and M_{h2}, say, the maximum deflection should be found by elastic theory for the span, of uniform uncracked flexural stiffness E_aI_1.

Yielding of steel

In continuous beams built unpropped, with steel beams in Class 1 or 2, it is possible that serviceability loading may cause yielding at internal supports. This is permitted for beams in buildings, but it causes additional deflection, which should be allowed for. ***Clause 9.3.1(8)*** provides a method. Its reference to *clause 9.3.1(6)* excludes its use in combination with the method in *clause 9.3.1(7)*. The bending moments at internal supports are found by elastic analysis, with allowance for effects of cracking and without redistribution. The two values given in the clause for factors f_2 correspond to different checks. The first is for dead load only: wet concrete on a steel beam.

Clause 9.3.1(8)

The second check, for the beam in service, will be more critical after creep has occurred, so the long-term modular ratio should be used. For the load combination for this check, *clause 9.3.1(8)*, on calculation of deflections, takes one by way of *clause 9.3.1(6)* to *clause 7.4.2.3(2)*, on the effects of cracking of concrete. It says use the characteristic combination.

The UK's National Annex to the previous EN 1994-1-1 said that the criteria should be quantified for each project, and recommended that 'for function and damage to structural and non-structural elements' the characteristic combination should be used. The load additional to that for the first check acts on the composite beam. For each analysis, appropriate assumptions are needed for the adjacent spans, on their loading and on the state of construction.

Yielding of welded mesh

Clause 7.5.1(6) says that welded mesh should not be included in the effective section unless it has been shown to have 'sufficient ductility', which is not defined. In the UK, Class B or Class C reinforcement is normally used (Simms and Hughes, 2011). Where welded mesh has been omitted in analyses for ULS, it is simplest to omit it also for SLS. If calculations find that it reaches yield at ULS, the risk of fracture is such that it should be excluded.

Shrinkage

In principle, shrinkage effects appear in all load combinations. For SLS, *clause 7.4.2.2(7)* refers to *clause 9*, where ***clause 9.3.1(9)*** enables effects of shrinkage on deflections of beams to be ignored for span/depth ratios up to 20. In more slender beams, shrinkage deflections are significantly reduced by provision of continuity at supports. *Clause 10.8.2* gives new rules for shrinkage in composite slabs. Example 9.1 gives results of calculations.

Clause 9.3.1(9)

Temperature

Clause 9.3.1 does not refer to temperature. *Clause 7.4.2.5(2)* requires account to be taken of temperature effects at ULS in some situations (e.g. where allowance for lateral-torsional buckling is made). For buildings, neither Ψ_0 nor Ψ_1 are given as zero in Table A.1.7 of EN 1990 (nor in the UK's National Annex to the previous EN 1990 of 2004) so if temperature effects are relevant at ULS, their effect on deflections should be considered.

Large openings in webs

Openings provided in the webs of composite beams, usually for services, can increase the deflection. There are design rules in *clause D.6* and further comment there. Approximate values for the percentage increase caused by each opening are available for rectangular and circular shapes (Lawson and Hicks, 2011). Increases of 1% or less are ignored – for example, a rectangular opening at any location in the span with a depth half that of the steel beam and a width of 2.5 times its depth has been found to increase deflection by about 1.5% if unstiffened, and by less if stiffened. The increase for a larger opening can exceed 5%.

9.3.2 Vibration

Limits to vibration in buildings are material-independent. Composite floor systems are lighter and have less inherent damping than their equivalents in reinforced concrete. During their design, dynamic behaviour should be checked against the criteria in clause A.1.8.3 of EN 1990 referred to from ***clause 9.3.2(1).*** These are general, and advise that the lowest natural frequency of vibration of the structure or member should be kept above a value to be agreed with the client and/or the relevant authority. No values are given for limiting frequencies. However, limiting values of the dynamic response can be given in the national annex or through guidance, such as that given by ISO (ISO 10137, 2017), which requires the consideration of the damping ratio for the floor. *Clause 9.3.2(1)*

More specific guidance can be found in the extensive literature on this subject (BS 6472-1, 2008; Smith *et al.*, 2009). These sources refer to several criteria that are likely to be specific to the individual project, and, with other aspects, should be agreed 'by relevant parties'. A Note to clause 9.2(2) of EN 1993-1-1 says that limits to vibration of floors can be set by the national annex. The UK's National Annex is expected to refer to 'specialised literature', such as the design guide cited above.

9.4 Cracking of concrete

9.4.1 General

It has been found (Randl and Johnson, 1982) that for composite beams in hogging bending, the long-established British methods for control of crack width were unreliable for initial cracks, which were wider than predicted. Before this, it had been found for reinforced concrete that the appropriate theoretical model for cracking caused by restraint of imposed deformation was different from that for cracking caused by applied loading. This has led to the presentation of design rules for control of cracking as two distinct procedures, both of which are needed

- for minimum reinforcement, in *clause 9.4.2*, for all cross-sections that could be subjected to significant tension by imposed deformations (e.g. by effects of shrinkage, which cause higher stresses than in reinforced concrete, because of restraint from the steel beam)
- for reinforcement to control cracking due to direct loading (*clause 9.4.3*).

The rules given in EN 1994-1-1 are based on an extensive and quite complex theory, supported by testing on hogging regions of composite beams (Johnson and Allison, 1983). Much of the original literature is in German. *Clause 9.4* does not refer to cracking in columns, which is unlikely, except perhaps in unbraced composite frames.

A detailed account of the theory has been published in English (Hanswille, 1997; Johnson, 2003), with comparisons with results of tests on composite beams, additional to those used originally. The paper includes derivations of some of the formulae given in *clause 9.4*, comments on their scope and underlying assumptions, and procedures for estimating the mean width and spacing of cracks. These are tedious, and so are not in EN 1994-1-1. Its methods are simple. *Tables 9.1* and *9.2* give maximum diameters and spacings of reinforcing bars for four design crack widths, $w_{lim,cal}$: 0.15, 0.2, 0.3 and 0.4 mm. Symbol $w_{lim,cal}$ is defined as 'design crack width'.

Concrete in tension in a composite beam or slab for a building will usually be in exposure class XC3, (EN 1992-1-1, Table 6.1) for which Table 9.2 of EN 1992-1-1 gives the recommended limit for durability as $w_{\text{lim,cal}} = 0.3\ k_{\text{surf}}$ mm, unless a national annex gives different values. This limit for $w_{\text{lim,cal}}$ is not for values measured on site, but for values $w_{\text{k,cal}}$ found from clause 9.2.3 of EN 1992-1-1, 'Refined control of cracking'. The factor k_{surf} is a function of the concrete cover, with values between 1.0 and 1.5 that depend on $c_{\text{min,dur}}$, a factor related to exposure, defined in clause 6.5.2.2 of EN 1992-1-1. For buildings, past practice has been to assume $w_{\text{lim,cal}} = 0.3$ mm – that is, to assume $k_{\text{surf}} = 1$. This is a conservative simplification that avoids the determination of k_{surf}.

For spaces with low or very low air humidity, class XC1 to Table 6.1 of EN 1992-1-1, Table 9.2 gives no limit for durability and Table 9.1 gives 0.4 mm for appearance, but that is expected to be replaced by 0.3 mm in the UK's National Annex 'unless there are specific requirements for appearance'. The limits are more severe for prestressed members, which are not discussed further. The severe environment for a floor of a multi-storey car park is discussed in Chapter 6.

Tables 9.1 and *9.2* are for reinforcement that is 'weldable ribbed and indented' steel bars, as defined in 5.2.1 of EN 1992-1-1. The use of other steel is outside the scope of Eurocode 4.

Clause 9.4.1(2)

Clause 9.4.1(2) says that an 'estimation' of crack width 'may be obtained by appropriate modification of the method in EN 1992 1-1:2023, 9.2.3'. It is expected than an amendment to EN 1994-1-1:2026 will provide a new informative *Annex K* giving rules based on EN 1992-1-1: 9.2.3 in a form applicable to composite beams.

Clause 9.4.1(3)

They are intended to take account of the following differences between the behaviours of composite beams and reinforced concrete T-beams. The steel member in a composite beam does not shrink or creep and has much greater flexural stiffness than the reinforcement in a concrete beam. Also, the steel member is attached to the concrete flange by connectors which, after initial slip, apply discrete forces, whereas in reinforced concrete there is monolithic connection. ***Clause 9.4.1(3)*** refers to the simplified methods developed for composite members.

Uncontrolled cracking

Clause 9.2.1(2) of EN 1992-1-1 (referred to from *clause 9.4.1(1)*) permits cracking of uncontrolled width in some circumstances – for example, beams designed as simply-supported, with a concrete top flange that is continuous over 'simple' beam-to-column connections. These are flexible in bending, and rotate about a point that cannot be predicted, as its position depends on tolerances and methods of erection of the steelwork. It may then be impossible to predict the widths of cracks. Where the environment is dry and the concrete surface is concealed by a flexible finish, such as carpeting, crack widths exceeding 0.4 mm may be acceptable.

Clause 9.4.1(4)

Even so, the minimum reinforcement required (for other reasons) by EN 1992-1-1 may be inadequate to prevent the fracture of small-diameter bars near internal supports, or the formation of very wide cracks. Minimum areas greater than those of EN 1992-1-1 are therefore specified in ***clause 9.4.1(4)*** and, for composite slabs, in *clause 10.8.1(2)*.

Table 9.1. Use of steel fabric as minimum reinforcement, to *clause 7.4.1(4)*

Bar size and spacing	Cross-sectional area, mm^2 per m width	Maximum thickness of slab, h_c, mm	
		Unpropped, 0.2%	Propped, 0.4%
6 mm, 200 mm	142	71	–
7 mm, 200 mm	193	96	48
8 mm, 200 mm	252	126	63

The maximum thicknesses of slab that can be reinforced by one layer of standard welded fabric, according to these rules, are given in Table 9.1. For composite slabs, the relevant thickness is that above the net depth of the steel sheeting, for parallel sheeting, and above the gross depth, for transverse sheeting.

In ground-bearing slabs, tensile strain is sometimes concentrated into cracks at defined locations by the use of shallow saw cuts. This process should be used with caution above composite beams. Its accuracy may not be sufficient to avoid damage to mesh reinforcement or shear connectors (Dougherty, 2023).

The maximum spacing of flexural reinforcement, given in Table 12.2 of EN 1992-1-1 as an NDP, is $3h$, but not exceeding 400 mm, where h is the total depth of the slab. This rule is for solid slabs. It is not intended for slabs formed with profiled steel sheeting, for which a rule is given in *clause 10.2.1(5)*: spacing not exceeding $2h_{cs}$ and 350 mm in both directions, where h_{cs} is the overall thickness of the slab.

9.4.2 Minimum reinforcement

This clause, also about uncontrolled cracking, is more general than *clause 9.4.1*, and applies to hogging moment regions of continuous composite beams. The stress σ_s immediately after initial cracking can be taken as f_{sk}. It is assumed that the curvature of the steel beam does not change, so all of the tensile force in the concrete just before cracking is transferred to the reinforcement, of area A_s, given by *Formula 9.4* in ***clause 9.4.2(1)***. *Clause 9.4.2(1)*

$$A_s\sigma_s = k_{is}k_c k\ A_{ct}\ f_{ct,\,eff} \tag{9.4}$$

The three correction factors are based on calibration work (Roik *et al.*, 1989). Factor k_{is} allows for the reduction of tension in the concrete slab caused by cracking and is given as 0.9. Factor k is for 'non-uniform self-equilibrating stresses'. These arise from primary shrinkage and temperature effects, which cause curvature of the composite member. Slip of the shear connection also causes curvature and reduces the tensile force in the slab. Its value is given as 0.8.

Factor k_c takes account of the non-uniform stress distribution in the concrete assumed to crack, as given by *Formula 9.5*. Its calculation can often be omitted, because it is less than 1.0 only where $z_0 < 1.2\ h_c$. Where there is a composite slab spanning onto the beam, the thickness of the concrete flange h_c is found using the gross depth of the sheeting. Then, the depth of the 'uncracked' neutral axis below the centre of the thickness h_c normally exceeds 1.2 h_c, giving $k_c = 1$. For sheeting parallel to the beam, h_c is found using the net depth of the sheeting.

The method of *clause 9.4.2(1)* is not intended for the control of early thermal cracking, which can occur in concrete a few days old, if the temperature rise caused by heat of hydration is excessive. The flanges of composite beams are usually too thin for this to occur. It would not be correct, therefore, to assume a very low value for $f_{ct,eff}$. The suggested value, 3 N/mm^2, was probably rounded from the mean 28 day tensile strength of grade C30/37 concrete, given in EN 1992-1-1 as 2.9 N/mm^2 – the value used as the basis for the optional correction given in ***clause 9.4.2(2)***. The difference between 2.9 and 3.0 is obviously negligible. If there is good reason to assume a value for $f_{ct,eff}$ such that the correction is not negligible, it is best used by assuming a standard bar diameter ϕ, calculating ϕ^*, and then finding σ_s by interpolation in *Table 9.1*. *Clause 9.4.2(2)*

With factors k as given above and $f_{sk} = 500$ MPa, $f_{ct,eff} = 3$ MPa, *Formula 9.4* gives $A_s/A_{ct} \geq 0.43\%$, which corresponds to the rule in Table 9.1 for propped construction.

Reinforcement for encasement of a steel web

Clause 9.4.2(6) gives a minimum value of A_s/A_{ct} for encasement of the type shown in *Figure 8.11*. As k_{is} is not referred to here, the value in *clause 9.4.2(1)*, 0.9, should be used. Detailing of this reinforcement may be determined by the requirements of *clause 8.3.3* for vertical shear and of EN 1994-1-2 for fire resistance. *Clause 9.4.2(6)*

9.4.3 Control of cracking due to direct loading

Clause 9.4.3(1) refers to *Tables 9.1* and *9.2*. The only data needed for them are the tensile stress in the reinforcement, σ_s, and the design crack width. ***Clause 9.4.3(2)*** specifies elastic global analysis to *clause 7*, allowing for the effects of cracking and tension stiffening. The preceding comments on global analysis for deformations apply also to this analysis for bending moments in regions with concrete in tension. *Clause 9.4.3(1)* *Clause 9.4.3(2)*

Both for solid and composite slabs, the design crack width and thickness of the concrete flange, h_c, will be known. It will be evident whether there should be one layer of reinforcement or two. Two layers will often consist of bars of the same size and spacing, which satisfies *clause 9.4.2(3)*.

Table 9.1 gives the maximum bar diameter. Figure 9.1 gives an example of its results, assuming two similar layers of bars at spacing s, with $k_c = 1$ and $f_{ct,eff} = 3.0$ N/mm^2. *Formula 9.4* then gives, for a slab of breadth b, the design rule

$$(\pi\phi^2 / 4)\ (2\ b / s) \geq 0.72 \times 3\ bh_c / \sigma_s$$

$$\text{Hence,} \quad h_c s \leq 0.727\phi^2 \sigma_s \qquad \text{(X9.3)}$$

For each bar diameter and a given crack width, *Table 9.1* gives $\phi^2\sigma_s$, so the product $h_c s$ is known. This leads to curves of bar spacing for chosen slab thicknesses, which can of course also be read as slab thicknesses for given bar spacing. Figure 9.1 can be used for slabs with one layer of bars by halving the slab thickness.

Figure 9.1. Bar spacing and maximum diameter for reinforcement in two equal layers, for $w_{lim,cal} = 0.3$ mm and $f_{ct,eff} = 3.0$ N/mm^2

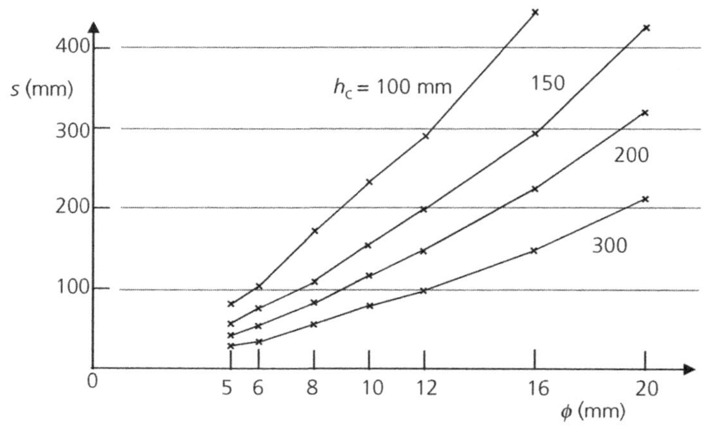

In practice, crack width is usually checked for reinforcement designed for an ultimate limit state, so A_s, ϕ and s have been chosen and σ_s can be found. If *Table 9.1* shows that this diameter ϕ is too high, smaller bars at closer spacing are needed. An alternative is to use *Table 9.2* to check the bar spacing. If it is too wide, the bar diameter is reduced, leading to closer spacing.

Tension stiffening

Clause 9.4.3(3) ***Clause 9.4.3(3)*** gives *Formula 9.7* for the stress σ_s that includes a term for the effect of tension stiffening. The wording 'may' means that other methods may be used, not that the correction is optional. Tension stiffening is now explained.

It was thought at one time that, for a given tensile strain at the level of the reinforcement, the total extension must be the extension of the concrete plus the width of the cracks, so that allowing for the former reduced the latter. The true behaviour is more complex.

The upper part of Figure 9.2 shows a single crack in a reinforced concrete member with a central reinforcing bar resisting tensile force, N. At the crack, the force, N, causes strain $\varepsilon_{s2} = N/A_sE_s$ in the bar, and the strain in the concrete is the free shrinkage strain ε_{cs}, which is negative, as shown. There is a transmission length L_e each side of the crack, within which there is transfer of shear between the bar and the concrete. Outside this length, the strain in both the steel and the concrete is ε_{s1}, and the stress in the concrete is fractionally below its tensile strength. Within the length $2L_e$, the curves $\varepsilon_s(x)$ and $\varepsilon_c(x)$ give the strains in the two materials, with mean strains ε_{sm} in the bar and ε_{cm} in the concrete.

The graph is assumed to represent the typical behaviour of a reinforcing bar in a cracked concrete flange of a composite beam, in a region of constant bending moment such that the crack spacing is $2L_e$.

Figure 9.2. Strain distributions near a crack in a reinforced concrete tension member

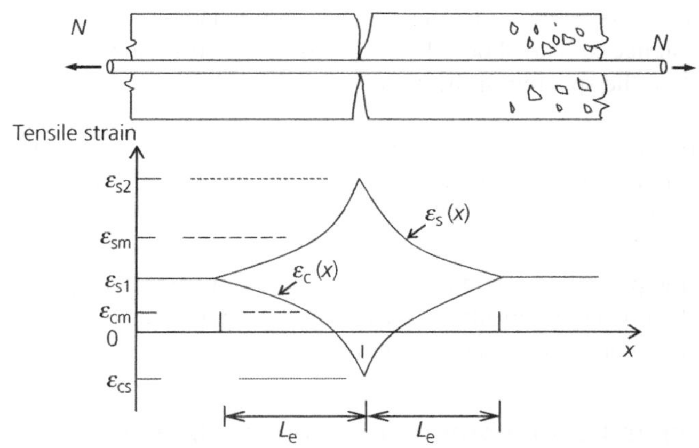

The curvature of the steel beam is determined by the mean stiffness of the slab, not the fully cracked stiffness, and is compatible with the mean longitudinal strain in the reinforcement, ε_{sm}.

Midway between the cracks, the strain is the cracking strain of the concrete, corresponding to a stress less than 30 N/mm^2 in the bar. Its peak strain, at the crack, is much greater than ε_{sm}, but less than the yield strain, if crack widths are not to exceed 0.4 mm. The crack width corresponds to this higher strain, not to the strain ε_{sm} that is compatible with the curvature, so a correction to the strain is needed. It is presented in *clause 9.4.3(3)* as a correction to the stress $\sigma_{s,0}$ because that is easily calculated, and *Tables 9.1* and *9.2* are based on stress. The strain correction cannot be shown in Figure 9.2 because the stress $\sigma_{s,0}$ is calculated using the 'fully cracked' stiffness, and so relates to a curvature greater than the true curvature. The derivation of the correction (Hanswille, 1997; Johnson, 2003) takes account of crack spacings less than $2L_e$, the bond properties of reinforcement, and other factors omitted from this simplified outline.

The property α_{st} (*Formula 9.9*) is independent of the modular ratio unless the 'cracked' composite cross-section includes concrete in compression. For simplicity, α_{st} may conservatively be taken as 1.0, because $AI > A_a I_a$.

Clause 9.4.3(4) says that for buildings, the quasi-permanent combination should 'typically' be used. Except for storage areas, the values of factors ψ_2 for floor loads in buildings are typically 0.3 or 0.6. The bending moments will then be much less than for the ultimate limit state, especially for cross-sections in Class 1 or 2 in beams built unpropped. There is no need to reduce the extent of the cracked regions below that assumed for global analysis, so the new bending moments for the composite members can be found by scaling values found for ultimate loadings. *Clause 9.4.3(4)*

Cracking in regions of biaxial tension

Background to the new ***clause 9.4.3(5)*** is available (Johnson, 2011). This type of cracking occurs where cantilever cross-girders are supported within the hogging region of continuous main girders. The Avonmouth viaduct in the UK is an example. It is less likely in buildings because beams are often designed as simply-supported. *Clause 9.4.3(5)*

Influence of profiled sheeting on the control of cracking

The references to profiled steel sheeting in *clause 9.4* are in *clause 9.4.1(4)*, '*...no account should be taken of any profiled steel sheeting*'; and in the definition of A and I in 9.4.3(3), '*...neglecting... profiled sheeting, if any*'. The effects of the use of profiled sheeting for a slab that forms the top flange of a continuous composite beam are as follows.

- There is no need for control of crack widths at the lower surface of the slab.
- Where the sheeting spans in the transverse direction, there is no evidence that it contributes to the control of cracks at the top surface of the slab.
- Where the sheeting spans parallel to the beam, it is likely to be discontinuous at an internal support, and so ineffective for crack-width control.

General comments on *clause 9.4*

Where unpropped construction is used for a continuous beam, the design loading for checking cracking is usually much less than that for the ultimate limit state, so that the quantity of reinforcement provided for resistance to load should be sufficient to control cracking. The main use of *clause 9.4.3* is then to check that the spacing of the bars is not too wide.

Where propped construction is used, the disparity between the design loadings for the two limit states is smaller. If cracks are to be controlled to 0.3 mm or 0.2 mm, a check to *clause 9.4.3* is more likely to influence the reinforcement required.

For beams in frames, the preceding comments apply where semi-rigid or rigid connections are used. Where floors have brittle finishes or an adverse environment, simple beam-to-column joints should not be used, because effective control of crack width may not be possible.

Example 9.1: two-span beam (continued) – serviceability limit state

Details of this beam are shown in Figure 8.27. All of the design data and calculations for the ultimate limit state are given in Example 8.6. For data and results required here, reference should be made to

- Table 8.3, for characteristic loads per unit length
- Table 8.4, for elastic properties of the cross-sections at the internal support (B in Figure 8.27(c)) and at mid-span
- Table 8.5 for bending moments at support B for uniform loading on both spans
- Figure 8.32 for bending-moment diagrams for design ultimate loadings, excluding the effects of shrinkage.

The secondary effects of shrinkage are significant in this beam and cause a hogging bending moment at support B of 117 kN m (Example 8.6). *Clause 9.3.1(9)* does not permit shrinkage to be ignored for serviceability checks on this beam, because it does not refer to lightweight-aggregate concrete, which is used here.

Stresses

From *clause 9.2.2(1)*, there are no limitations on stress; but stresses in the steel beam need to be calculated, because if yielding occurs under service loads, account should be taken of the resulting increase in deflections from *clause 9.3.1(8)*.

Table 9.2. Hogging bending moments at support B and stresses in steel bottom flange, for the characteristic load combination and cracked cross-section

Loading	Permanent (on steel beam)	Permanent (on composite beam)	Variable	Shrinkage
w, kN/m	5.78	1.2	17.5	–
Modular ratio	–	20.6	20.6	29.2
$10^{-6}\ I_{y,B}$, mm^4	337	445	445	445
$M_{Ek,B}$, kN m	104	18	260	117
$10^{-6}\ W_{a,bot}$, mm^3	1.50	1.72	1.72	1.72
$\sigma_{a,bot}$, N/mm^2	69	10	151	68

Yielding is irreversible, so, as explained under *clause 9.3.1(8)*, it should be checked for the characteristic load combination. The maximum stress in the steel beam occurs in the bottom flange at support B. Results for variable load on both spans and 15% of each span cracked are given in Table 9.2. The permanent load, other than floor finishes, is assumed to act on the steel beam alone.

The modular ratio is irrelevant, as the concrete is fully cracked. The relevant section properties are shown; I is a second moment of area and $W_{a,bot}$ is a section modulus. The total bottom-flange compressive stress is

$$\sigma_{\mathrm{Ek,bot,a}} = 298\ \mathrm{N/mm^2} (= 0.84 f_y)$$

so no allowance is needed for yielding.

Deflections

The maximum deflection of span AB of the beam will occur at about 4.8 m from A (40% of the span), when variable load acts on span AB only. The additional deflection caused by slip of the shear connection can be ignored, if all three conditions of *clause 9.3.1(4)* are satisfied.

One of these is that either the degree of shear connection is at least 0.5 (not satisfied here) or that the shear per connector is less than P_{Rd}. Here, $P_{\mathrm{Rd}} = 50.8$ kN. Elastic analysis of the uncracked section at 15% of the span from the internal support finds $P_{\mathrm{Ed}} = 38$ kN per connector, so no account need be taken of slip.

Calculated maximum deflections in span AB, with 15% of each span assumed to be cracked, are given in Table 9.3. The frequent combination is used, for which $\psi_1 = 0.7$, so the variable loading is $0.7 \times 17.5 = 12.3$ kN/m. Hand calculations for this beam with non-uniform cross-section are too long to give here.

Table 9.3. Deflections at 4.8 m from support A, for the frequent combination

Load	Modular ratio	Deflection (mm)
Dead, on steel beam	–	9
Dead and imposed on composite beam	20.6	15
Shrinkage	29.2	8

Shrinkage deflections in continuous beams are often negligible, because the primary and secondary deflections are in opposite directions. The option of neglecting shrinkage in *clause 7.4.2.2(7)* is for ULS only. Here the beams have a span/depth ratio exceeding 20. The shrinkage is high, and predicted shrinkage deflection is not negligible. It would be 15 mm if the spans were simply-supported. Here, it is reduced by the secondary bending moment at support B, to about 8 mm.

The functioning of the floor may depend on its maximum deflection relative to the supporting columns. The total deflection of the composite slab assumed here, if cast unpropped, is about 9 mm for the frequent combination. Example 10.1 shows that this is increased by about 3 mm by shrinkage. This is relative to the supporting beams, and increases the maximum floor deflection from 32 to 44 mm (span/273).

As explained above, EN 1990 leaves the designer to decide what deflection is 'excessive', based on the type of structure involved and its intended use. It is likely to be excessive here, especially if partitions are to be provided. It could be reduced by the use of propped construction for the beams or for both beams and slabs.

In practice, deflections of continuous beams are rarely troublesome. The beam here is an extreme example, chosen partly to illustrate the use of lightweight-aggregate concrete and the effects of unpropped construction.

Control of crack width

Clause 9.4 applies to reinforced concrete that forms part of a composite member. In the beam considered here, the relevant cracks are those near support B caused by hogging bending of the beam, and cracks along the beam caused by hogging bending of the composite slab that the beam supports. The latter are treated in Example 10.1 on a composite slab.

Formula 9.4 for minimum reinforcement includes factor k_c, which is calculated for the uncracked composite section, using $n_0 = 10.3$. For both effective widths, 1.6 m and 2.5 m, k_c=1. A value is required for the strength of the concrete when cracks first occur. As unpropped construction is used, there is at first little load on the composite member, so from *clause 9.4.2(1)*, conservatively, $f_{\mathrm{ct,eff}} = 3.0$ N/mm^2.

Assuming that 10-mm bars are used for the minimum reinforcement in regions of sagging bending, *Table 9.1* gives $\sigma_s = 320$ N/mm^2. Then, from *Formula 9.4*,

$$100\, A_s / A_{ct} \geq 100 \times 0.9 \times 1 \times 0.8 \times 3.0 / 320 = 0.675\%$$

One layer of 10-mm bars at 125-mm spacing provides 628 mm^2/m, which is **0.78%.**

In regions of hogging bending, the 12-mm bars at 150-mm spacing used in Example 8.6 provide 1131 × 100/(1600 × 80) = **0.88%.** This is sufficient because *Table 9.1* gives

$$\sigma_s = 280 \text{ N/mm}^2 \text{ and minimum reinforcement is } 0.675 \times 320/280 = \mathbf{0.77\%}.$$

Clause 7.5.1(5) also sets a limit, as a condition for the use of plastic resistance moments with bars in tension. For this concrete, $f_{lctm} = 2.31$ N/mm^2, and $f_{sk} = 500$ N/mm^2. Hence, from *Formula 7.6* with $k_c = 1.0$,

$$\mathbf{100}\, \rho_s \geq 100 \times (355/235)(2.31/500) = \mathbf{0.70\%} \qquad \text{(X9.4)}$$

This is also satisfied.

Cracking due to direct loading

Only the most critical cross-section, at support B, will be considered. *Clause 9.4.3(4)* permits the use of the quasi-permanent combination, for which the variable loading is $\psi_2 q_k$, with

$\psi_2 = 0.6$ (an NDP), from clause A.1.6.3(1) of EN 1990.

From Table 9.2, the bending moment at B that stresses the reinforcement is

$$M_{E,qp,B} = 18 + 260 \times 0.6 + 117 = 291 \text{ kN m}$$

It is at first assumed that the 12-mm reinforcement used in Example 8.6 will be satisfactory. The neutral axis for the cracked section is 33 mm above the centre of the steel section (Table 8.4), which in turn is 355 mm below the top of the slab, so the neutral axis is 322 mm below the top of the slab. For the cracked composite section,

$$10^{-6} I = 445 \text{ mm}^4, \text{ so the section modulus for reinforcement at depth 30 mm is}$$
$$10^{-6} W_s = 445/(322-30) = 1.52 \text{ mm}^3$$

Hence, from *clause 9.4.3(3)*, for *Formula 9.7*, $\sigma_{s,o} = 291/1.52 = 191$ N/mm^2

The correction for tension stiffening, *Formula 9.7*, is now calculated, with $\rho_s = 0.00884$ (above). Using values obtained in Example 8.6,

$$\alpha_{st} = AI/(A_a I_a) = (9880 + 1131) \times 445/(9880 \times 337) = 1.47$$

From *Formula* 9.8 with f_{ctm} from Example 8.6

$$\Delta\sigma_s = 0.4 \times 2.31/(1.47 \times 0.00884) = \mathbf{71\ N/mm^2} \qquad \text{(X9.5)}$$

From *Formula 9.7*, $\sigma_s = 191 + 71 = 262$ N/mm^2

The characteristic crack width is to be limited to 0.3 mm. *Table 9.1* gives the maximum bar diameter as 12 mm. From *Table 9.2*, the bar spacing should not exceed 150 mm. The use of **12-mm bars at 150 mm** spacing at support B satisfies both conditions. They extend over a width of 1.6 m. Over the remaining flange width of each beam, 0.9 m, the tensile force in the slab will be lower, due to shear lag, and reduction in bar diameter to 10 mm could be considered. The rules proposed in *Annex K,* referred to under *clause 9.4.1(2),* find the crack width here to exceed 0.4 mm.

Serviceability, if the spans were designed as simply supported

It is found in Example 8.6 that if each span is designed as simply supported, all checks for ULS are satisfied using the same cross-section and method of construction. The mid-span deflections under the loading used above are as follows, assuming that the span between supporting columns is 11.76 m

- from loading 5.76 kN/m on the steel beam: 20.3 mm
- from loading 13.5 kN/m on the composite beam, with $n = 20.6$: 19.3 mm
- from primary shrinkage, with $n = 29.2$: 15.4 mm

The total deflection is high, at 55 mm, or span/218, and there will be further deflection of the composite slabs relative to the beams. Performance could be improved in several ways, such as the use of propped construction or precambering the steel beams. The deflection due to imposed loading is $20.3 \times 12.3/13.5 = 18.5$ mm, or span/636, which is not excessive.

The connections to the columns would probably be similar to those designed in Example 11.1, though the thickness of the end plates could be reduced from 12 mm to 10 mm. This reduction, and neglecting the top reinforcement, roughly halves the initial stiffness of each connection, but it remains well within the range defined in EN 1993-1-8 as 'semi-rigid', and would slightly reduce the deflections calculated above.

It could be difficult to control crack widths at supports to 0.3 mm, but this may not be required. *Clause 9.4.1(4)* says that where '*the control of crack width is of no interest*' the longitudinal reinforcement for unpropped construction should be at least 0.2% of the area of concrete. This requires 160 mm^2/m, so a 200-mm mesh of 7-mm bars could be used (192 mm^2/m). They could fracture under the design ultimate loads, but their presence would not be relied on in the design of the connection.

Roger P Johnson and Stephen J Hicks
ISBN 978-1-83662-921-4
https://doi.org/10.1108/978-1-83662-920-720253012

CHAPTER 10
Composite slabs with profiled steel sheeting for buildings

This chapter corresponds to Section 10 of EN 1994-1-1, which has the following clauses.

- Scope *Clause 10.1*
- Detailing provisions *Clause 10.2*
- Actions and action effects *Clause 10.3*
- Analysis for internal forces and moments *Clause 10.4*
- Verification of profiled steel sheeting as shuttering for ultimate limit states *Clause 10.5*
- Verification of profiled steel sheeting as shuttering for serviceability limit states *Clause 10.6*
- Verification of composite slabs for the ultimate limit states *Clause 10.7*
- Verification of composite slabs for serviceability limit states *Clause 10.8*

10.1 General

Composite slabs are an efficient form of floor construction, but they are less robust than continuous concrete slabs with two-way reinforcement. Their orthotropic structural properties and staged construction lead to relatively complex rules for design. Some verifications are done by manufacturers of the sheeting and embodied in proprietary software and tables of limiting spans and load levels. A designer using these aids should make further verifications, but may not be fully aware of the assumptions that underlie the tables.

New entrants to this competitive market should become familiar not just with rules in the Eurocodes, but with the extensive guidance provided by the industry – for example, in the UK, by the Steel Construction Institute and other organisations.

Potential situations that could lead to unsatisfactory or unsafe performance include the following

- under-prediction of the loads that may be applied during construction
- as-built layouts of profiled sheets that differ from those assumed during design
- substitution of similar depth sheets that may not necessarily have the same structural performance as those specified
- the addition of significant line loads (e.g. from partitions) during the life of the building
- the application of repeated travelling loads, as from fork-lift trucks (see *clause 10.1.1(3)*)
- unintended location of shear connectors on the 'unfavourable' side of troughs in profiled sheeting, which can reduce their slip capacity and shear resistance.

This list is mainly for designers. Problems that could arise during construction (e.g. unsatisfactory through-deck welding of studs) are outside the scope of this guide. Further guidance on design and construction is available (Dougherty, 2023).

Scope

The form of construction and the scope of *clause 10* are defined in ***clause 10.1.1.*** The shape of the steel profile, with ribs running in one direction, and its action as tensile reinforcement for the finished floor result in a system that effectively spans in one direction only. The slab can also act as the concrete flange of a composite beam spanning in any direction relative to that of the ribs. Provision is made for this in the rules for design of beams in *clauses 7, 8* and *9*. *Clause 10.1.1*

Symbols

Eurocode 4 uses many symbols for dimensions of composite slabs, as shown in *Figure 10.2* and in Figure 10.1, where G is the centre of area of the sheeting and 'plastic n.a.' is its plastic neutral axis.

Figure 10.1. Symbols for composite slabs with trapezoidal sheeting

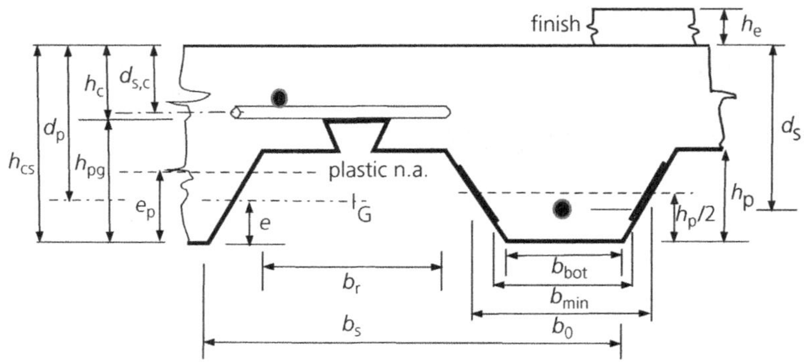

Clause 10.1.1(2) ***Clause 10.1.1(2)*** limits the scope to sheeting with 'narrowly spaced webs', meaning sheeting with the ratio of rib width to rib spacing, b_r/b_s, less than the specified value. This ratio is an important property of a composite slab. If the troughs are too narrow, there may be insufficient resistance to vertical shear. If the rib is too wide, the ability of the slab to spread loads across several webs may be inadequate, especially if the thickness of the slab above the sheeting is minimised to save weight.

The maximum value of b_r/b_s is given as 0.65, subject to a Note. Such a wide range of profiles is in use that this route for a higher value to be used for a particular sheeting is provided, involving testing. Such sheets are used in shallow-floor construction – for example, ArcelorMittal profile Cofraplus 220 has $b_r/b_s = 584/750$ (mm) $= 0.78$.

The design methods for composite slabs given in *clause 10* are based on test procedures described in *clause B.3*. The test to failure is under static loading. Thus if dynamic effects are expected, the detailed design for the particular project must ensure that the integrity of the composite action is maintained
Clause 10.1.1(3) (***clauses 10.1.1(3)*** and ***10.1.1(4)***).
Clause 10.1.1(4)
Clause 10.1.1(5) Guidance on the degree of lateral restraint provided to steel beams (***clause 10.1.1(5)***) is available in EN 1993-1-3 and elsewhere (Gardner, 2011). Inverted U-frame action relies also on flexural restraint. This subject is covered in comments on *clause 8.4.2*.

Because of the wide range of profiles in use, resistance to longitudinal shear has always been based on tests. The *m-k* test method is no longer supported, so profiles with shear properties found in this way are now outside the scope of Eurocode 4. 'Ductile' failure is defined in *clause 10.7.2(3)*, and is a requirement for the use of partial shear connection.

Types of shear connection

As for other types of composite member, bond is not accepted as a reliable method of shear connection
Clause 10.1.2.1(1) (***clause 10.1.2.1(1)***). Sheeting without local embossments is permitted where the profile shape is re-entrant (*Figure 10.1(2)*). Here, the distinction between 'frictional interlock' and 'bond' is, in effect, that the former is what remains after the 25 cycles of loading specified in *clause B.4.4(3)*.

The quality of mechanical interlock is sensitive to the height or depth of the small local deformations of the sheeting, so tight tolerances (*clause B.4.3(2)*) should be maintained on these during manufacture with occasional checking on site and reference to the essential tolerances given in EN 1090-4. (The tolerances on the depth of embossments are –0.5/+1.0 mm.) These two types of interlock can be
Clause 10.1.2.2(1) insufficient to provide full shear connection, as defined in ***clause 10.1.2.2(1)***. They can be augmented by anchorages at the ends of each sheet, as shown in *Figure 10.1*, or design can be based on partial shear connection.

10.2 Detailing provisions

Clause 10.2.1(1) The limits to thickness of composite slabs given in ***clauses 10.2.1(1)*** and ***10.2.1(2)*** are based on sat-
Clause 10.2.1(2) isfactory experience of floors with these dimensions. No limits are given for the depth of the profiled sheeting. Its minimum depth will be governed by deflection. For a slab acting compositely with a

beam, the minimum depths are increased (*clause 10.2.1(2)*) to suit the detailing rules for stud connectors, such as the length of stud that extends above the sheeting, and the concrete cover. A slab used as a diaphragm is treated similarly.

Where a slab spans onto a hogging moment region of a composite beam, the reinforcement transverse to the span of the slab is governed by the rules for the flange of the beam (e.g. *Table 9.1*), and is likely to exceed the amount given in ***clause 10.2.1(4).*** The minimum bearing lengths for sheeting (***clause 10.2.3***) are based on accepted practice. *Clause 10.2.1(4)* *Clause 10.2.3*

10.3 Actions and action effects

Profiled sheeting

Where props are used for profiled sheeting (***clause 10.3.1(2)***), care should be taken to set these at the correct level, taking account of any expected deflection of the surface that supports them. *Clause 10.3.1(2)*

Normally, to increase the speed of construction, profiled sheeting is not propped. It then has to support the weight of the wet concrete and the construction loading. Its design is likely to be governed by either this situation or the check on the deflection of the finished floor.

For the serviceability limit state, the deflection of the sheeting at the time when the concrete hardens is important, for use when checking the total deflection of the floor in service. The construction load and any extra loading from heaping of concrete are not present at this time, so the deflection is from permanent load only.

Clause 10.3.2(1) refers to 'ponding', and ***clause 10.3.2(2)*** gives a condition for its effects to be ignored. Where profiled sheeting is continuous over several supports, this check should be made using the most critical arrangement of imposed load. There is further comment under *clause 10.6(2)*. *Clause 10.3.2(1)* *Clause 10.3.2(2)*

Composite slab

The resistances of composite slabs are determined by plastic theory or by results from tests in which all the loading is resisted by the composite section (*clause B.4.3(6)*). This permits design checks for the ultimate limit state to be made on this basis (***clause 10.3.3(2)***). *Clause 10.3.3(2)*

10.4 Analysis for internal forces and moments

Profiled steel sheeting

Clause 10.4.1(1) refers to EN 1993-1-3 (BS EN 1993-1-3:2024), which gives guidance on global analysis of continuous members of light-gauge steel, including the use of finite-element methods. ***Clause 10.4.1(2)*** rules out plastic redistribution where propping is used, but not where unpropped sheeting extends over more than one span, as is usual. Subsequent flexure over a permanent support will be in the same direction (hogging) as during construction, whereas at the location of a prop it will be in the opposite direction. Any redistribution should satisfy clause 8.1.4.1(9) of EN 1993-1-3, which requires it to be justified by results from tests. *Clause 10.4.1(1)* *Clause 10.4.1(2)*

Elastic global analysis can be used because a safe lower bound to the ultimate resistance is obtained. Elastic moments calculated for uniform stiffness are normally greatest at internal supports. The reduction in stiffness due to parts of the cross-section yielding in compression will be greatest in these regions, which will cause redistribution of moment from the supports to mid-span. Redistribution should not be used in analyses for serviceability loads because of its uncertain effect on deflections.

In the UK, partial-plastic global analysis of continuous double-span decks has traditionally been used in design by UK manufacturers to satisfy the rules in clause 8.1.11 of EN 1993-1-3 for combined bending and reaction from a support. This approach can provide up to about 30% redistribution of moment from the internal support. Larger amounts of redistribution are sometimes used in current design, based on product-specific double-span tests according to EN 1993-1-3.

Composite slab

As the steel sheets are normally continuous over more than one span, and the concrete is cast over this length without joints, the composite slab is in reality continuous. If elastic global analysis is used based on the uncracked stiffness, the resulting moments at internal supports are high. To resist these moments may require heavy reinforcement. This can be avoided by designing the slab as a series of

simply-supported spans (*clause 10.4.2(5)*), provided that crack-width control is not a problem. Other
Clause 10.4.2(3) approaches that reduce the quantity of hogging reinforcement needed are given in ***clauses 10.4.2(3)***
Clause 10.4.2(4) and ***(4)***.

Numerical and experimental research on continuous slabs has been reported (Stark and Brekelmans, 1990). With typical relative values of resistance moment at internal supports and at mid-span, the maximum design loads calculated by elastic analysis with limited redistribution were found to be less than those obtained by treating each span as simply supported. This arises because the resistance to sagging moment is not fully used.

If the slab is to be treated as continuous, plastic analysis is more advantageous. The studies showed that no check on rotation capacity need be made provided the conditions given in *clause 10.4.2(4)* are satisfied. Where reliance is placed on the contribution of sheeting to resistance to hogging bending, designers should consider whether the layout of the sheets might be modified during construction.

In the UK, composite slabs are often designed as simply supported in normal conditions. However, in fire conditions, the slab is treated as continuous to enable plastic analysis to be used. The reinforcement over the supporting beams should then be placed as close to the top of the slab as possible.

Effective width for concentrated point and line loads

The ability of composite slabs to carry masonry walls or other heavy local loads is limited. The rules
Clause 10.4.3 of ***clause 10.4.3*** for the effective widths b_{m}, b_{em} and b_{ev} are important in practice. They are based on a mixture of simplified analysis, test data and experience (BS 5950-4, 1994). The effective widths depend on the ratio between the longitudinal and transverse flexural stiffnesses of the slab, and are lower than those for solid reinforced concrete slabs.

Clause 10.4.3(2) The rule for effective width of slab in ***clause 10.4.3(2)*** is related to its effective thickness. Where there is a small top rib, the depth h_{c} in *Formula 10.1* is as shown in Figure 10.2, not as in *Figure 10.4* of EN 1994-1-1.

The previous clauses on the use of nominal transverse reinforcement have been supplemented by
Clause 10.4.3(5) the more detailed ***clauses 10.4.3(5)*** and ***(6)***. The quantity of this reinforcement (*paragraph (5)*) is as
Clause 10.4.3(6) before. Its use is permitted without detailed calculation of the local effects of local imposed loading up to the limits in *paragraph (4)*. The limits are as before where the thickness of the concrete above the gross depth of the sheeting is small. For thicker topping, there is a more generous rule for concentrated loads. The limit for distributed load is unchanged at 5.0 γ_{Q} kN/m^2. There is a new method for designing the transverse reinforcement in the slab. Examples 10.1 and 10.2 are relevant.

A rule for line loading has been added, assuming the most adverse direction for the line: parallel to the span. The rules for concentrated loads are not generous, and should not be assumed to apply for the 'largely repetitive' loads to which *clause 10.1.1(3)* refers, such as the wheels of fork-lift trucks and mobile elevated working platforms.

Formulae 10.5 and *10.6* include the uncracked second moments of area I_{cross} and I_{long}. The sheeting is included, of course, in I_{long} but not in I_{cross}. It is simpler to use the average thickness of the concrete than to allow for its irregular thickness. The contribution of reinforcement is usually negligible and can be omitted.

10.5 and 10.6 Verification of profiled steel sheeting as shuttering

Clause 10.5(1) The design checks before composite action is established are done to EN 1993-1-3. ***Clause 10.5(1)*** refers to the loss of effective cross-section that may be caused by deep deformations of the sheeting. This loss and the effects of local buckling are both difficult to determine theoretically.

Design recommendations provided by manufacturers are based in part on results of loading tests on the sheeting concerned. These data should include the nominal cross-sectional area of the sheeting (A_{p}), the positions of the elastic and plastic neutral axes of the profile, denoted e and e_{p} respectively (*Figure 10.7*), and the effective cross-section (A_{pe}). This is normally slightly less than A_{p} because the embossed parts of the cross-section (provided to improve shear connection) may reduce stiffness and resistance to longitudinal force.

The maximum deflection of $L/180$ given in the Note to ***clause 10.6(2)*** is accepted good practice. It is expected to be confirmed in the UK's National Annex with the addition of upper limits: 20 mm where the loads from ponding are ignored (see *clause 9.3.2(2)*) and 30 mm where they are included. *Clause 10.6(2)*

10.7 Verification of composite slabs for the ultimate limit states

10.7.1 Design criterion

No comment is needed.

10.7.2 Bending resistance

The rules in *clause 10.7.2* are based on research (Stark and Brekelmans, 1990). From ***clause 10.7.2(1)***, partial shear connection is linked with a reduction in the compressive force in the concrete. This is consistent with its definition in *clause 3.1.27*, because the bending resistance of a composite slab to sagging bending is almost proportional to this force. The method is not used for hogging bending. *Clause 10.7.2(1)*

Resistance to sagging bending

To avoid the risk of sudden failure in longitudinal shear, the use of partial shear connection is limited in ***clause 10.7.2(2)*** to slabs with 'ductile' behaviour, which is defined in ***clause 10.7.2(3)*** in terms of behaviour in tests. For plain sheeting, the ultimate shear resistance is not significantly greater than that for initial slip. Such behaviour is not 'ductile'. For profiles with deformations such as embossments, the expected ultimate behaviour involves a combination of friction and mechanical interlock after initial slip, giving a relationship between load and deflection that is ductile for most profiles. *Clause 10.7.2(2)* *Clause 10.7.2(3)*

The degree of partial shear connection, η, is not a useful concept for composite slabs, as it varies along the length of the beam, as shown in Example 10.1. There is no equivalent to the rules for minimum degree of shear connection in beams. From Figure 10.7, it is seen that if $\tau_{u,Rd}$ is gradually reduced to zero, the curve AB for M_{Rd} becomes more horizontal, the location of shear failure, point D, approaches mid-span, and the bending resistance reduces to that of the sheeting, a ductile mode.

The compressive force N_c in a width b_c of the slab at any cross-section is usually the force N_p transferred from the sheeting by shear failure at the surface of the sheeting, at a shear stress averaged over the length of the relevant shear span, L_x, which is the distance from the cross-section to the nearest support. The design stress used in the method, $\tau_{u,Rd}$, is an equivalent value for a horizontal plane, not the stress at the surface of the sheeting. Where longitudinal reinforcement is provided in the troughs, it is assumed to be at yield, and a force $A_s f_{sd}$ is added to N_p, *clause 10.7.2(8)*.

The characteristic shear strength, τ_{Rk}, is found from tests defined in *clause B.4*, on which comments are given in Chapter 13 of this guide, with a worked example. The specification for the tests involves a compromise between the need to explore interactions between the many relevant parameters, and limiting the cost of testing to a level such that the use of new profiles is not discouraged. A partial factor γ_{VS} is applied to obtain $\tau_{u,Rd}$, which is used in *clause 10.7.2(8)*. The value 1.25 is recommended in *clause 4.4.1.2(6)* for this factor. This is expected to be confirmed in the UK's National Annex.

Comment on ***clause 10.7.2(4)*** on hogging bending is given under *clause 10.7.2(13)*. ***Clause 10.7.2(5)*** refers to the effective cross-sectional area of sheeting, A_{pe}. Allowance for its deformations should be made in accordance with EN 1993-1-3, or based on testing. No guidance on this testing is given. It could be based on a standard test in *Annex B* where longitudinal failure was observed but the degree of shear connection was at least 1.0 when A_{pe} was used. Test results are also influenced by local buckling within the flat parts of the steel profile, and by the enhanced yield strength at cold-formed corners. *Clause 10.7.2(4)* *Clause 10.7.2(5)*

For a composite slab in sagging bending, tests can be done in which the shear span is long enough, or the end anchorage is sufficient, for flexural failure to occur. If the strengths of the materials are known, the effective area of the sheeting, when in tension, can be calculated from the moment resisted. An estimate made by reducing the total area A_p by half of the embossed areas is sometimes used.

Bending resistances of composite slabs are based on rectangular stress-blocks (***clause 10.7.2(7)*** to *clause 10.7.2(8)*), as shown in Figure 10.2. The rules in *paragraph (7)* for resistance to sagging bending make provision for partial shear connection, and for the yield force N_s from any reinforcement of area A_s placed in troughs within the width of slab b_c to which the calculations apply. These bars (*clause 10.7.2(11)*) can improve resistance to fire and are sometimes provided in deep decking of the *Clause 10.7.2(7)*

Figure 10.2. Longitudinal forces in composite slab with partial shear connection, at failure in longitudinal shear

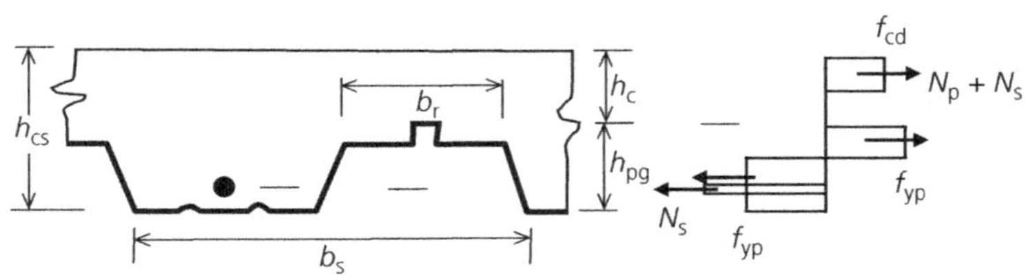

type used in slim-floor systems (Johnson and Shepherd, 2013). (Dimension h_{cs} in Figure 10.2 (from *clause 3.2.2*) is shown as h in *Figure 10.8*.)

For an assumed flexural failure at a cross-section at a distance L_x from the nearest support, the compressive force N_c in a slab of breadth b_c is the sum of the force N_p transferred from the sheeting at mean shear stress $\tau_{u,Rd}$ and the force N_s from the 'lower reinforcement' at yield (*Formula 10.9*), transferred by shear stress in the concrete. The derivation of the design shear strength $\tau_{u,Rd}$ takes account of the difference between the perimeter of a cross-section of sheeting and its overall breadth. The shear span L_x is reduced where hogging bending is present because partial connection is permitted for continuous composite slabs only where the concrete slab is in compression, as for continuous beams.

The analytical model for partial shear connection is similar to that used for composite beams, *clause 8.2.1.3*, and has been verified for slabs by full-scale tests (Bode and Sauerborn, 1991).

Clause 10.7.2(8) ***Clause 10.7.2(8)*** gives two limits to the force N_p. They are that it cannot exceed the force based on the stress $\tau_{u,Rd}$ (*Formula 10.8*) and that it (N_p) plus N_s cannot exceed the resistance of the concrete to compression (*Formula 10.7*). The depth of the stress block, z_c, is defined in *clause 3.3.2*. It cannot exceed the effective thickness of the slab, $h_{cs} - h_{pg}$. This definition shows that where there is a top rib, the compression zone in the slab cannot extend to the main top surface of the sheeting. Whichever limit governs, the sheeting may not be fully stressed in tension. Then, the stress distribution is as shown in Figure 10.2.

Usually, the degree of shear connection, and hence the force N_p, will be decided first, with $\tau_{u,Rd}$ found from tests. If the tension in the sheeting, $N_p - N_s$, is less than N_{pl} for the sheeting ($= A_{pe}f_{yp,d}$), part of its cross-section is in compression and there is a second neutral axis. The geometry of cross-sections of sheeting can be complicated, but there is no need to find the position of this axis, because M_{Rd} for the
Clause 10.7.2(9) cross-section can be found by the approximate method of ***clause 10.7.2(9)*** (Stark and Brekelmans, 1990).

Formula 10.11 is an interpolation between two values. If $N_p = A_{pe} f_{yp,d}$, it gives $z_p = h - 0.5\ z_c - e$, which is correct because the sheeting is fully plastic in tension. If $N_p = 0$, it gives $z_p = h - 0.5\ z_c - e_p$, which is correct because the sheeting is fully plastic in bending.

Where the neutral axis is above the sheeting (*Figure 10.6*) the line of action of the tensile force in the sheeting is known, and the resistance $M_{pl,Rd}$ is given by *Formula 10.14* with $M_{pr} = 0$ and z_s from *Formula 10.12*.

Where there is a neutral axis within the sheeting (*Figure 10.7*), for simplicity, the tensile stress block is split into two parts. The upper part of the block balances the compression N_p in the slab, with a line of action at distance between e and e_p above the base, as given by *Formula 10.10*. The lower part balances the compression in the sheeting, and provides a bending resistance M_{pr}, given by *Formula 10.13*. The formula is obviously correct where $N_p = 0$ and the tension in the lower part of the block equals the compression in the sheeting. It is also correct where $N_p = A_{pe} f_{yp,d}$ and $M_{pr} = 0$, and provides interpolation between these limits.

Summary: resistance to flexure by the partial connection method

To satisfy *clause 10.7.2* it is necessary to show that throughout the span (coordinate x) the curve of design bending moment, M_{Ed} (x), nowhere lies above the curve of design resistance, $M_{pl,Rd}$ (x), which is a function of the degree of shear connection, $\eta(x)$. These curves (Figure 10.7) are used in Example 10.1. Initially, $\tau_{u,Rd}$ and M_{pa} are known, and the route for any assumed value L_x is as follows, assuming that $z_{pl}/h_{cs} \leq 0.2$.

- Find N_p from *Formula 10.8*.
- Find N_s from *Formula 10.9*.
- Find z_c from *Formula 10.7* (but not exceeding $h_{cs} - h_{pg}$).
- Find z_p from *Formula 10.11* and z_s from *Formula 10.12*.
- Find M_{pr} from *Formula 10.13* and $M_{pl,Rd}$ from *Formula 10.14*.
- Repeat with a new L_x to obtain the curve $M_{pl,Rd}(x)$.

This curve is compared with the curve $M_{Ed}(x)$, calculated from the loading.

Further notes on sagging bending

In design of reinforced concrete beams, the compressive strain in concrete is limited by the use of the β factors given in *clause 8.2.1.2*. This is to prevent premature crushing of the concrete before the reinforcement yields. The corresponding rule for composite slabs is given in ***clause 10.7.2(11)***. For $z_{pl}/h_{cs} > 0.2$, its *Formulae 10.15* give values β_d which are about 2% higher than those for S355 steel in *Figure 8.3*. *Clause 10.7.2(11)*

Where a composite slab is supported on secondary beams, part of it will lie within the effective width of a primary beam. Their directions of span are parallel. Care must then be taken to avoid assuming in design that an area of concrete slab is acting as the compression flange of both the primary beam and the composite slab. *Clause 10.7.2* does not refer to this.

Clause 10.7.2(12) is a reminder that reinforcement in troughs must be fully anchored at a cross-section where it is assumed to be at yield in tension. Insufficient anchorage is likely if large-dimeter bars are used in slabs with relatively short spans (e.g. 2.5 m). *Clause 10.7.2(12)*

Resistance to hogging bending

For a composite slab in hogging bending, the sheeting may not be continuous, *clause 10.7.2(4)*, so its contribution to bending resistance is usually ignored, ***clause 10.7.2(13)***. Where sheeting is continuous, its compressive strength may be reduced by local buckling, and can be found conservatively from EN 1993-1-5. Tests on hogging regions of continuous composite slabs have been reported (Guo and Bailey, 2007). *Clause 10.7.2(13)*

10.7.3 Longitudinal shear for slabs without end anchorage

A proportion of the vertical reaction R_{Ed} from a supporting beam is transferred through the sheeting to the slab above it. Frictional restraint to longitudinal slip in this region increases the compressive force in the concrete slab. This increase may only be allowed for when $\tau_{u,Rd}$ is found using *Formula B.9* in *Annex B* rather than *Formula B.8*.

A note to ***clause 10.7.3(2)*** gives a recommended value, 0.6, for the coefficient of friction, and permits alternatives from a national annex or from testing. *Clause 10.7.3(2)*

10.7.4 Longitudinal shear for slabs with end anchorage

Clause 10.7.4 refers to two of the types of end anchorage defined in *clause 10.1.2.1*. The anchorage provided by studs is ductile. It is preferable to that provided by deformed ribs of re-entrant profiles, where poor compaction of concrete may occur. *Clause 10.7.4*

In the partial-connection method, the anchorage force is used as a contribution to the total force N_c (***clause 10.7.4(2)***). For through-deck-welded studs it can be calculated, to ***clause 10.7.4(3)***. The model is based on the weld collar pulling through the end of the sheeting. The design anchorage resistance is the lesser of that given by *clause 8.6.9.1* (on resistance of studs in troughs of parallel sheeting) and $P_{pb,Rd}$, the bearing resistance of the sheet, from *Formula 10.16*. Even where 16-mm studs are used in lightweight concrete, the bearing resistance (typically about 18 kN) will be the lower of the two. *Clause 10.7.4(2)* *Clause 10.7.4(3)*

Where anchorage forces are available on both sides of the steel flange, it and the sheeting provide continuous resistance to tension across the steel beam. Some or all of this resistance can be deemed to be effective as transverse reinforcement (*clause 8.6.11.4(4)*), but that part of it must be deducted from its contribution to N_c. Resistance $P_{pb,Rd}$ cannot serve both purposes.

Studs used as end anchorage of sheeting also resist longitudinal shear from the beam. They are biaxially loaded, so *clauses 8.6.8.3(1)* and *8.6.9.1(5)* apply. A summary of the checks required is now given, for a stud located in a trough of transverse sheeting, as shown in Figure 8.34. Its shear resistances in the trough are $k_\ell P_{Rd}$ and $k_t P_{Rd}$ (*clause 8.6.9*) and from *Formula 8.23* its shank shear resistance is

$$P_{s,Rd} = (0.8 f_u \pi d^2 / 4) / \gamma_V$$

The bearing resistance of the sheeting $P_{pb,Rd}$ (*clause 10.7.4(3)*) will be lower than the resistances of the stud. Suppose a designer wishes to allocate resistances as follows

- N_1, in the direction transverse to the beam, for end anchorage of the sheeting
- N_2 as a contribution to transverse reinforcement
- N_3 as the longitudinal shear force on the stud from the composite beam.

The values chosen for N_1 to N_3 should satisfy three conditions. For the sheeting,

$$N_1 + N_2 \leq P_{pb,Rd}$$

For shank failure of the stud between the sheeting and the beam,

$$N_2^2 + N_3^2 \leq P_{s,Rd}^2$$

For failure of the stud in the trough,

$$(N_1 / k_\ell P_{Rd})^2 + (N_3 / k_t P_{Rd})^2 \leq 1$$

Clause 10.7.4(4) ***Clause 10.7.4(4)*** refers to a European Technical Product Specification (if available) for the contribution from deformed ribs, because end anchorage is not used in standard test specimens.

10.7.5 Vertical shear

This quite complex new clause increases design resistance to vertical shear by adding a contribution from the concrete slab (Schmeckebier and Kurz, 2023). ***Clause 10.7.5(1)*** refers to EN 1992-1-1, where resistance to vertical shear depends on the effective depth d of the section.

Clause 10.7.5(1)

Symbol d_q is defined in *clause 3.2.2* as the 'effective depth of the composite slab', and in *clause 10.7.5(6)* as 'distance from the centroid of the tension forces to the extreme fibre of the concrete in compression'. The 'distance from the top surface to the centroid of the profile' (symbol d_p, from *clause 3.2.2*) is not used, because d_q depends on any reinforcement in the troughs, and on the anchorage of the tension forces in the sheeting and reinforcement. Anchorage may be hard to calculate, especially where the shear connection is partial, with two neutral axes. Sheeting that is neither continuous over the supporting beam nor anchored to it should not be assumed to be effective for resistance to vertical shear. Anchorage is used in Example 10.1.

Symbol d_s, used in *clause 10.7.5(6)*, is defined in *clause 3.2.2* as 'depth to steel reinforcement in tension' and is shown in *Figure 10.7*.

Figure 10.3 shows a cross-section of part of a steel I-section beam supporting one end of a composite slab. It uses the symbols of *Figure 10.9* for the external forces. Surface AB represents a crack, and compression acts on surface BC of the concrete. The diagonal strut shown by the dashed arrow is an internal force. These forces give the resistance to vertical shear at a cross-section distant d_q from the face of a support. They are calculated in Example 10.1.

Clause 10.7.5(2) *Formula 10.18* in ***clause 10.7.5(2)*** shows that the shear resistance $V_{v,Rd}$ is the sum of five terms, which are discussed in turn.

Figure 10.3. Components of resistance to vertical shear of a composite slab

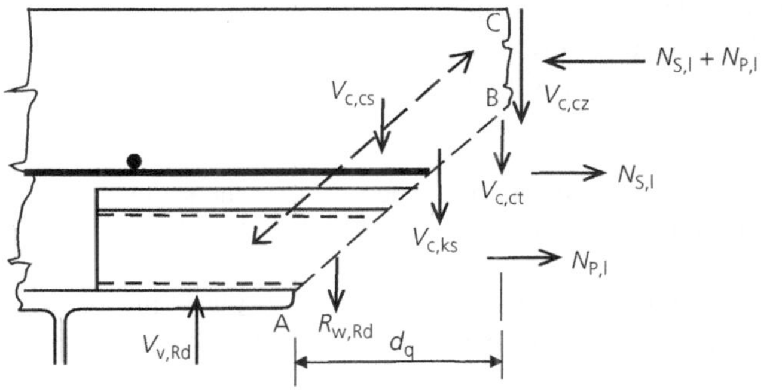

Symbol $R_{w,Rd}$ is a resistance to transverse forces based on yielding, not buckling, defined in clause 8.1.6 of EN 1993-1-3. *Formula 10.18* gives it as a lower limit to resistance $V_{v,Rd}$. In practice, $R_{w,Rd}$ is sometimes determined from results of tests, and its value is accepted as the shear resistance. Clause 8.1.6.3 of EN 1993-1-3 gives resistances both for sheeting continuous across a support and for end bearing. Its Formula 8.37 is applicable only where the sheeting extends at least 40 mm beyond the end of its bearing length. This may not occur in composite slabs, and then Eurocode 3 gives no value for $R_{w,Rd}$. Eurocode 4 refers, as an alternative, to a European Technical Product Specification. At present, manufacturers of sheeting provide loading tables for composite slabs, using resistances of sheeting to vertical shear based on tests. It is expected that they will provide revised tables, based on these clauses.

Formula 10.18 for resistance $V_{v,Rd}$ includes the term 0.5 $R_{w,Rd}$, so the other terms, which need extensive calculation, can be omitted if $V_{v,Ed}$ can be kept below 0.5 $R_{w,Rd}$. These other terms, for the concrete slab, are now discussed.

Resistance $V_{c,cz}$, from *Formula 10.19*, is about 6% of the anchored tension in the sheeting and any reinforcement in its troughs. Anchorage at an end support is difficult, so this resistance may be low. Resistance $V_{c,cs}$ is higher: the proportion is about 21%.

Resistance $V_{c,ks}$, *Formula 10.20*, depends on a dimension h_{pc} which refers to the 'lowest shear connection device' in a trough of sheeting. This will not exceed half the depth h_p of a trough. The resistance is proportional to the tensile strength of the concrete, and is also relatively low.

Resistance $V_{c,ct}$, *Formula 10.22* in ***clause 10.7.5(6)***, based on fracture energy in the crack propagation zone, is low for lightweight-aggregate concrete, but may not be negligible for normal-weight concrete. *Clause 10.7.5(6)*

Clause 10.7.5(7) gives an alternative method for resistance to vertical shear. Its Notes give two alternatives for the extent to which the resistances of the concrete slab and the steel sheeting can be added (*Formula 10.23*). These depend on the method of construction *Clause 10.7.5(7)*

- for unpropped construction, with $V_{b.e.,Rd}$ taken as $\leq V_{g,Ed}$ and $\leq V_{b,Rd}$, then $k_v = 1$, where $V_{g,Ed}$ is the loading applied direct to the steel sheeting, and $V_{b,Rd}$ is a shear resistance that takes account of shear buckling (clause 8.1.5 of EN 1993-1-3)
- for propped construction, where $V_{g,Ed} = 0$ and $V_{b.e.,Rd}$ is taken as $V_{b,Rd}$, then $k_v = 0.5$.

This method is used in Example 10.1. A similar method is given in ***clause 10.7.5(8)*** for regions in hogging bending. *Clause 10.7.5(8)*

Clause 10.7.5(9) refers to the interactions between normal and shear stress in sheeting that arise from hogging bending where two-span sheetings are designed as continuous. *Clause 10.7.5(9)*

10.7.6 Punching shear

Figure 10.11 shows a rectangular area of slab with sides labelled 'surfaces'. With sheeting spanning in the direction shown, shear resistance of surfaces labelled 2 are calculated as for a reinforced concrete beam of depth h_c using clause 8.4.3 of Eurocode 2, and those for surfaces 3 as for a composite slab of effective depth d_q. The shear resistance (see Example 10.2) is the sum of these two results, *clause 10.7.6(1)* to *(3)*. The width of the 'loaded area' labelled 1 is presumably the actual size of a concentrated load, b_p, plus 45° spread through the floor finish of thickness h_e, as shown in *Figure 10.4*. Clause 8.4.3 of EN 1992-1-1 gives the shear resistance as a stress, so one needs to know the depth of slab on which this stress is assumed to act. For Eurocode 4, it is assumed to be the depth to the transverse reinforcement.

Clause 10.7.6(1)
Clause 10.7.6(2)
Clause 10.7.6(3)

Formula 10.26 in *clause 10.7.6(3)* for the resistance $V_{pt,Rd}$ is the same as *Formula 10.18* for vertical shear, except for the omission of the term $V_{c,cs}$ as explained in its Note.

Wheel loading from fork-lift trucks and mobile elevated work platforms should always be checked for punching shear. Additional slab reinforcement may be required (Steel Construction Institute, 2011).

10.8 Verification of composite slabs for serviceability limit states

10.8.1 Control of cracking of concrete

Clause 10.8.1(1) **Clause 10.8.1(1)** refers to EN 1992-1-1, where, from clause 9.2.1(2)

> Cracks may be permitted to form without any attempt to control their width, provided that they do not impair the functioning of the structure.

Clause 10.8.1(2) For this situation, the provisions for minimum reinforcement above internal supports of composite slabs (***clause 10.8.1(2)***) are the same as those for beams in *clause 9.4.1(4)*, where further comment is given. Crack widths should always be controlled above supports of slabs subjected to travelling loads. The methods of EN 1992-1-1 are applicable, neglecting the presence of the sheeting. The '*area of the concrete above the ribs*' in this clause can be based on a thickness that extends only to a top rib, if present.

10.8.2 Deflection

Clause 10.8.2(1) The reference in ***clause 10.8.2(1)*** to EN 1990 is to the basic criteria for verifying deformations. The rest of the clause is as in the previous EN 1994-1-1 except as follows.

Clause 10.8.2(3) For global analysis of the composite slab, ***clause 10.8.2(3)*** refers to *clause 7*, but it is not stated whether all the references to continuous beams apply. Probably, they may be assumed to apply where relevant, unless in conflict with a provision in *clause 10* – for example, *clause 7.4.4(8)* on limited redistribution of moments at SLS probably does apply to continuous slabs, because *clause 10.4.2(1)* allows redistribution for ULS verifications.

Clause 10.8.2(4) The rules in ***clause 10.8.2(4)*** permit calculation of deflections to be omitted if both of two conditions are satisfied. Previously, the first condition referred to limits to the ratio of span to effective depth given in EN 1992-1-1. These were 20 for a simply-supported slab, 26 for an external span of a continuous slab, and 30 for an internal span. They have been replaced by the single ratio of 26, with a new definition of effective depth.

Deflection from shrinkage

Clause 10.8.2(5) ***Clause 10.8.2(5)*** is new. Research has found (Al-Deen *et al.*, 2015) that shrinkage of concrete causes sagging curvature of composite slabs because drying occurs from the top surface only. Formulae are given for estimating the resulting deflection. For normal-weight concrete, they are consistent with the assumption that the free shrinkage strain varies from 400×10^{-6} at the top surface of the slab to zero at the bottom. The resulting curvature depends on the ratios of the depths and the cross-sectional areas of the concrete and the sheeting, so the formulae are approximations. The deflection is small, but may not be negligible. Taking a slab of thickness 120 mm and span 3 m as an example, the deflection is 4.5 mm.

Clause 10.8.2(7) The second condition of *clause 10.8.2(4)*, in ***clause 10.8.2(7)***, applies to 'external' spans only. This presumably includes simply-supported spans, which are not referred to. It relates to the initial slip load found in tests – information which may not be available to the designer. Two-point loading is used for testing, so the 'design service load' must be converted to point loading. This can be done by assuming that each point load equals the end reaction for a simply-supported span under service loading.

Testing by the manufacturer of the sheeting is likely to have established whether the shear behaviour is ductile, to *clause 10.7.2(3)*, but the rule in *clause 10.8.2(7)* requires load to be measured at a different end slip, 0.5 mm, which may not have been done. This load can be quite variable, and should presumably be taken as the mean from all the tests on the profile used in which longitudinal shear failure occurred. Where this second condition is not satisfied, ***clause 10.8.2(8)*** may lead to the use of end anchorage, for which the tied-arch method of calculation given in ***clause 10.8.2(9)*** is available. *Clause 10.8.2(8)* *Clause 10.8.2(9)*

Tied-arch model

The use of this model seems likely to arise only where

- a high proportion of the shear connection is provided by welded studs
- its amount has been established by calculation to *clause 10.7.4*
- no relevant test data are available.

If the end anchorage is provided by deformed ribs, then experimental verification is essential, and the slip behaviour will be known.

In the tied-arch model, the whole of the shear connection is provided by end anchorage. Accurate calculation of deflection is difficult, but the arch will be so shallow that the following simplified method can be used.

Figure 10.4. Tied-arch model for deflection of a composite slab with an end anchorage

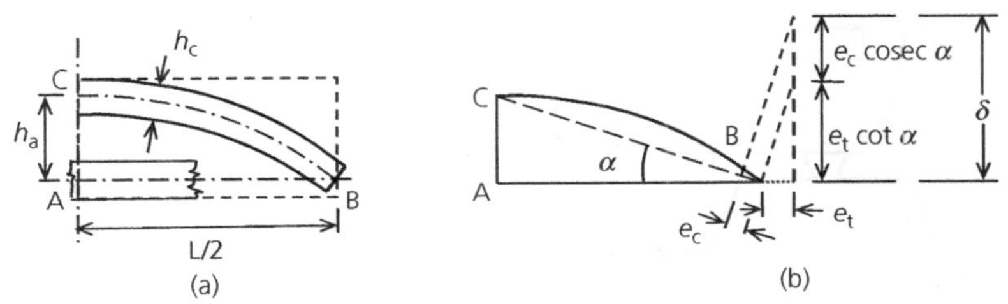

The tie AB shown in Figure 10.4(a) consists of the effective area of the steel sheeting, with its centroid coincident at each support with that of the arch. The thickness h_c of the arch rib must be such that at the relevant bending moment, its compressive stress at mid-span is not excessive. There is interaction between its assumed thickness h_c and the lever arm h_a, so a little trial-and-error is needed. For a given loading, the longitudinal forces at mid-span are now known, and the strains ε_c and ε_t in the concrete and steel members can be found, taking appropriate account of creep. In Figure 10.4(b), curve CB represents the arch rib. The ratio L/h_a is unlikely to be less than 20, so both the curve length and the chord length CB can be taken as $L/2$, and

$$\text{cosec}\ \alpha \approx \cot \alpha = L/(2\ h_a)$$

The changes in length of the members shown in the figure are

$$e_t = \varepsilon_t L/2 \text{ and } e_c = \varepsilon_c L/2$$

so the deflection δ is given by

$$\delta/\text{L} = (\varepsilon_t + \varepsilon_c)[L/(4\ h_a)] \qquad \text{(X10.1)}$$

with both strains taken as positive numbers.

This method can be over-conservative. The shortening of the arch, including creep, usually far exceeds the extension of the sheeting, so a low compressive stress should be assumed, even though this reduces the angle α. The simply-supported composite slab of Example 10.1 provides an illustration. Its span is 2.93 m. Assuming a compressive stress of 3.5 N/mm^2 in the concrete leads to a deflection from service loading of 14 mm; much greater than the 4.0 mm found later.

Taken as a whole, *clause 10.8.2* can be difficult to satisfy. Designers need information from manufacturers of sheeting that may not be available and will need interpretation, as tests can reproduce only a few of the sets of dimensions, materials and so on that are used in practice.

Example 10.1: two-span continuous composite slab

Data

Details of the geometry assumed for this composite slab are shown in Figure 10.5(a). There is no reinforcement in the troughs of the sheeting. The properties of the concrete and reinforcement are as used in the worked example on a two-span beam in Chapters 8 and 9, where further information on them is given. The spacing of the supporting composite beams is wider, 3.0 m rather than 2.5 m. Their top-flange width is 190 mm.

The different cross-section assumed for the sheeting satisfies the condition of *clause 10.1.1(2)* for 'narrowly spaced webs'. It provides shear connection by embossments, in accordance with *clause 10.1.2.1(a)*. Its overall depth h_{pg} is 70 mm and the slab thickness h_c is 60 mm. Some of the design data used here have been taken from the sheeting manufacturer's brochure.

Figure 10.5. Composite slab: (a) dimensions; (b) forces for $M_{pl,Rd}$, sagging, with full shear connection; (c) second moment of area, sagging; (d) second moment of area, hogging

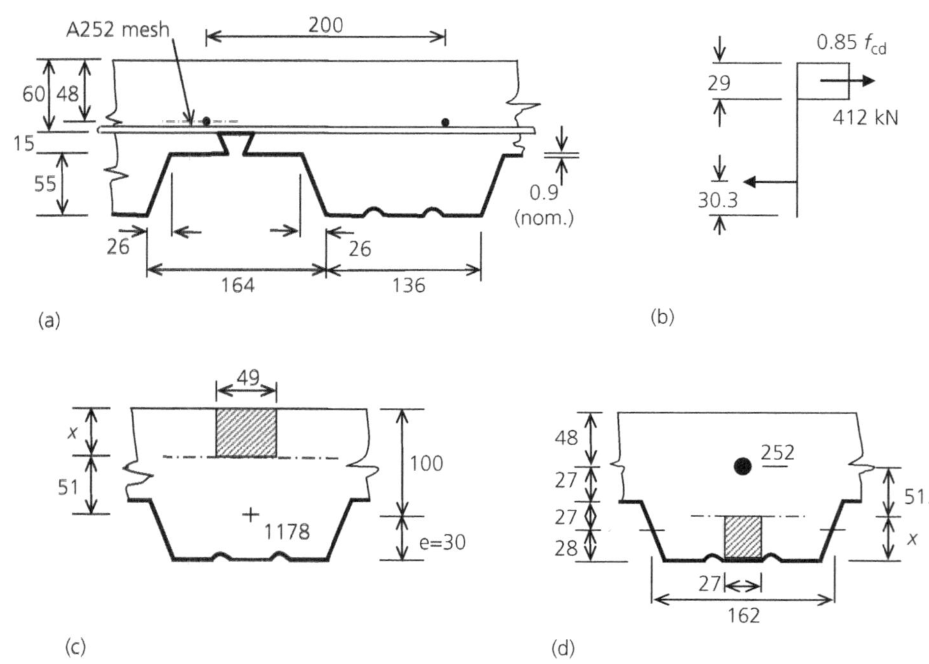

For simplicity, it is assumed that the transverse reinforcement in regions where the supporting beams resist hogging bending is determined by resistance to lateral-distortional buckling, as in Example 8.6: $A_s = 565$ mm^2/m (12-mm bars at 200 mm spacing). For other regions, it is assumed initially that A252 mesh is provided ($A_s = 252$ mm^2/m), resting on the sheeting. This has 8-mm bars at 200 mm spacing, both ways. These details satisfy the detailing rules of *clause 10.2.1*.

At the outset, two decisions have to be made that affect the verification of a composite slab

- whether construction will be unpropped or propped
- whether the spans are modelled as simply supported or continuous.

Here, it will be assumed that unpropped construction will be used wherever possible. The sheeting is provided in 6 m lengths, so no 3-m span can have sheeting continuous at both ends. For a building 9 m wide, there will also be spans where 3 m lengths of sheeting are used, so several end-of-span conditions will be considered. Some of the design checks on a composite slab are usually satisfied by wide margins. These checks will be simplified here by making conservative assumptions.

Properties of materials and profiled sheeting

- Lightweight-aggregate concrete: grade LC25/30. These properties are explained in Example 8.6: $E_{cm} = 20.4$ GPa; $f_{cm} = 27.1$ N/mm^2; $f_{cd} = 14.2$ N/mm^2 for plastic resistances; short-term modular ratio $n_0 = 10.3$; $f_{ctm} = 2.31$ MPa; design density 1950 kg/m^3, including reinforcement and sheeting.
- Creep is allowed for by using $n = 20.6$ for all loading (*clause 7.4.2.2(11)*).
- Reinforcement: $f_{sk} = 500$ N/mm^2; $f_{sd} = 435$ N/mm^2.
- Profiled sheeting: nominal thickness including zinc coating, 0.9 mm; bare-metal thickness 0.86 mm (for zinc coating of 275 g/m^2); area, $A_p = 1178$ mm^2/m; weight, 0.10 kN/m^2; second moment of area, $10^{-6}\, I_{y,p} = 0.548$ mm^4/m; plastic neutral axis, $e_p = 33$ mm above bottom of section (notation as in *Figure 10.7*); centre of area, $e = 30.3$ mm above bottom of section; $E_a = 210$ kN/mm^2; yield strength, $f_{yk,p} = 355$ N/mm^2; $\gamma_{M,p} = 1.0$; plastic moment of resistance in hogging and sagging bending, taking account of the effect of embossments (*clause 10.5(1)*), $M_{pa} = 6.18$ kN m/m.
- Composite slab: 130 mm thick; volume of concrete, 0.105 m^3/m^2.

The value $f_{cd} = 14.2$ N/mm^2 used here includes the factor $k_{tc}\, \eta_{cc}$ as 0.85, which is the maximum allowed by *clause 5.1(3)* for plastic resistances.

Loading for the profiled sheeting

- The dead weight of the floor is $g_{k1} = 1.95 \times 9.81 \times 0.105 = 2.01$ kN/m^2.
- From Table A.1 of EN 1991-1-1 (BSI, 2025) for fresh concrete the weight density should be increased by 1 kN/m^3, so for initial loading on the sheeting, $g_{k1} = 2.01 \times 20.5/19.5 = 2.11$ kN/m^2.
- Construction loading during casting of concrete, including dynamic and heaping effects, is assumed to be $q_k = 1.0$ kN/m^2 for all stages of construction.

Loading for the composite slab

- Floor finish (permanent): $g_{k2} = 0.48$ kN/m^2.
- Variable load (including partitions, services, etc.): $q_k = 7.0$ kN/m^2.
- The loads per unit area are summarised in Table 10.1, using $\gamma_{F,g} = 1.35$, $\gamma_{F,q} = 1.5$.

Table 10.1. Loadings per unit area of composite slab (kN/m^2)

Type of load	Characteristic, maximum	Characteristic, minimum	Ultimate, maximum	Ultimate, minimum
During concreting:				
sheeting and concrete	2.11	0.10	2.85	0.13
imposed load	1.00	0	1.50	0
Total	3.11	0.10	4.35	0.13
For composite slab				
slab and floor finish	2.01 + 0.48	2.49	3.36	3.36
imposed load	7.00	0	10.5	0
Total	9.49	2.49	13.9	3.36
For deflection of composite slab after construction (frequent)	$0.7 \times 7 + 0.48 = 5.38$	0.48	not applicable	not applicable

Verification of the sheeting

The width of bearing of a sheeting on the steel top flange can range from 50 mm (the minimum, *clause 10.2.3(2)*) to about 95 mm, as shown in Figure 10.6(a). Assuming that the line of action of the reaction is at the centre of the bearing length, the lengths of simple spans could range from 2.86 m to 2.905 m, plus the tolerance on beam spacing. Conservatively, the value 2.93 m will be used for both sheeting and slabs, for bending moments and longitudinal shear and all combinations of end conditions. It corresponds to a member continuous over two spans, with 50 mm bearing length at each simple support, Figure 10.6(b). For vertical shear, the span is taken as 3.0 m to give the correct loads for the supporting beams.

Deflections are critical only for single spans or continuous sheeting that is sensitive to ponding effects, and 2.86 m is used. During construction, load from the sheeting causes transverse bending in the steel flanges. It can usually be ignored because longitudinal stress in the flanges is quite low at this stage.

Figure 10.6. Effective spans for slabs on beams at 3 m spacing: (a) bearing lengths; (b) slab continuous over two spans (dimensions in mm)

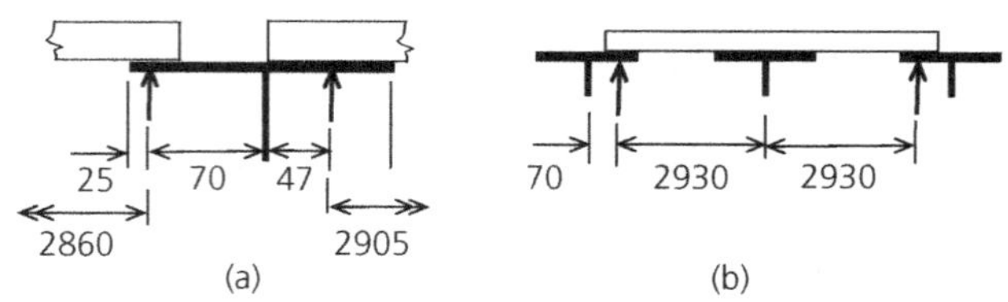

Ultimate limit state

For the sheeting during concreting, with a simply supported span of 2.93 m,

$$M_{\mathrm{Ed}} = 4.35 \times 2.93^2/8 = 4.67 \text{ kN m/m}$$

This is 76% of M_{pa}, the bending resistance of the sheeting.

Excluding effects of continuity, for vertical shear: $\boldsymbol{V_{\mathrm{Ed}} = 4.35 \times 1.5 = 6.52}$ **kN/m**

There are 6.7 webs of depth 61 mm per metre width of sheeting, at an angle $\cos^{-1}$ 55/61 to the vertical. Their shear resistance is governed by buckling, and is found to be sufficient, as a quite long calculation to clause 8.1.6.3 of EN 1993-1-3 gives $\boldsymbol{V_{\mathrm{b,Rd}} = 16}$ **kN/m**.

Deflection during construction

A note to *clause 10.6(2)* recommends that the deflection should not exceed span/180, an NDP. The worst case is where a span is simply-supported. Then,

$$\delta = 5wL^4/(384\, E_{\mathrm{a}} I_{\mathrm{y}}) = 5 \times 2.01 \times 2.86^4 \times 1000/(384 \times 210 \times 0.548) = 15.2 \text{ mm}$$

For fresh concrete, this is increased to 15.2 × 20.5/19.5 = 16.0 mm.

From *clause 10.3.2(2)*, allowance should be made for ponding if the deflection exceeds one-tenth of the slab thickness (13 mm), as it does here. The specified thickness of the additional concrete is 0.7δ. Its weight, for fresh concrete, is

$$0.7 \times 0.016 \times 20.5 = 0.23 \text{ kN/m}^2$$

This increases the deflection to 16 × 2.34/2.11 = **17.7 mm**, which is span/161, and so exceeds span/180. It follows that where the sheeting is not continuous at either end of a 3 m span, propping should be used during construction.

The effects of continuity at one end of unpropped spans are now considered. The most adverse condition occurs with wet concrete on one span and not on the other. Assuming that no construction load is present, the loadings on the spans are 2.11 kN/m^2 and 0.10 kN/m^2. From elastic analysis, the **final deflection** is not excessive: 13.4 × 2.01/2.11 = **12.7 mm**, with a small reduction when the other span is cast.

Verification of the composite slab

Plastic resistance moment with full shear connection

For sagging bending (Figure 10.5(c)), the plastic neutral axis is likely to be above the sheeting, so *Figure 10.6* for stress distribution applies. The tensile force in the sheeting, when at yield, and assuming that the gross decking area can be used, is

$$F_{y,p} = A_p f_{ypd} = 1178 \times 0.355 = 418 \text{ kN/m} \qquad \text{(X10.2)}$$

The depth of slab in compression with full shear connection is

$$z_{pl} = 418/14.2 = 29 \text{ mm} \qquad \text{(X10.3)}$$

The lever arm is 130 – 30.3 – 29/2 = 85 mm

$$\text{so } \boldsymbol{M_{pl,Rd}} = 418 \times 0.085 = \mathbf{35\ kN\ m/m} \qquad \text{(X10.4)}$$

Second moments of area

For sagging bending, the transformed width of concrete is 1000/20.6 = 49 mm/m. For an elastic neutral axis at depth x (Figure 10.5(c)), first moments of area gives

$$1178(100 - x) = 49\, x^2/2$$

whence $x = 49$ mm. Hence

$$\mathbf{10^{-6} \boldsymbol{I_y}} = 0.548 + 1178 \times 0.051^2 + 49 \times 0.49^3/3 = \mathbf{5.53\ mm^4/m} \text{ (sagging)}$$

For hogging bending, each trough in the sheeting has a mean width of 162 mm. There is one trough per 300 mm, so the transformed width of concrete in compression is

$$49 \times 162/300 = 27 \text{ mm/m}$$

The mesh with area 252 mm^2/m is 82 mm above the base, and the sheeting is not included, as it may not be continuous. For a neutral axis at height x above the bottom of the slab (Figure 10.5(d)),

$$252(82 - x) = 1178(x - 30.3) + 27x(x/2)$$

whence $x = 30.6$ mm. Thus, the neutral axis almost coincides with the centre of area of the sheeting, and

$$\mathbf{10^{-6} \boldsymbol{I_y}} = 0.548 + (252 \times 51.4^2 + 27 \times 30.6^3/3) \times 10^{-6} = \mathbf{1.47\ mm^4/m} \text{ (hogging)}$$

Deflection

This is considered first, as it sometimes governs the design. *Clause 10.8.2(4)* gives conditions under which a check on deflection may be omitted. One of these is $L/d_q \leq 26$. The depth d_q to the centroid of the sheeting is 100 mm. The ratio is 2.86/0.1 = 28.6, and the condition is not satisfied.

Deflection is a reversible limit state, for which a note to clause 8.4.3.3(1) of EN 1990 recommends use of the frequent combination. The ψ_1 factor for this combination depends on the floor loading category. From Table A1.7 in EN 1990, it ranges from 0.5 to 0.9, and is here taken as 0.7, the value for 'congregation' areas. From Table 10.1, the maximum and minimum loadings are 5.38 kN/m^2 and 0.48 kN/m^2. For a simply-supported loaded span of 2.93 m, with $I_y = 5.53 \times 10^6$ mm^4/m, the deflection is 4.0 mm. A more accurate calculation for the two-span slab, assuming 15% of each span to be cracked, and with 0.48 kN/m on the other span, gives the deflection of the fully-loaded span as 3.5 mm.

Clause 10.8.2(5) gives the additional deflection from shrinkage for a continuous slab as

$$0.1\, \varepsilon_{sh} L^2/h. \text{ Here, } \varepsilon_{sh} = 500 \times 10^{-6}, L = 2.93 \text{ m and } h = 0.13 \text{ m, giving } \delta_{sh} = 3.3 \text{ mm.}$$

Hence, the total deflection of the slab, if cast unpropped, is

$$\boldsymbol{\delta_{total}} = 12.7 + 3.5 + 3.3 = \mathbf{19.5\ mm}$$

which is span/150.This value is not the total deflection, as it is relative to the levels of the supporting beams, and is so high that in most situations, propped construction should be used for the slab designed here.

Ultimate limit states for the composite slab

Resistance to bending by the partial-connection method

From Table 10.1, the maximum loading is 13.9 kN/m². For a simply-supported span,

$$\boldsymbol{M_{\mathrm{Ed}}} = 13.9 \times 2.93^2/8 = \mathbf{14.9\ kN\ m/m} \text{ at mid-span}$$

The method given after the comments on *clause 10.7.2(9)* is used. Example 13.1 finds from tests one to four on slabs with the sheeting used here that its design shear strength is

$$\tau_{\mathrm{u,Rd}} = 0.144 \text{ MPa} \qquad \text{(X10.5)}$$

Specific conditions for the use of test results in design are given in *clause B.4.1(4)*. They relate to the thickness and strength of the sheeting and the strength of the concrete. The method used for the value found for $\tau_{\mathrm{u,Rd}}$ is now related to these conditions.

- The thickness of the sheeting including coating, 0.9 mm, is the same as in the tests.
- Test results with mean concrete strength f_{cm} may be used where the specified strength f_{ck} is not less than $0.8 f_{\mathrm{cm}}$. Here $f_{\mathrm{ck}} = 25$ N/mm² and $f_{\mathrm{cm}} = 26.4$ N/mm² (converted from cube results), which is acceptable.
- The concrete density in the tests, measured 1.5 h after mixing, was 1944 kg/m³. This takes no account of loss of moisture. The design density, $\leq$ 1800 kg/m³, is lower, but acceptable because the more important check on compressive strength is satisfied.
- The specified yield strength of the sheeting, 350 N/mm², is below the yield strength in the tests, as required.

The thickness of the slab was 170 mm for tests one to four and 120 mm for tests five to eight. The design shear strength used here, 0.144 MPa, is calculated from the tests on the 170-mm slabs. The tests on 120-mm slabs gave a higher value for $\tau_{\mathrm{u,Rd}}$. It is not used for the 130-mm slabs in this example, for reasons explained in Example 13.1.

Design partial-interaction diagram

The two curves shown in Figure 10.7 are now constructed, with $x = 0$ at the left-hand support. For M_{Ed} (L_{x}), the design ultimate load for the slab from Table 10.1 is 13.9 kN/m², assumed to act on the composite member. The verification for flexure used simply-supported spans of 2.93 m, which gives $M_{\mathrm{Ed,max}}$ as 14.9 kN m/m, at mid-span. The parabolic bending-moment diagram is plotted, for a half span, as 0C in Figure 10.7.

Figure 10.7. Design partial-interaction diagram for a simply-supported span

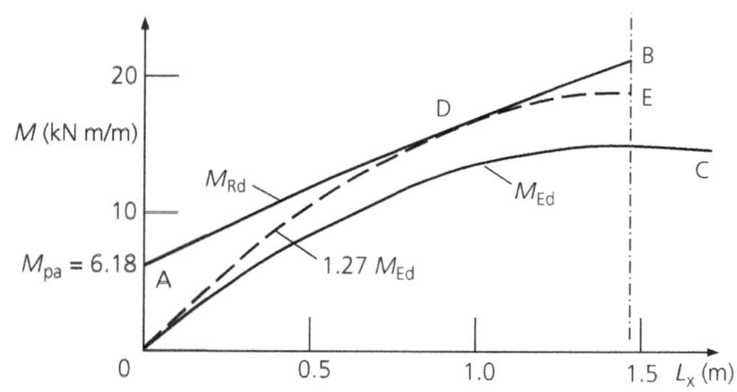

From *Formula 10.8* the compressive force per metre width of slab at distance L_{x} from an end support, with L_{x} in m units, is

$$N_{\mathrm{p}} = \tau_{\mathrm{u,Rd}} b_{\mathrm{c}} L_{\mathrm{x}} = 0.144 \times 1000\ L_{\mathrm{x}} = 144\ L_{\mathrm{x}} \text{ kN/m}$$

For full interaction, $N_{\mathrm{c,f}} = A_{\mathrm{p}} f_{\mathrm{yp,d}} = 412$ kN/m

The length of shear span needed for full interaction is therefore $L_{\mathrm{s,f}} = 412/144 = 2.86$ m

This exceeds span/2 (1.47 m), so full interaction is not achieved in a span of this length.

The depth of the full-interaction stress block in the slab x_{pl} was found above to be 29 mm.

From *Formula 10.11* with z_p in mm,

$$z_p = 130 - 29/2 - 33 + (33 - 30.3) \times (144\ L_x/412) = 82.5 + 0.944\ L_x$$

The reduced bending resistance of the composite slab is given by *Formula (10.13)*, with $M_{pa} = 6.18$ kN m/m,

$$M_{pr} = 1.25 \times 6.18\ (1 - 144\ L_x/412) = 7.72 - 2.7\ L_x \leq 6.18 \text{ kN m/m}$$

so this result applies where $L_x \geq (7.72 - 6.18)/2.7 = 0.570$ m

From *Formula 10.14*, the plastic resistance for $0.57 \leq x \leq 1.47$ m in kN m units is

$$M_{pl,Rd} = M_{pr} + N_p z_p = 7.72 - 2.7\ L_x + 0.144\ L_x(82.5 + 0.944\ L_x)$$
$$= 7.72 + 9.11\ L_x + 0.151\ L_x^2$$

For $x < 0.57$ m,

$$M_{Rd} = 6.18 + 0.144\ L_x(82.5 + 0.944\ L_x) = 6.18 + 11.8L_x + 0.151\ L_x^2$$

The curve $M_{Rd}(L_x)$ is plotted as AB in Figure 10.7. It lies above the curve 0C for M_{Ed} at all cross-sections, showing that there is sufficient resistance to longitudinal shear.

The loading for failure in longitudinal shear is found as follows. Curve 0C is scaled up until it touches curve AB. The scale factor is found to be 1.27 (curve 0DE), with contact at $x = 1.0$ m, where $M_{Rd} = 17.0$ kN m/m. Shear failure thus occurs along a length of 1.0 m adjacent to an end support. The degree of shear connection is $N_p/N_{c,f}$. It is $144 \times 1.0/412 = 0.35$ at this cross-section, and higher elsewhere.

Curve AB gives the bending resistance at mid-span as 21.4 kN m/m, which is 1.4 times M_{Ed} at that section; but mid-span is not the critical cross-section. It is at $x = 1.0$ m, where $M_{Rd}/M_{Ed} = 1.27$. This shows that for this check there is no need to make use of the reduction in sagging M_{Ed} from continuity. Where this is necessary, *clause 10.4.2* on global analysis and *clause 10.7.2(4)* on resistance to hogging bending are applicable.

The calculations summarised in this example include some provisions of EN 1994-1-1 that are unlikely to be needed for routine design. They are relevant to the preparation of design charts or tables for composite slabs using sheeting of a particular thickness and profile. These are normally prepared by specialists working on behalf of the manufacturer.

Resistance to vertical shear

From *clause 10.7.5(1)*, shear resistance is checked at a distance d_q from the face of the support, which is 95 mm from the web of the supporting beam. Where shear connection is partial, as here, the line of action of the tension in the profiled sheeting, and hence the value for d_q, is unclear. Here, the centroid of the steel section (the value for full shear connection) is used, which is slightly conservative.

Hence, $d_q = h_{cs} - e = 130 - 30.3 = 100$ mm

The shear span is $1.5 - 0.095 - 0.10 = 1.305$ m. Calculations are done for a width of slab equal to its wavelength, 0.3 m here. Hence,

$$V_{Ed} = 1.305 \times 0.3 \times 13.9 = \mathbf{5.44\ kN} \text{ per 0.3 m}$$

In *clause 10.7.5, Formula 10.17* for $V_{v,Rd}$, the resistance to vertical shear, requires calculation of five separate shear resistances, starting with the lower limit, $R_{w,Rd}$.

Shear resistance of the sheeting

For $R_{w,Rd}$, clause 8.1.6 of EN 1993-1-3 is needed, but appears not to apply at a simple support, for reasons given in the comments on *clause 10.7.5*. In practice, test results may be used. Tests on a similar profile (ComFlor 60) gave $R_{w,Rd}$ as about 13 kN per 0.3 m. Half of this exceeds V_{Ed}, so it is likely that a resistance based on tests will be sufficient. Calculations for the other terms in *Formula 10.18* are now given.

Shear resistance of the concrete compression zone, $V_{c,cz}$

Clause 10.7.5(3) requires the anchorage forces in the sheeting $N_{p,1}$ and mesh reinforcement $N_{s,1}$. From the minimum bearing length of *clause 10.2.3(2)* for sheeting, $L_{bs} = 50$ mm. It is assumed that the reaction is at 25 mm from the tip of the steel flange, and that the concrete slab continues to the centre of the steel flange, so that $L_{bc} = 95$ mm.

For the sheeting, $\tau_{u,Rd}$ was found excluding the effect of the support reaction, so for its anchorage length of 50 mm, $N_{p,1} = 0.144 \times 50 \times 0.3 = \mathbf{2.16\ kN}$ per 0.3 m.

As there may be no cross-wires of the mesh reinforcement within the anchorage length, the contribution $N_{s,1}$ from its longitudinal bars (8-mm bars every 200 mm) is neglected.

With $f_{ctm} = 2.31$ N/mm^2, *Formula 10.19* gives

$$V_{c,cz} = 0.67\ (2.31/25)(0 + 2.16) = \mathbf{0.13\ kN}\ \text{per 0.3 m}$$

Shear resistance inducing spalling of concrete

Formula 10.20 for $V_{c,ks}$ requires the '*distance of the centroid of the lowest shear connection device to the lower flange*'. For trapezoidal sheeting, this dimension, h_{pc}, is assumed to be 5 mm. At this height above the base, from Figure 10.5(a), $b_{min} = 141$ mm. Hence,

$$V_{c,ks} = 1.64\ b_{min} h_{pc}\ f_{ctm} = 1.64 \times 0.141 \times 5 \times 2.31 = \mathbf{2.67\ kN}\ \text{per 0.3 m}$$

Shear resistance due to the resistance of the compression strut

Using values for $N_{p,1}$ and $N_{s,1}$ found above, *Formula 10.21* gives

$$V_{c,cs} = 0.21(2.16 + 16.4) = \mathbf{3.90\ kN}\ \text{per 0.3 m}$$

Shear resistance of the concrete in the crack propagation zone

Values needed for *clause 10.7.5(6)* include $b_0 = 162$ mm, $f_{ctm} = 2.31$ N/mm^2 and $E_{cm} = 20.4$ GPa. The main tension force is from the sheeting, so $d_s = 100$ mm. It is assumed that normal-weight sand is used for the concrete. From Table M.1 of EN 1992-1-1, $\eta_{lw,fct} = 0.89$, and

$$G_f = 24 + 16 \times 0.89 \times 2.31 = 56.9\ \text{N/m}$$

From *Formula 10.22*,

$$V_{c,ct} = 0.09(56.9 \times 20\ 400/2.31) \times 0.162(100/270)/1000 = \mathbf{2.71\ kN}\ \text{per 0.3 m}$$

Sum of shear resistances from the concrete slab

From *Formula 10.18* with $\gamma_C = 1.5$,

$$V_{v,Rd}\,(\text{slab}) = 0.5/1.5(0.13 + 2.67 + 3.90 + 2.71) = \mathbf{3.14\ kN\ per\ 0.3\ m}$$

This contribution is much less than half of the 13 kN per 0.3 m from the profiled sheeting, so from *Formula 10.18* it has no effect on the result, which is $V_{v,Rd} =$ **13 kN per 0.3 m.**

Trials using seven UK deck types have found that $V_{c,cz}$ contributes less than 15% of $V_{v,Rd}$(slab), so this term could be omitted from preliminary checks.

Alternative method of *clause 10.7.5(7)*, Note 2 for resistance to vertical shear, $V_{v,Rd}$

From *Formulae 10.23* and *10.24*,

$$V_{v,Rd} = \tau_{c,Rd}(100\ b_{min}/b_s)z_{cs} + k_v V_{b,e,Rd}$$

Strength $\tau_{c,Rd}$ is found from Eurocode 2 to be 0.82 MPa.

Width b_{min} at 5 mm above the base of a trough of sheeting is found, above, to be 141 mm, and b_s, the wavelength of the sheeting, is 300 mm. The lever arm z_{cs} is 100 mm. Note 2 to this clause gives $k_v = 0.5$, and from this Note, $V_{b,e,Rd}$ is taken as the buckling resistance of the sheeting, 16 kN/m. The formula then gives

$V_{v,Rd} = 46.5$ kN/m, or **14.0 kN per 0.3 m**.

This simplified method gives in this case a resistance similar to that of the sheeting alone, estimated from tests. Both are well above $\boldsymbol{V_{Ed}}$ **at 5.44 kN** per 0.3 m.

Serviceability limit state – cracking

Clause 10.8.1(1) is for 'continuous' slabs. This slab can be assumed to satisfy the condition of *clause 10.8.1(2)*: to have been designed as simply-supported in accordance with *clause 10.4.2(5)*. That clause refers to *clause 10.8.1(2)* on nominal reinforcement at internal supports which, for unpropped construction, should be at least 0.2% of the area of concrete above the ribs.

The mean concrete thickness above the main top flange, 75 mm, is relevant, so the area required is 150 mm^2/m, and the A252 mesh used here is sufficient. However, if any spans are constructed propped, to reduce deflections, the area required is doubled, and A252 mesh is not sufficient.

Example 10.2: resistance to wheel load

The data are as in Example 10.1 except that

- the composite slab is simply-supported over a span of 3.0 m
- it is assumed that its imposed loading, $q_{Ek} = 7$ kN/m^2 included 1.5 kN/m^2 for partitions, which is replaced by a wheel load F_{Ek} $(= F_{Ed}/1.5)$ treated as a point load with a square contact area, $b_p = 50$ mm, travelling across the 3-m span.

Resistances to a load of this type from bending, vertical shear and punching shear are found, using *clauses 10.4.3* (effective width) and *10.7.6* (punching shear). It will be shown that punching shear does not govern because both bending and shear resistances are reached at lower values of F_{Ed}. For simplicity, no other loading is included.

The thickness of finishes, h_e, is taken as 30 mm. The thickness of concrete, Figure 10.5 is $h_c = 60$ mm. From *clause 10.4.3*, for global analysis the load is distributed over an effective width of sheeting (*Formula 10.1*):

$$b_m = b_p + 2(h_c + h_e) = 50 + 2(60 + 30) = 230 \text{ mm}$$

The bending resistance varies across the span (Figure 10.7), and the moment envelope for a travelling point load is parabolic, so the critical cross-section for bending is at $L_p = 1.0$ m, as found above for distributed loading. Hence, $M_{Rd} = 17.0$ kN m/m.

For longitudinal bending of the composite slab, from *Formula 10.2* the effective width is

$$b_{em} = 0.23 + 2 \times 1.0(1 - 1.0/3.0) = 1.56 \text{ m} \quad \text{and} \quad M_{Rd} = 17 \times 1.56 = 26.5 \text{ kN m}$$

The point load that gives $M_{Ed} = M_{Rd}$ at this location is $\boldsymbol{F_{Ed}}$ **= 39.7 kN**

For transverse bending, F_{Ed} exceeds the limit of *clause 10.4.3(4)* for the use of nominal reinforcement. It is found from the resistance to *clause 10.4.3(6)* from the reinforcement provided, which is 8-mm mesh at 200 mm. It gives the slab a resistance to sagging bending, $m_{Rd} = 5.68$ kN m/m. *Clause 10.4.3(7)* links it to a limit to F_{Ed} using the ratio $(I_{cross}/I_{long})^{1/4}$ for the composite slab, which is 0.65. From *Formula 10.5* for transverse bending with $m_{Ed} = m_{Rd}$,

$\boldsymbol{F_{Ed, slab}} \leq 5.68/(0.3 \times 0.65) =$ **29.1 kN**

For vertical shear, the load is assumed to be close to a support, with $L_p = 0.15$ m, so that $V_{Ed} = 0.95\,F_{Ed}$. From *clause 10.4.3(3)*, the ratio $h_p/h_{cs} = 70/130 = 0.54$, less than 0.6, so from *Formula 10.4*, the effective width for resistance to vertical shear (but not punching shear) is

$$b_{ev} = 0.23 + 0.15(1 - 0.15/3.0) = 0.37 \text{ m}$$

The resistance of the composite slab to vertical shear was estimated (above) to be 13 kN per 0.3 m, so $V_{Rd} = 13 \times 0.37/0.3 = 16.0$ kN.

Hence, $\boldsymbol{F_{Ed}} \leq 16/0.95 = \mathbf{16.9\ kN}$

Punching shear

Punching is now checked to *clause 10.7.6*, where *Figure 10.11* shows a so-called 'loaded area' at the top surface of the slab. It is assumed that this can allow for the 45° dispersion through finishes, so may be greater than the actual loaded area, 50 mm square. The 0.48 kN/m^2 load from finishes is equivalent to 30 mm of lightweight concrete, which is assumed to provide 45° dispersion, giving the loaded area on the slab, 110 mm square.

The length of the failure surface for the *deck slab*, AB in Figure 10.8, depends on the depth d_q for the composite slab (100 mm, found above), because it depends on the model for failure of the *composite slab*. Similarly, the length BC of the failure surface of the *composite slab* depends on the dimension $d_{s,c}$ for the deck slab (which may also be shown in *Figure 10.11* as d_{cs} or d_{sc}). For a slab, it is the mean of the depths of the reinforcement in both directions.

Here, the longitudinal reinforcement is irrelevant, so $d_{s,c}$ is taken as the depth to the transverse reinforcement, 56 mm (Figure 10.5(a)). Hence, the effective width for resistance of the composite slab (AD plus BC in Figure 10.8) is $2(110 + 56) = 332$ mm.

From Example 10.1, it is assumed that the shear resistance of the composite slab is at least half of the value 13 kN per 0.3 m found there. Hence, the resistance from lengths AD and BC is

$$V_{\mathbf{Rd,comp}} \geq (13/2)(0.332/0.3) = \mathbf{7.2\ kN}$$

For resistance from the concrete slab, *clause 10.7.6*, 'Punching shear', refers to clause 8.4.3 of EN 1992-1-1 on punching shear. That method relies on reinforcement in both directions. On a plane such as AB in Figure 10.8 only transverse reinforcement may be effective, so the rules of clause 8.2.2 of EN 1992-1-1 will be used. The design shear stress is given in Formula 8.27:

$$\tau_{Rd,c} = (0.66/\gamma_V)[100\,\rho_l f_{ck}(d_{dg}/d)]^{1/3}$$

Figure 10.8. Dimensions: punching shear from point load

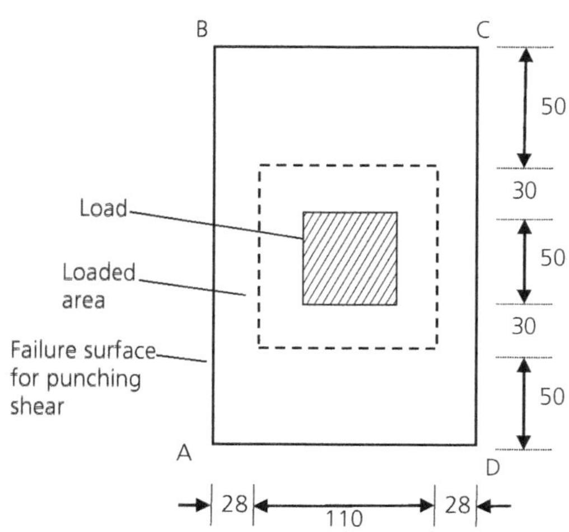

Here, $d_{dg} = 56$ mm, $100\ \rho_l = 252/(1000 \times 56) = 0.45$, $f_{ck} = 23$ MPa, $d = 56$ mm, $\gamma_V = 1.4$, whence $\tau_{Rd,c} = 0.90$ MPa.

The breadth over which this shear stress acts, b_w in EN 1992, is $2(110 + 100) = 420$ mm from Figure 10.8.

Corresponding to clause 8.2.1(3) of EN 1992-1-1, $V_{Rd} = \tau_{Rd,c}\ b_w z$, and $z = 0.9\ d_v = 50.4$ mm, so that

$$V_{Rd,conc} = 0.90 \times 0.42 \times 50.4 = 19.1 \text{ kN}$$

Adding the 7.2 kN found above, the failure load for punching shear should not exceed $\boldsymbol{F_{Rd}} = \mathbf{26.3\ kN}$, with the current transverse reinforcement. This is higher than the value found above and does not govern the design. Therefore, the maximum point load is $\boldsymbol{F_{Ed}} = \mathbf{16.9\ kN}$, determined by resistance of the composite slab to vertical shear.

Roger P Johnson and Stephen J Hicks
ISBN 978-1-83662-921-4
https://doi.org/10.1108/978-1-83662-920-720253013

CHAPTER 11
Composite joints in frames for buildings

This chapter corresponds to *Section 11* of EN 1994-1-1, which has the following clauses

- Scope *Clause 11.1*
- Analysis, modelling and classification *Clause 11.2*
- Design methods *Clause 11.3*
- Resistance of components *Clause 11.4*

11.1 Scope

Clause 11 is based on research on beam-to-column and beam-to-beam joints of the types used in steel and composite frames for buildings, so its scope has been limited to 'frames for buildings'. The definition of composite joints to which ***clause 11.1(1)*** refers includes joints between composite and reinforced concrete members. These could occur, for example, in a tower block with a concrete core and composite floors. No application rules are given for them, but EN 1992-4, 'Design of fastenings for use in concrete' may be useful (BSI, 2018). Other references are Henderson (2017) and Kuhlmann *et al.* (2014). *Clause 11.1(1)*

As implied by ***clause 11.2.1(1)***, *clause 11* and *Annex A* are essentially extensions to the Eurocode for joints between steel members, EN 1993-1-8 (BSI, 2024). It is over 200 pages long. It is impracticable here to provide explanations of its clauses that are used in Examples 11.1 and 12.1. The cross-references to it in Eurocode 4 reflect a major change in its layout: much of its detail has been moved into 66 pages of annexes. Those used here are from *Annex A*, which defines structural properties of basic components of connections, such as stiffness and resistance, and *Annex B*, which covers their use in design. There are only minor changes from the previous Eurocode 4, except for the new clause number (11 was previously 8). *Clause 11.2.1(1)*

It is assumed that the reader will be familiar with the provisions in EN 1993-1-8 for the steel members that *clause 11* considers in detail. These are I- and H-sections, which may have concrete-encased webs. Plate girders are not excluded.

The application rules of EN 1994-1-1 are limited to composite joints at simple supports and to those in hogging bending where reinforcement is in tension and the lower part of the steel section is in compression (*Figure 11.1*). There are no application rules for joints where the axes of the members connected do not intersect, or do so at angles other than 90°; but the basic approach is more general than the procedures prescribed in detail, and is capable of application in a wide range of situations.

EN 1993-1-8 specifies the majority of the calculations needed for composite joints. The worked examples and much of the comment in the present guide are limited to a single type of joint – the double-sided configuration shown in Figure 11.9, with an end plate and an uncased column, as shown on the left of *Figure 11.1*. Further commentary on the analysis and design of this joint will be found in Example 11.1, Chapter 12 and Example 12.1. Guidance on joints where the steel beam supports hollow-core precast slabs is available (Lam and Fu, 2006).

The word 'connection' (as used, for example, in *clause 11.2.2*) means the set of components that connect a member to another member – for example, an end plate, its bolts and a column flange. Thus, a 'connection' is a distinct part of a 'joint'. Design rules for joints are based on the 'component' concept. Each component of a joint – bolts, end plates, flanges and webs of the members joined, slab reinforcement and so on – contributes to its flexibility and resistance to bending and shear. These contributions are evaluated separately. Flexibility depends mainly on the least stiff component, and the resistance of the joint depends on that of its weakest component.

No alternative to this method has been found, despite the extensive calculations required, only a few of which will relate to yielding or failure of the joint, and so will govern its resistance. Most components are found to be stiffer or stronger than necessary. Preparation of tables in advance is, therefore, possible for the stiffness and resistance of a wide range of steel and composite joints in accordance with EN 1993 and EN 1994.

The design rules have changed little since ENV 1994-1-1 appeared in 1994. Much background information and some tables of resistances were published in the 1990s, as discussed below. Since then, the industry in the UK has shown little inclination to adopt the new methods. For buildings, 'simple' (nominally pinned) beam-to-column joints are preferred, with resistance to bending that is assumed to be zero. Their design is covered in Annex C of EN 1993-1-8 for steelwork, but not in EN 1994. Tabulated data for non-composite joints is available (Brown and Iles, 2013; Moreno, 2014).

Detailed guidance on the Eurocode methods for composite joints appeared in 1998, in the context of design to British codes (Couchman and Way, 1998). An explanation of the provisions and approximations in the Eurocodes, with worked examples, was then prepared (ECCS TC11, 1999). It refers to the draft codes as they were in 1998, so some differences, mainly in symbols, will be found between it and the published EN Eurocodes. With over 200 pages, it provides much broader coverage than is possible here. An account is available of the use of semi-continuous joints in practice (Huber, 2001). Some partial factors used in this material may differ from those in the relevant current national annex.

11.2 Analysis, modelling and classification

Clause 11.2.1(1)

Clause 11.2.1(1) refers to clause 7 of EN 1993-1-8, which covers the same subjects as *clause 11.2*. Table 7.1 in EN 1993-1-8 defines the links between the three types of global analysis, elastic, rigid-plastic and elastic-plastic, and the types of models used for joints. This enables the designer to determine whether the stiffness of the joint, its resistance or both properties are relevant to the analysis. Joints in steelwork are classified by stiffness as rigid, nominally pinned or semi-rigid; and by resistance as full-strength, nominally pinned or partial-strength. This classification relates the property of the joint (stiffness or resistance) to that of the connected member, normally taken as the beam.

The classification applies also to composite joints. The rotational stiffness of a joint (bending moment per unit rotation) is reduced below the initial elastic stiffness, $S_{j,ini}$, when the applied bending moment, $M_{j,Ed}$, exceeds two-thirds of the resistance $M_{j,Rd}$ of the joint, to allow for inelastic behaviour. As $M_{j,Ed}$ is initially unknown, clause 7.1.2 of EN 1993-1-8 defines a reduced stiffness, $S_{j,ini}/\eta$, which may be used in global analyses for all values of $M_{j,Ed}$. It gives values for the stiffness modification coefficient η, ranging from 2.0 to 3.5, that depend on the type of joint. This η should not be confused with same symbol used for degree of shear connection in *clause 8*.

Clause 11.2.2(1)

Clause 11.2.2(1) gives η_j for a contact-plate connection, shown in the right of *Figure 11.1* and in Figure 11.5(b), because they are not included in EN 1993-1-8. This value is for elastic global analysis. There is further comment on stiffness of joints under *clause 11.3.3*.

Clause 11.2.3(2)

The classification of a composite joint may depend on the direction of the bending moment (sagging or hogging). This is less likely in a steelwork joint, and so is referred to in ***clause 11.2.3(2)***. From *clause 11.2.3(3)*, I_b should be the value for the uncracked composite section at mid-span, using the short-term modular ratio. The stiffer the beam is assumed to be, the less likely it is that the joint can be classified as rigid. The span L_b is between centres of supporting columns.

The stiffness class depends on the ratio of the initial elastic stiffness of the joint, $S_{j,ini}$, to that of the connected beam, $E_a I_b / L_b$. This is shown in Figure 11.2(a), where the ratio 8 is for frames with bracing defined in EN 1993-1-8, 7.3.2.5. Most beam-to-column joints are in a region of hogging bending, where the only difference from an all-steel joint is the stiffness added by slab reinforcement continuous across the joint, mentioned in *clause 11.3.3(1)*.

Choice in the calculation of beam stiffness is permitted by the use of 'may' in ***clause 11.2.3(3)*** – for example, a representative value of modular ratio could be used. Account could also be taken of cracked and uncracked lengths within the beam in accordance with *clause 7.4.2.3*, but the additional calculation would not normally be worthwhile. *Clause 11.2.3(3)*

Figure 11.1. Model for a two-span beam in a frame

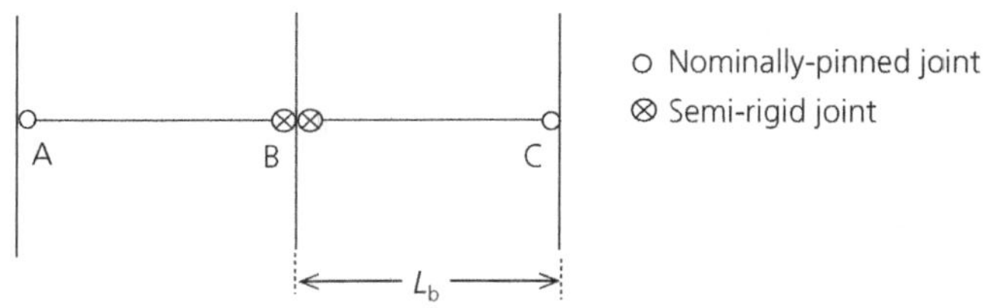

Outline of the modelling of joints for global analysis

In global analysis, nominally pinned joints are represented by pins, and semi-rigid joints by rotational springs, as shown in Figure 11.1 for a two-span beam of uniform depth, supported by three columns in a braced frame. Joints to external columns are usually designed as nominally pinned to reduce bending moments in the columns. The use of partial-strength semi-rigid joints at point B, rather than nominally pinned joints, involves much more calculation but has advantages in design, namely

- possible reduction in the section sizes for beams
- reduction in the deflection of beams
- reduction in crack widths near support B.

In comparison with full-strength rigid joints, the advantages are

- beams less susceptible to lateral-torsional buckling
- simpler construction and a significant reduction in cost
- lower bending moments in columns.

For global analysis, the precise location of each beam-to-column connection has to be decided. Some 'simple' connections are treated in the UK as nominally pinned, based on experience, even where the rules of *clause 11* would class them as semi-rigid. The bending moments found in columns are strongly influenced by the assumed location of these pins, which has been controversial and is not mentioned in EN 1994-1-1. Some 'complementary information' on this was issued in the UK in 2005, supported by its structural steelwork industry and those of France, Germany, Spain and Sweden. It located pins at 100 mm from the adjacent face of the supporting column, which can be either a flange or the web. It is available from the Steel Construction Institute in the UK.

Figure 11.2. Moment-rotation relationships for joints in braced frames

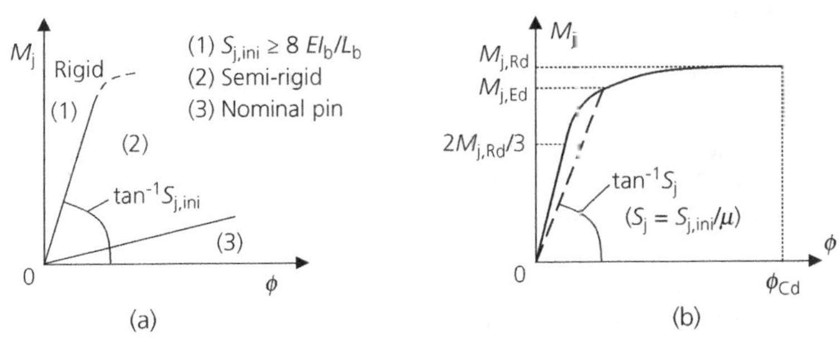

Figure 11.3. Model for an internal beam-to-column joint

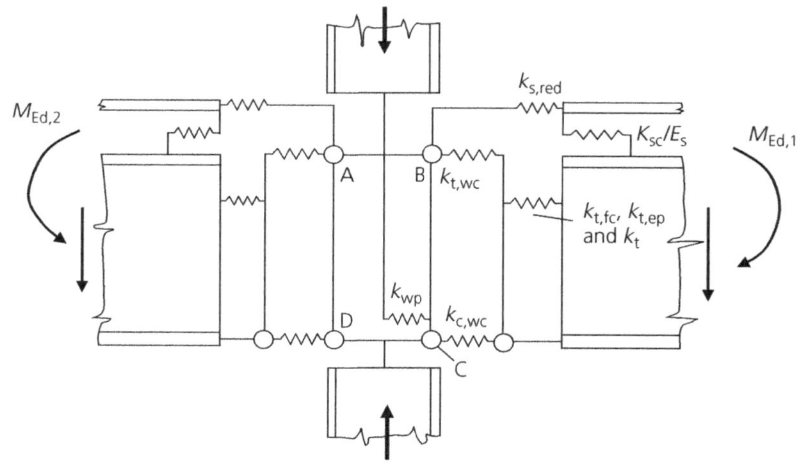

The initial stiffness of a joint is assembled from the stiffnesses of its components, represented by elastic springs. Those for an end-plate joint with a single row of bolts in tension, between beams of equal depth, are shown in Figure 11.3, in which all elements except springs and pins are rigid. The notation for the spring stiffnesses k_i here and in Examples 11.1 and 12.1 is as in clause 8 and Annexes A and B of EN 1993-1-8 as follows. The name is followed by the number of the formula in Annex A for its value.

k_{wp}	stiffness of column web panel in shear (A.13) (A.14)
$k_{c,wc}$	transverse compression of column web (A.26) (A.27)
$k_{t,wc}$	extension of column web in transverse tension (A.32)
$k_{t,fc}$	bending of column flange, caused by tension from a single row of bolts (A.35)
$k_{t,ep}$	bending of end plate, caused by tension from a single row of bolts (A.36)
k_t	extension of bolts in tension, for a single row of bolts (A.40) (A.41)

Stiffnesses in EN 1994-1-1, but not in EN 1993-1-8, are

$k_{s,r}$	extension of reinforcement (defined in *clause A.3.1.1* of *Annex A*)
K_{sc}/E_s	slip of shear connection (defined in *clause A.4(2)* of *Annex A*).

Coefficients for concrete-encased columns are given in *clauses A.3.3.1* and *A.3.3.2*.

Each 'spring' has a finite resistance, governed by yield or buckling of the steel. The design method ensures that non-ductile modes, such as fracture of bolts, do not govern. Springs are now referred to by their property k_i. For the tension region, the weakest of springs $k_{t,fc}$, $k_{t,wc}$, k_t and $k_{t,ep}$ is found. This resistance, together with that of the tension reinforcement, is compared with the compressive resistance of spring $k_{c,wc}$. The product of the lower of these resistances and the effective lever arm gives the plastic bending resistance of the joint. The resistance can be increased by strengthening the weakest link – for example, by the addition of column-web stiffeners if $k_{t,wc}$ or $k_{c,wc}$ governs.

Where the beams are of unequal depth, or $M_{Ed,1} \neq M_{Ed,2}$ (Figure 11.3), rotation at the joint is increased by shearing deformation of the column web. For beams of equal depth, this is the area ABCD in Figure 11.3. Its deformation is represented by the spring of stiffness k_{wp}.

The notation $M_{Ed,1}$ and so on with $M_{Ed,1} \geq M_{Ed,2}$ is as in *Table A.1* and replaces the more general notation of EN 1993-1-8: $M_{j,b2,Ed}$ and so on. Depending on the out-of-balance moment $|M_{Ed,1} - M_{Ed,2}|$, the column web panel may govern the resistance of the joint. It is assumed here that these bending moments have positive values when hogging, and that $M_{Ed,1} \geq M_{Ed,2}$.

The stiffnesses of joints depend on the *transformation parameter* β, defined in clause 7.2.3 of EN 1993-1-8. It allows for the effect on shear in the web panel of unequal vertical shear forces from the beams on either side of a column. However, it is defined in terms of the beam bending moments at the joints: $\beta = (M_{Ed,1} - M_{Ed,2})/M_{Ed,1}$, so for beams of the type shown in Figure 11.1, $0 \geq \beta \geq 1$.

The values M_{Ed} depend on joint stiffnesses, and hence on β, so a simpler method is needed. This is the 'approximate' method of clause 7.2.3(3), which is for use where the internal action effects are 'proportional throughout the analysis' (as assumed here). For the joints discussed here, it gives $\beta = 0$ where $M_{Ed,1} = M_{Ed,2}$. Otherwise, $\beta \approx 1$. These values are used in Example 12.1, and results based on the design bending moments are also given. Clause 7.2.3 of EN 1993-1-8 is worded 'may' rather than 'should'. Where $M_{Ed,1}$ and $M_{Ed,2}$ are only slightly different, designers may prefer to use a guessed value such as 0.7, rather than 1.0. There is further comment in Chapter 12 and in Example 12.1.

The parameter β is also used in the methods of EN 1993-1-8 for finding the stiffness of a web panel in shear, the resistance of the column web to transverse tension and transverse compression, and resistance of end-plate bolts to tension where there is more than one row. Hence, all of these values may depend on whether the two bending moments M_{Ed} are equal or different.

11.3 Design methods

Clause 11.3.1(1)

Clause 11.3.1(1) refers to clause 8 of EN 1993-1-8, which with its Annexes A and B defines the 'basic components' of a steelwork joint, their resistances and their elastic stiffnesses, and shows how these are assembled to obtain the resistances, rotational stiffness and rotation capacity of complete joints.

A composite joint has these additional components

- longitudinal slab reinforcement in tension
- concrete encasement, where present, of the column web
- steel contact plates, if used (not covered in EN 1993-1-8)
- slip of shear connection, allowed for by modifying the stiffness of the reinforcement.

Combination of stiffness coefficients and calculation of $S_{j,ini}$

All the properties of components given in, or cross-referenced from, EN 1994-1-1 satisfy the conditions of *clause 11.3.1* for the basis of the information. This important clause needs detailed explanation. It concerns the application to composite joints of Formulae B.6 to B.9 of EN 1993-1-8. They are explained here as applied to a steel end-plate joint with bolts in tension. Where there are several rows, each with a stiffness $k_{eff,r}$, their combined stiffness in tension k_{eq} is found taking account of the different levels of the rows. Here, a single row is assumed.

The contributions of each tensile component to the stiffness of a joint depends on its 'lever arm' above the centre of compression, usually at bottom-flange level.

The symbols k are stiffnesses, so the flexibilities are $1/k$. The flexibility in tension of a bolted end-plate connection, $1/k_{eq}$, is the sum of the flexibilities of four components in series: the bolts ($1/k_t$), a T-stub that represents the end-plate ($1/k_{t,ep}$), bending of the column flange under tension from the bolts ($1/k_{t,fc}$) and extension of the column web ($1/k_{t,wc}$). The notation is that of Table B.3 and clause B.4.2.1 of EN 1993-1-8. The total flexibility is given by

$$1/k_{eq} = 1/k_t + 1/k_{t,ep} + 1/k_{t,fc} + 1/k_{t,wc} = \Sigma(1/k_i) \qquad \text{(B.8) in EN 1993-1-8}$$

showing that the stiffnesses act in parallel, as shown in Figure 11.4. The total flexibility is greater than that of any component, and the total stiffness is less than any of the four k's. It acts at a lever arm z_1 above the centre of compression, the bottom flange.

Composite end-plate joint

Suppose now that there is reinforcement in the slab, attached to the steel beam by shear connection with its own stiffness. Their combined stiffness, $k_{s,red}$, equivalent to stiffness $k_{eff,2}$ in EN 1993-1-8 for a second row of bolts, acts with lever arm z_2. Although in series with the bolts, it cannot be added directly because their lever arms are different. Symbol k_{eq} now includes the reinforcement. The addition is

$$k_{eq}z_{eq} = k_{eff,1}z_1 + k_{eff,2}z_2 \qquad \text{(B.7) of EN 1993-1-8}$$

with the effective lever arm z_{eq} given by

$$z_{eq} = (k_{eff,1}z_1^2 + k_{eff,2}z_2^2)/(k_{eff,1}z_1 + k_{eff,2}z_2) \qquad \text{(B.9) of EN 1993-1-8}$$

The z^2 terms arise because for a given joint rotation about the bottom flange, the tension in a bolt is kz, with moment kz^2 about the bottom flange. Formula B.9 in EN 1993-1-8 is for two rows of bolts at different levels. Here it will be used for one row of bolts and the reinforcement in tension, with some symbols different, Equation X12.1.

In clause B.4.1(4) of EN 1993-1-8 the rotational stiffness $S_{j,ini}$ is related to the spring stiffnesses k_i by

$$S_{j,ini} = Ez_{eq}^2 / [\Sigma_i(1/k_i)] \qquad \text{(B.6) of EN 1993-1-8}$$

Symbol E in EN 1993-1-8 corresponds to E_a in EN 1994-1-1. It may seem strange that the stiffnesses k_i have the dimension of length, not force per unit deformation. This is because $S_{j,ini}$ is a bending moment per unit rotation and has the correct dimensions when each k is defined as a length.

Comparison with Formula B.8 above shows that the $1/k_i$ terms in Formula B.6 are flexibilities in series. A row of bolts combined with reinforcement provide stiffnesses in series, as shown by Formula B.7.

For a steel joint with one row of bolts in tension, there are other stiffnesses in parallel: $k_{c,wc}$ for the column web in compression and, for unequal end moments, k_{wp} for the web plate in shear. Both have the lever arm z_{eq}, as they relate to the total tensile force across the joint, as shown in Figure 11.4.

Figure 11.4. Assembly of stiffnesses for a composite joint

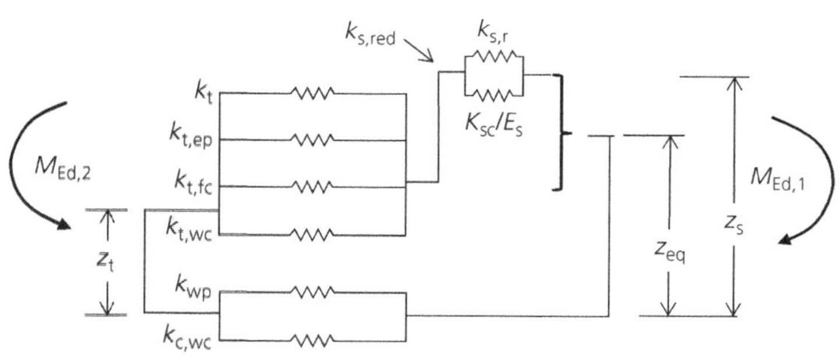

The stiffness $k_{s,red}$ from the reinforcement in the slab, with a lever arm z_s greater than that of the bolts, has two components: $k_{s,r}$ for the reinforcing bars and K_{sc}/E_s to allow for slip of the shear connection. They act in parallel. The effective stiffness is defined by $k_{s,red} = k_{slip}\,k_{s,r}$, with k_{slip} defined in *clause A.4(2)* as a function of both $k_{s,r}$ and K_{sc}/E_s. This is why *Formula A.5* for k_{slip} is unlike the usual formula for stiffnesses in parallel.

Stiffness $k_{s,red}$ acts in series with the stiffness of the bolt row or rows. The combination has an effective stiffness k_{eq} and an effective lever arm z_{eq}, intermediate between z_s and z_t. These stiffnesses are defined in *clauses A.3.1.1* and *A.4*, and are further explained in Chapter 12 of this book, on *Annex A*.

Clause 11.3.2(1) The additional stiffness from the reinforcement has no influence on resistance to vertical shear, which is fully covered by EN 1993-1-8: B.3.4 (***clause 11.3.2(1)***).

Clause 11.3.2(2) ***Clause 11.3.2(2)*** is for 'a composite joint with full shear connection'. In framed structures for buildings, composite joints other than nominal pins are almost always in regions of hogging bending, for which full shear connection is required. This does not apply to a nominally pinned joint, as these are often adjacent to a region of sagging bending with partial shear connection.

The use of Informative *Annex A* for finding rotational stiffnesses is expected to be permitted by the UK's National Annex. It satisfies ***clause 11.3.3(1)*** and is illustrated in Example 12.1. *Clause 11.3.3(1)*

The coefficient ψ is given in ***clause 11.3.3(2)*** for contact-plate joints, because it is not in Table 7.5 of EN 1993-1-8. It is used in clause 7.2.6(4) of EN 1993-1-8 to define the shape of the moment-rotation curve for the joint at bending moments $M_{j,Ed}$ that exceed $2M_{j,Rd}/3$, as follows. Let $S_{j,ini}$ be the stiffness at low bending moments. *Clause 11.3.3(2)*

For $2M_{j,Rd}/3 < M_{j,Ed} \leq M_{j,Rd}$, the stiffness is $S_j = S_{j,ini}/\mu$ (X11.1)

(Figure 11.2(b)), where $\mu = (1.5\,M_{j,Ed}/M_{j,Rd})^{\psi}$ (X11.2)

Rotation capacity

The rotation capacity of composite joints, ϕ_{Cd} in Figure 11.2(b), has been extensively researched by Anderson and Aribert (2000) and Bose and Hughes (1995). There are many relevant parameters. Analytical prediction is still difficult, and there are as yet no design rules sufficiently well established to be included in EN 1994-1-1.

The reference in ***clause 11.3.4(2)*** to 'experience' is relevant to the use of so-called 'simple' joints, and makes design possible without calculating either the available rotation capacity of a joint or the rotation required. Further guidance is available (ECCS, 1999). *Clause 11.3.4(2)*

11.4 Resistance of components

This clause supplements clause 8.2 and the annexes of EN 1993-1-8. The effective width of concrete flange in tension is the same at a joint as for the adjacent beam (***clause 11.4.2.1(1)***). *Clause 11.4.2.1(1)*

Clause 11.4.2.1(4) applies at an external column with a partial- or full-strength joint. The tensile force in the bars that pass the column must be transferred to it – for example, by being looped round it. This applies also at internal columns where there is a change in the tension in the bars (*Figure 11.2* and Example 11.1). *Clause 11.4.2.1(4)*

Research has found (Anderson *et al.*, 2000) that for external columns the use of a truss model of the type shown in *Figure 11.2* may not prevent shear failure of the slab on the exterior side of the column. Its conclusion is that $A_{s,t}$, the area of transverse reinforcement assumed to resist the force F_{tq} in *Figure 11.2* should be not less than A_s, the area of longitudinal reinforcement to be anchored behind the column.

Clauses 11.4.2.2(1) and ***11.4.3(1)*** permit the same 45° spread of force in a contact plate as used in EN 1993-1-8 for an end plate. For the flange and root radius of the column, the term $5(t_{fc} + s)$ in Formula A.16 of EN 1993-1-8 shows that the force is assumed in EN 1993-1-8 to spread at $\tan^{-1} 2.5$ (68°). Where the compressive force relied on in design exceeds the resistance of the steel bottom flange, the length of the contact plate should allow for this (Figure 11.5(b)). *Clause 11.4.2.2(1)* *Clause 11.4.3(1)*

Figure 11.5. Beam-to-column end-plate joints: (a) showing top bolts only; (b) detail of contact plate between beam bottom flange and column

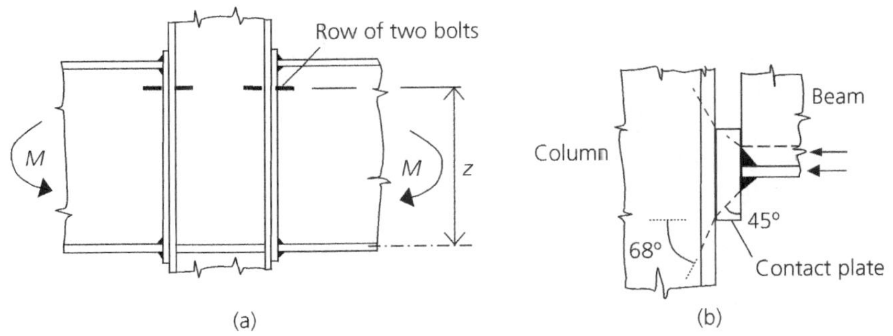

Column web panel in shear

Clause 11.4.4.1(2) The model used in ***clause 11.4.4.1(2)*** is illustrated in Figure 11.6, which shows an elevation of the concrete encasement of a column web. Its width is $h - 2t_f$ (column depth less flange thicknesses) and depth z, the lever arm between the resultant horizontal forces from the beam. A shear force V is transferred through the encasement, which is of thickness $b_c - t_w$ (column width less web thickness). The concrete strut ABDEFG has width $0.8(h - 2t_f)\cos\theta$, where

$\tan\theta = (h - 2t_f)/z$, so its area is

$$A_{vc} = 0.8\,(b_c - t_w)(h - 2t_f)\cos\theta \qquad (11.2)$$

Its compressive strength is $0.85\nu f_{cd}$, giving the force C in Figure 11.6. For horizontal equilibrium at B and F, $C\sin\theta = V$. These are *Formulae 11.1* to *11.3* in *clause 11.4.4.1.*

Figure 11.6. Strut model for the shear resistance of the concrete encasement to a column web

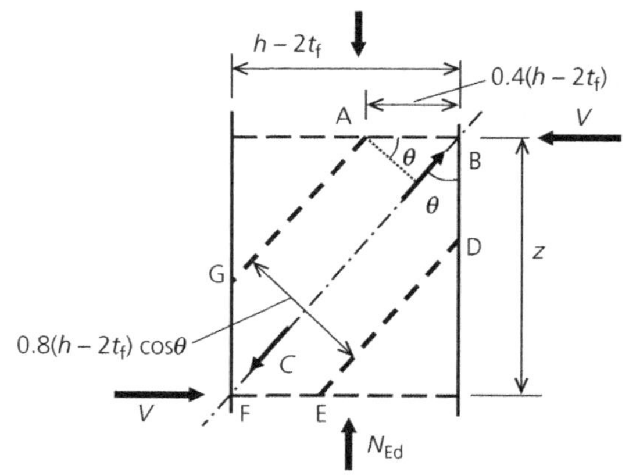

Clause 11.4.4.1(3) The shear strength of concrete is modified by longitudinal compression in the column. This is allowed for by the factor ν in ***clause 11.4.4.1(3)***, which ranges from 0.55 for zero axial compression to 1.1 for $N_{Ed} \geq 0.55\,N_{pl,Rd}$. For a fully-encased composite column, the area of concrete that contributes to $N_{pl,Rd}$ in *Formula 11.4* will be greater than that between the flanges, used in *clause 11.4.4.1(2)*. This presumably applies also to the '*contribution of the concrete encasement*' in *clause 11.4.4.2(3)*.

Clause 11.4.4.2(2) The contribution of concrete encasement to the resistance of a column web to 'transverse' (typically, horizontal) compression is given in ***clause 11.4.4.2(2).*** For an end-plate connection to a column flange, the depth of encasement assumed to resist compression, $t_{eff,c}$, is shown in Figure 11.7(a), with the 2.5:1 dispersion, referred to above, extending through the root radius r.

The horizontal compressive strength of the concrete is $0.85\,k_{wc,c}f_{cd}$, where $k_{wc,c}$ depends on the 'longitudinal' (typically, vertical) compressive stress in the column, $\sigma_{com,c,Ed}$, as shown in Figure 11.7(b).

Figure 11.7. Model for resistance to compression of the concrete encasement to a column web

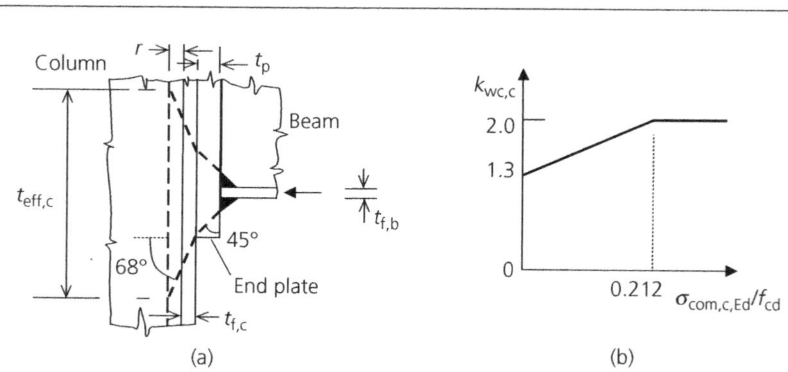

Example 11.1: semi-rigid end-plate joints in a two-span beam in a braced frame

In development work that followed the publication of ENV 1994-1-1, a set of application rules for composite joints was prepared, more detailed than those in *Section 8* and *Annex A* of EN 1994-1-1:2004. They were published as a model annex J (ECCS, 1999) and provide useful guidance in this example.

Figure 11.8. Model for a two-span beam ABC, with an internal column EBF

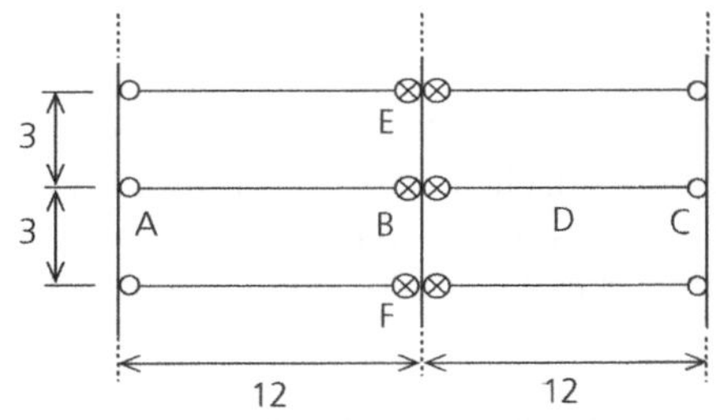

The design process

The flexural stiffness of a semi-rigid joint depends on the bending moment applied to it, and elastic global analysis including joint flexibility is quite complex. The initial design is based on design ultimate loading and resistances. Approximate global analyses are used that estimate or ignore rotation of the joints and make conservative assumptions about cracking of concrete and effects of shrinkage. These show that the details assumed for the joint provide sufficient resistance to the loading used in Example 8.6. Also, the mode of failure has high ductility because the mode of failure is yielding of the thin steel end plate.

Stiffness of joints is the subject of Chapter 12. These are determined in its Example 12.1 and used in global analyses in this Example 11.1 for serviceability loading that allow for joint rotation, creep, cracking and shrinkage. These quite complex calculations require understanding of EN 1993, 'Design of steel structures, Part 1-8: Joints'. They lead to predictions of deflections and crack widths.

Sequence of design and calculations

This list of steps may be useful for those who do not wish to follow the details.

1. Define the limited frame for global analyses (members ABC and EBF in Figure 11.8).
2. Design for ductile failure in bending by yield of reinforcement and flexure of the end plate.
3. Choose end plate thickness, bolt details and area of top reinforcement.
4. Use approximate methods to decide whether allowance need be made for yielding at the joint during construction. Result: no allowance needed.
5. Check whether vertical shear can be resisted by the bottom row of bolts only, leaving the top row fully available in tension.
6. For the construction stage (no reinforcement), find the resistances in tension and the weakest link in the tension chain. It is combined end-plate bending and bolt yield, with $F_{T,2,Rd} = 216$ kN.
7. Find the resistances of the column flange and web to transverse tension and compression, as they may govern. Hence find the resistance of the steel joint during unpropped construction: $M_{j,Rd,steel} = 82$ kN m.
8. For the composite joint, add the tension force from the reinforcement at yield. Using its lever arm, find $M_{j,Rd,comp} = 366$ kN m for $M_{Ed,1} = M_{Ed,2}$ (imposed loading on beams AB and BC), reducing to 285 kN m for imposed loading on BC only ($M_{Ed,1} > M_{Ed,2}$).
9. Calculate stiffnesses of all components and combine them to find stiffnesses of the steel and composite joints for both types of loading (see Example 12.1). Do global analyses of the limited frame (Figure 11.13). Check that resistances are sufficient for both 'max-max' and 'min-max' loadings on the two spans.
10. From analyses with serviceability loads, find deflections. Consider effects of shrinkage. Review crack-width control and classification of the joints.

Data

The subject of Examples 8.6 and 9.1 is a two-span beam ABC continuous over its central support (see Figures 8.27 to 8.32). It is now assumed that this beam is one of several similar beams in a multi-storey braced frame (Figure 11.8). Its joints with the external columns are nominal pins. The spans of the composite-slab floors are 2.5 m, as before. For simplicity, in the work on beams AB and BC, column EBF will be treated as fixed in position and direction at nodes E and F. These beams are now partly discontinuous, being attached to the column at B by the partial-strength end-plate joints shown in Figure 11.9, which also gives dimensions of the column section. Its other properties are as follows: HEB 240 steel section, $A_a = 10\ 600\ \text{mm}^2$, $f_y = 355\ \text{N/mm}^2$, $10^{-6} I_y = 112.6\ \text{mm}^4$.

The end plates are of mild steel, $f_y = 275\ \text{N/mm}^2$, and relatively thin, 12 mm, to provide the plastic behaviour required. They are attached to the beam by fillet welds of size (throat thickness) 10 mm to the flanges, and 8-mm welds to the web. They are each attached to the column by four Grade 8:8 M20 bolts with the properties $f_{ub} = 800\ \text{N/mm}^2$, $f_{yb} = 640\ \text{N/mm}^2$ and net area at root of thread $A_{s,b} = 245\ \text{mm}^2$ per bolt.

Figure 11.9. Details of the beam-to-column end-plate joints

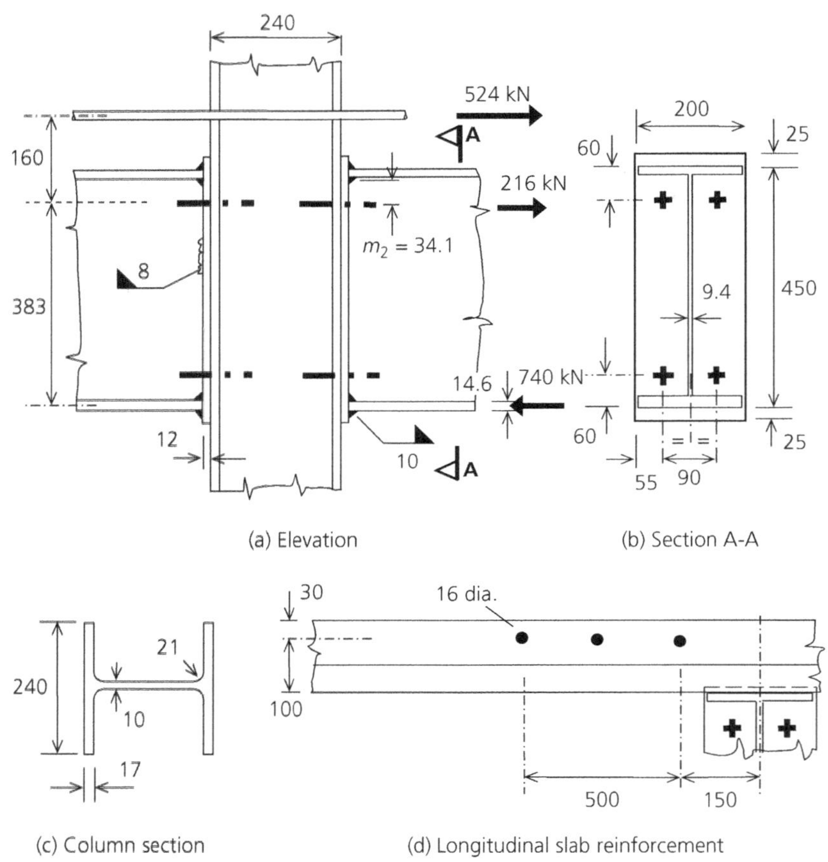

Longitudinal reinforcement at support B

The only other change from the data used in Examples 8.6 and 9.1 on geometry, materials and loadings concerns the reinforcement in the slab. Partial-strength joints need rotation capacity. Its value cannot be found at this stage, but it is known to increase with both the diameter of the reinforcing bars in the slab and the area of reinforcement provided. However, the amount of top reinforcement should be limited, so that the whole of the compressive force across the joint can be resisted by the beam bottom flange and the unstiffened column web.

Detailed guidance is available (Couchman and Way, 1998). For a steel beam of depth 450 mm in S355 steel, the recommended minimum areas are 3000 mm^2 for bars with 5% elongation and 860 mm^2 for bars with 10% elongation. The recommended maximum amount depends on the size of the column and the details of the bolts in tension, and is about 1200 mm^2 for this example. The recommended bar diameters are 16 mm and 20 mm, with the comment 'smaller diameter bars are generally less ductile'. For these reasons, the previous reinforcement (ten 12-mm bars, $A_s = 1131$ mm^2) is replaced by six 16-mm hot-formed bars (minimum elongation 10%): $A_s = 1206$ mm^2, with $f_{sk} = 500$ N/mm^2. This increases the second moment of area of the cracked composite section to $10^{-6}I_y = 451$ mm^4.

For effective width, from *clause 11.4.2.1(1)*, the value for the beam (1.60 m) should be used. The bars have to miss the column and could be spaced as shown in Figure 11.9(d). This spacing is checked later for crack-width control.

Classification and resistance of the joints

It is assumed, initially, that flexural failure of a joint will occur in a ductile manner by yielding of the reinforcement in tension and the end-plate or column flange in bending, and that at bottom-flange level the compressive resistance of the column web will be sufficient. As the spans are equal, it is unlikely that shear of the column web will be critical.

The joint is expected to be 'partial-strength'. This can be checked by comparing the tension resistance of the top two bolts, from Tables 5.4 and 5.9 of EN 1993-1-8, with the force to yield the beam top flange

$$\mathbf{F_{t,Rd,bolts}} = 2(0.9 f_{ub} A_s / \gamma_{M2}) = 2 \times 0.9 \times 0.8 \times 245 / 1.25 = \mathbf{282\ kN} \qquad (X11.3)$$

$$F_{Rd,flange} = b_f t_f f_{yd} = 190 \times 14.6 \times 0.355 = 985 \text{ kN}$$

Thus, the resistance moment $M_{j,Rd}$ for the joint will be much less than $M_{pl,Rd}$ for the beam, and lateral buckling of the beam will be less critical than before. There is no need to find the stiffness of the joint at this stage, because it is clearly either 'rigid' or 'semi-rigid'. Either type can be treated as 'semi-rigid'.

Tables in Appendix B of Couchman and Way (1998) enable a rough check to be made on this initial design, without much calculation. They give resistances $M_{j,Rd}$ in terms of the cross-section of the steel beam, its yield strength, the thickness and grade of the end plate, the number and size of bolts in tension, and the area of reinforcement. Even though the beam used here is an IPE section, it can be deduced that $M_{j,Rd}$ is about 400 kN m. It will be found later to be 366 kN m.

Approximate global analysis

For both spans fully loaded, it was found in Example 8.6 that M_{Ed} at B was 527 kN m from loading (see Figure 8.32) plus 117 kN m from shrinkage. The flexural stiffness of the joint is not yet known, but it will be between zero and 'fully rigid'. If fully rigid, the joint would obviously be 'plastic' under ultimate loading, and there would then be no secondary shrinkage moment. At mid-span, for the total load of 35.7 kN/m (see Table 8.3), the sagging bending moment from loading would be approximately

$$35.7 \times 12^2 / 8 - 400 / 2 = 443 \text{ kN m}$$

The extent to which this is reduced by the secondary effect of shrinkage depends on the rotation of the joint, unknown at this stage.

If the joint acts as a pin, there would be no reduction from shrinkage and the mid-span moment would be $35.7 \times 12^2/8 = 643$ kN m.

It is recommended in Couchman and Way (1998) that mid-span resistances should be taken as 0.85 $M_{pl,Rd}$ to limit the rotation required at the joints. From Example 8.6, $M_{pl,Rd}$ with full shear connection is 1043 kN m, so the bending resistance of the beam is obviously sufficient.

Vertical shear

For $M_{Ed} = 400$ kN m at B, the vertical shear at B is

$$\mathbf{F_{v,Ed,B}} = 35.7 \times 6 + 400 / 12 = \mathbf{247\ kN}$$

The shear resistance of the four M20 bolts is now found, using Table 5.9 of EN 1993-1-8. Two of the bolts are likely to be at yield in tension. From this table, the shear applied to these bolts must satisfy

$$F_{v,Ed}/F_{v,Rd} + F_{t,Ed}/(1.4\,F_{t,Rd}) \le 1.0 \quad \text{(X11.4)}$$

The net shear area of each bolt is $A_{s,b} = 245$ mm^2, so from the table,

$$F_{v,Rd} = 0.6 f_{ub} A_{s,b}/\gamma_{M2} = 0.6 \times 800 \times 0.245 / 1.25 = 94.1 \text{ kN}$$

From Equation (X11.4) with $F_{t,Ed} = F_{t,Rd}$, $F_{v,Ed} \le (1 - 1/1.4)\,F_{v,Rd} = 27$ kN

$$\text{For four bolts, } \boldsymbol{F_{v,Rd}} = 2(94.1 + 27) = \mathbf{242\ kN} \quad \text{(X11.5)}$$

This shows that it may be necessary to add a second pair of bolts in the compression region of the joint.

Bending resistance of the joint, excluding reinforcement

Unpropped construction was used in Example 8.6. From Table 8.3, the design ultimate load for the steel beam is 7.8 kN/m. For the construction phase, this is increased to 9.15 kN/m to allow for the higher density of fresh concrete and the imposed loading during construction.

For rigid joints at the internal support between two 12 m spans,

$$M_{Ed,B} = wL^2/8 = 9.15 \times 12^2 / 8 = 165 \text{ kN m}$$

The plastic resistance of the joint during construction is required. An upper limit is easily obtained. The lever arm from the top bolts to the centre of the bottom flange is

$$z_{bolts} = 450 - 60 - 7.3 = 383 \text{ mm} \quad \text{(X11.6)}$$

$$\text{The limit to the resistance is: } F_{t,Rd,bolts} z_{bolts} \le 282 \times 0.383 = 108 \text{ kN m} \quad \text{(X11.7)}$$

and is found (below) to be 82 kN m, governed by the T-stub. It is assumed that the flexibility of the joint will be sufficient to reduce the bending moment at support B to a level such that during construction, no permanent deflection arises from inelastic behaviour.

Resistance of T-stubs and bolts in tension

The calculation of the bending resistance consists of finding the 'weakest links' in both tension and compression. In tension, the column flange and the end plate are each modelled as T-stubs, and prying action may occur. Some of the dimensions required are shown in Figure 11.10. The dimensions m shown in Figure A.4 of EN 1993-1-8 overlap with 20% of the corner fillet or weld (Figure 8.3 of EN 1993-1-8). Thus, in Figure 11.10(a) for the end plate

$$m = 100 - 55 - 9.4/2 - 0.8 \times 8 \times 1.414 = 31.2 \text{ mm} \quad \text{(X11.8)}$$

It is evident from the geometry shown in Figure 11.10 that the end plate is weaker than the column flange, so its resistance is now found. Table 8.2 of EN 1993-1-8 gives three possible failure modes

(1) yielding of the end plate
(2) a combination of (1) and (3)
(3) failure of the bolts in tension.

Yield-line theory is used for bending of the plate. The theory is similar to the plastic hinge theory for steel frames, with hinges replaced by lines in the plate along which plastic rotation is assumed to occur. These form failure mechanisms, with deformations so much greater than those in the elastic range that the latter can be ignored in calculations. The mechanism that gives the lowest failure load depends on the geometry of the end plate and layout of the bolts. Many mechanisms are possible. Details are given by Couchman and Way (1998). The total length of the hinge lines is represented by an 'effective length', ℓ_{eff}. These are given in Table A.3 of EN 1993-1-8 (BSI, 2024).

The critical mechanism for this end plate will be either that shown in Figure 11.10(b) or a circular fan, for which the effective length is $\ell_{eff,cp} = 2\pi\, m$. For the non-circular pattern, $\ell_{eff,nc}$ in Table A.3 of EN 1993-1-8 is αm, with α given by Formula A.33 and depends on m, m_2 and e, which are shown in Figure 11.10 and in Figure A.10 of EN 1993-1-8. In Table 8.2 of clause 8.3.4 of EN 1993-1-8, dimension e is replaced by n (used below and in Figure 11.10).

Figure 11.10. Dimensions of T-stubs, and the yield line pattern

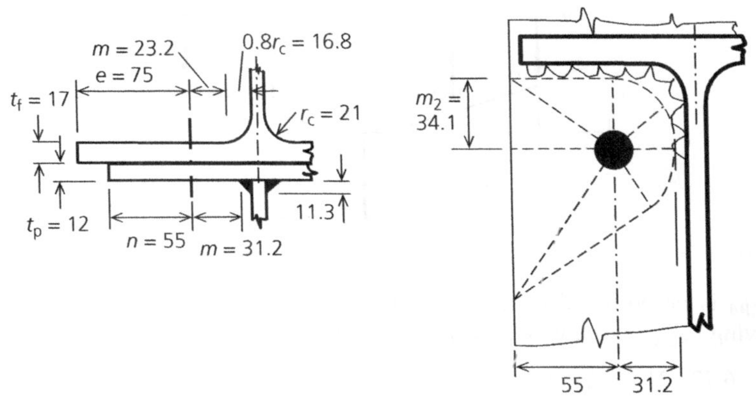

(a) Plan details of T-stub

(b) Yield-line pattern in end plate

With the 20% overlap referred to above, $m_2 = 60 - 14.6 - 0.8 \times 10 \times 1.414 = 34.1$ mm, m = 31.2 mm (above) and $e = 55$ mm. Hence, $\alpha = 6.77$, and $\ell_{\text{eff,nc}} = \alpha m = 6.77$ m.

As $6.77 > 2\pi$, the circular pattern governs for mode (1), and its perimeter is

$$\ell_{\text{eff,1}} = 2\pi m = 6.28 \times 31.2 = 196 \text{ mm}$$

The plastic resistance per unit length of plate is

$$m_{\text{pl,Rd}} = 0.25\, t_f^2 f_y / \gamma_{M0} = 0.25 \times 12^2 \times 0.275 / 1.0 = 9.90 \text{ kN m/m} \tag{X11.9}$$

(Notation m_{pl} is from yield-line theory, not to be confused with dimension m.)

For mode 1, yielding is confined to the plate. From Table 8.2 of EN 1993-1-8, equations for this mode, allowing for prying forces, are:

$$F_{\text{T,1,Rd}} = 4\, M_{\text{pl,1,Rd}} / m$$

$$M_{\text{pl,1,Rd}} = \ell_{\text{eff,1}} m_{\text{pl,Rd}} = 0.196 \times 9.9 = 1.94 \text{ kN m} \tag{X11.10}$$

From Equation X11.8 for m,

$$\mathbf{F_{T,1,Rd}} = 4 \times 1.94 / 0.0312 = \mathbf{249\ kN}$$

Yield-line theory gives an alternative method that is simpler here and does not need $\ell_{\text{eff,1}}$. For a circular fan at failure under a central point load W, with bending resistance the same for both sagging and hogging yield lines, $W = 4\,\pi\, m_{\text{pl}}$. Hence, assuming that $F_{\text{T,1,Rd}}$ above is for a row of two bolts,

$$F_{\text{T,1,Rd}} = 8\pi \times 9.9 = 249 \text{ kN}$$

The agreement with the result above confirms that forces from Table 8.2 of EN 1993-1-8 are per row, not per bolt.

Mode 2 is more complex. The equation in Table 8.2 for the tension resistance $F_{\text{T,2,Rd}}$ is

$$F_{\text{T,2,Rd}} = \left(2M_{\text{pl,2,Rd}} + n\sum F_{\text{t,Rd}}\right) / (m+n) \tag{X11.11}$$

It is explained with reference to Figure 11.11, where Q is the total prying force. Taking A and E as fixed points, the work done by the tensile force in the top flange is $F_{\text{T,2,Rd}}\,(m+n)\theta$. It equals the sum of the strain energy in bending the plate (which is $2M_{\text{pl,2,Rd}}\,\theta$ for a rotation θ) and in extending the bolts, $n\theta\, \Sigma F_{\text{T,2,Rd}}$.

Figure 11.11. Plan of a T-stub, showing failure mode 2

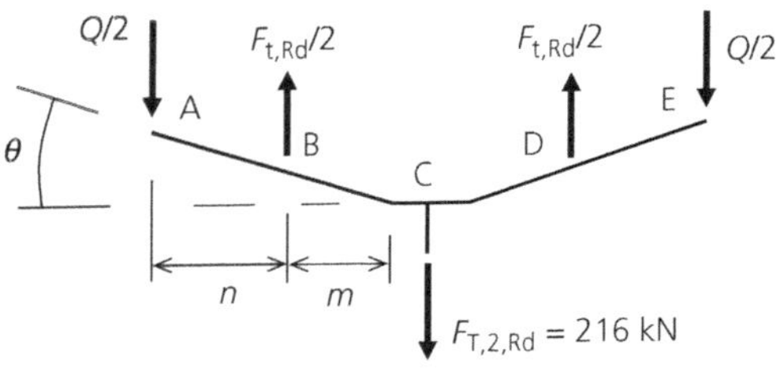

Resistance $M_{pl,2,Rd}$ is proportional to $\ell_{eff,2}$, the value for a non-circular pattern from Table A.3 of EN 1993-1-8, giving $\ell_{eff,nc} = \alpha m$, with $\alpha = 6.77$ (above).

$$\text{Hence, } \ell_{eff,nc} = 6.77 \times 31.2 = 211 \text{ mm} \quad \text{(X11.12)}$$

From Figure 8.9(a), n = 55 mm, but Table 8.2 of EN 1993-1-8 requires that n is the lesser of e_{min} (55 mm) and 1.25 m (39.0 mm); so $m + n = 31.2 + 39.0 = 70.2$ mm.

From Equation X11.10, $M_{pl,2,Rd} = \ell_{eff,2}\, m_{pl,Rd} = 0.211 \times 9.9 = 2.09$ kN m.

From Equation X11.3, the tensile resistance of a pair of bolts is $\Sigma F_{t,Rd,\text{ bolts}} = 282$ kN.

From Equation X11.11,

$$F_{T,2,Rd} = (2 \times 2.09 + 0.039 \times 282)/0.0702 = \mathbf{216\ kN} \quad \text{(X11.13)}$$

This is lower than $F_{T,1,Rd}$, and as $F_{T,3,Rd} = 282$ kN for failure of the bolts (mode 3), mode 2 governs. The prying force (Figure 11.11) is, therefore,

$$Q = F_{t,Rd} - F_{T,2,Rd} = 282 - 216 = 66 \text{ kN}$$

Beam web in tension

The equivalent T-stub in Figure 11.11 applies a tensile force of 216 kN to the web of the beam. Its resistance is given by Formula (A.39) in clause A.11.1 of EN 1993-1-8 as

$$F_{t,wb,Rd} = b_{eff,t,wb} t_{wb} f_{y,wb} / \gamma_{M0}$$

with $b_{eff,t,wb}$ taken as the effective length of the T-stub: $\ell_{eff,2} = 211$ mm. Hence,

$$F_{t,wb,Rd} = 211 \times 9.4 \times 0.355 / 1.0 = 704 \text{ kN} \quad \text{(X11.14)}$$

so this does not govern.

Column web in tension

The effective width of the column web in tension, clause A.6.1.1(1) of EN 1993-1-8, is

$$F_{t,wc,Rd} = \omega b_{eff,t,wc} t_{wc} f_{y,wc} / \gamma_{M0} \quad \text{(A.28) in EN 1993-1-8}$$

where ω is a reduction factor to allow for shear in the column web. For both spans fully loaded, the shear is zero, and $\omega = 1$. The column web is thicker than the beam web, so from result X11.14, its resistance does not govern.

Column web in transverse compression

The resistance of the column web to transverse compression is given by Formula A.15 in EN 1993-1-8

$$F_{c,wc,Rd} = \omega k_{wc} \rho b_{eff,c,wc}\, t_{wc} f_{y,wc} / \gamma_{M0} \quad \text{(X11.15)}$$

The factor ω depends on the ratio of the end moments. For $M_1 = M_2$ with $\beta = 0$, it is 1.0.

For $M_1 < M_2$ and β taken as 1.0, ω depends on the effective width of the web, given by

$$b_{eff,c,wc} = t_{fb} + 2\sqrt{(2)}\, a_p + 5(t_{fc} + s) + s_p$$

where a_p is the throat thickness of the bottom-flange welds, 10 mm here; t_{fb} is the thickness of the beam flange (14.6 mm); t_{fc} is the thickness of the column flange (17 mm); s_p allows for 45° dispersion through the end plate, and is 38.6 mm here; and s is the root radius of the column section ($s = r_c =$ 21 mm). Hence,

$$b_{eff,c,wc} = 14.6 + 28.3 + 5(17 + 21) + 38.6 = 271 \text{ mm}$$

From Formula A.19 in EN 1993-1-8, $\omega = 0.613$ for $\beta = 1$.

Factor ρ allows for buckling in transverse compression. It is given in clause A.5.1.1(7) of EN 1993-1-8 and needs a long calculation. The results are: $\rho = 0.901$ for $\beta = 0$ and $\rho = 1.0$ for $\beta = 1$. Where the vertical compressive stress exceeds $0.7f_{y,wc}$, factor k_{wc} is reduced below 1.0 (not applicable here).

From Formula X11.15,

$$\text{for } \beta = 0, \ \boldsymbol{F_{c,wc,Rd}} = 0.901 \times 271 \times 10 \times 0.355 / 1.0 = \mathbf{867\ kN}$$

$$\text{for } \beta = 1, \ \boldsymbol{F_{c,wc,Rd}} = 0.613 \times 1 \times 962 = \mathbf{590\ kN} \qquad \text{(X11.16)}$$

Resistances of joints to hogging bending

Bending resistance of the steel joint

For the steel beam, from Equation X11.13, $F_{T,Rd} = 216$ kN (< 590 kN). From Figure 11.9(a), the lever arm is 383 mm. The resistance for both beams fully loaded is governed by bending of the end plate T-stub, and

$$\boldsymbol{M_{j,Rd,steel}} = 216 \times 0.383 = \mathbf{82\ kN\ m} \qquad \text{(X11.17)}$$

From Example 8.6, the plastic bending resistance of the steel beam, an IPE 450 section, is 604 kN m. This exceeds seven times $M_{j,Rd,steel}$. Clause 7.3.3.2(2) of EN 1993-1-8 then permits this joint to be classified as 'nominally pinned' for the construction stage if it has 'sufficient rotation capacity'; that is, if it is 'ductile'.

The model for the critical mode 2 includes failure (not just yielding) of the top row of bolts in tension. However, the joint is closely based on a type given in Couchman and Way (1998). The 'ductile' behaviour of joints of this type is confirmed in ECCS TC11 (1999).

Resistance of the composite joint

For the composite joint, the reinforcement can be at yield in tension. Its resistance is

$$F_{t,s,Rd} = 1206 \times 0.500 / 1.15 = 524 \text{ kN}$$

This increases the total compressive force to

$$\boldsymbol{F_c} = 216 + 524 = \mathbf{740\ kN} \qquad \text{(X11.18)}$$

For $\beta = 1$, this exceeds $F_{c,wc,Rd} = 590$ kN, which governs, so only 374 kN (590 – 216) is available from reinforcement. Then,

$$\boldsymbol{M_{j,Rd}} = 82 + 374 \times 0.543 = \mathbf{285\ kN\ m}$$

For $\beta = 0$ the limit is 867 kN, so the reinforcement contributes 524 kN, giving

$$\boldsymbol{M_{j,Rd}} = 82 + 524 \times 0.543 = \mathbf{366\ kN\ m} \qquad \text{(X11.19)}$$

Thus, allowing for unequal moments by assuming $\beta = 1$ has a significant effect on the calculated resistance of the joint. In some cases a more competitive design could be obtained by following EN 1993-1-8, Formula 7.6. This gives separate values for β for the joints for beams 1 and 2. They are functions of $M_{j,Ed,1}$ and $M_{j,Ed,2}$, which depend on the stiffnesses $S_{j,ini}$, so design is iterative.

Check on vertical shear

For the maximum design beam load of 35.7 kN/m on both spans, with $M_{j,Ed} = M_{j,Rd} = 366$ kN m, the vertical shear V_{Ed} in each beam at B is 245 kN, which just exceeds the shear resistance found earlier, 242 kN.

What is the effect of shrinkage? This 'plastic' model for failure of the span is not associated with a particular rotation of the joint. The result $V_{Ed,B} = 245$ kN is not altered by shrinkage. Its effect is to increase the rotation of the joint.

It would probably be found from elastic-plastic global analysis that the vertical shear at B is reduced by the flexibility of the joints. If necessary, two extra M20 bolts could be added in the lower half of each end plate. This would have no effect on the preceding results for resistance to bending.

Comment on the design of the joint for maximum loading, ultimate limit state

The final design for maximum load on both spans shows that the failure mode is simple and ductile: flexure of the T-stub below the top flange of the beam and yielding of the slab reinforcement which contributes 78% of the total resistance to bending. The forces acting on the right-hand beam are shown in Figure 11.9(a). Some further checks are now explained.

Maximum load on span BC with minimum load on span AB

This loading causes rotation of the column at joint B, and its bending resistance causes an abrupt change in the tension in the slab reinforcement at B. Its anchorage within the slab is considered below, as it affects the detailing of transverse reinforcement. This loading also causes maximum shear in the column web, which reduces its resistance to transverse tension or compression. These are long calculations to EN 1993-1-8. The results are

- resistance of column web to compression, $F_{c,wc,Rd} = 590$ kN, previously 867 kN
- resistance of column web to shear, $V_{wp,Rd} = 629$ kN.

Both of these are more than sufficient, so this load arrangement does not govern the design of the joint.

Stiffness of the joints and rotation capacity, ultimate limit state

The initial stiffnesses of these joints are calculated in Example 12.1, using the simpler of the rules for the transformation parameter β given in EN 1993-1-8. They are given in Table 11.1. The joints are numbered 1 and 2, consistent with the notation M_1, M_2, with $M_1 \geq M_2$, used in Table 7.4 of EN 1993-1-8 (not intuitive, as 1 is on the right of 2 in diagrams).

Table 11.1. Initial stiffnesses of joints, $S_{j,ini}$ (kN m/mrad)

	No shear in column web	Joint 1 with shear	Joint 2 with shear
Steel joint	64.3	37.5	37.5
Composite joint, elastic	135	53.5	67.4

In some of the preceding global analyses, it was assumed that the joints would be rigid until their resistance $M_{j,Rd}$ was reached. For any further loading, they would act as hinges. This neglect of the elastic rotation of the joints at moments below $M_{j,Rd}$ leads to over-estimation of the hogging bending moments at the joints. The mid-span moments, therefore, exceed those calculated. This method is safe for the verification of the joints, and is appropriate where there is ample bending resistance at mid-span, as in this example.

There is little inelastic curvature in the regions of sagging moment, so the rotation at the joints is much less than that required for the development of mid-span plastic hinges. The typical joint details given in Couchman and Way (1998), which were used here, were shown to have adequate rotation capacity by a calculation method (Anderson *et al.*, 2000) supported by tests (*clause 11.3.4(3)*) so no further verification of rotation capacity is required. Joint details '*which experience has proved have adequate properties*' (*clause 11.3.4(2)*) may be used without further testing.

Analyses for ultimate limit states

Elastic global analyses are done for beams AB and BC in Figures 11.8 and 11.13, with simple joints at A and C and joints at B, analysed here, to a column EBF which is fixed in position and rotation at E and F. In accordance with Table 7.2 of EN 1993-1-8, the stiffnesses of the joints, S_j, are taken for all loadings as half the values in Table 11.1. Other data are as in this Example 11.1. Each span is assumed to be cracked for 15% of its length adjacent to support B.

Results

For maximum loads on both spans (35.7 kN/m), $M_{j,Ed} = 356$ kN m at both joints, just less than their resistance of 366 kN m. When the load on span AB is reduced to the minimum, 9.42 kN/m, the stiffnesses are different (Table 11.1) and $M_{j,Ed,1} = 212$ kN m, less than the resistance of 285 kN m. For the other joint, $M_{Ed,2} = 100$ kN m.

Anchorage of the force from the reinforcement

The change in tension in the reinforcement is estimated as follows. From the results above with maximum difference between the design ultimate loads, all acting on the composite beams, the difference between the hogging moments is 212 – 100 = 112 kN m.

For the joints, with $\boldsymbol{M_{j,Rd,comp}} = \mathbf{366}$ **kN m**, much higher than 212 kN m, the stresses are found by elastic theory. The second moment of area of the cracked section is $I_y = 451$ m^2 mm^2, and the neutral axis is 33 mm above the centre of the steel beam (Table 8.4). The stress in the reinforcement for a bending moment of 112 kN m is 70 N/mm^2, giving the change in tensile force in the bars as 85 kN; 42.5 kN on each side of the column.

The strut-and-tie model shown in Figure 11.12 requires transverse reinforcement to resist the force 42.5 kN that it shows, and the force depends on the directions chosen for the struts. It is assumed that the transverse reinforcement acts at distance 420 mm from the centre of the column (Figure 11.12). The distance AC is then 516 mm, the transverse tension is 30 kN and the compression in the strut is 52 kN. The width of the strut at the column can be up to about 240 $\sqrt{(2)} = 340$ mm, and the depth of concrete available is 80 mm. This gives the low mean compressive stress of $52/(80 \times 0.340) = 2$ N/mm^2, so the strut is obviously strong enough.

Figure 11.12. Strut-and-tie model for anchorage of unbalanced tension in slab reinforcement, showing the tension in transverse reinforcement

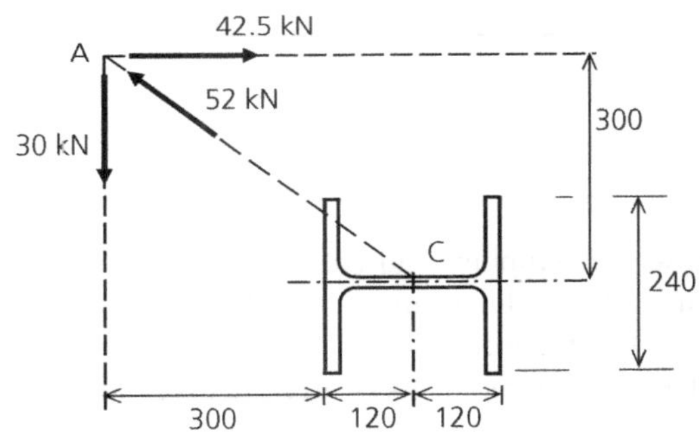

The existing transverse reinforcement (Example 8.6) is 12-mm bars at 200 mm spacing. Those above the beam supporting minimum load will not be fully stressed by transverse bending, and should easily resist an extra 30 kN, which is less than the design resistance of one 12-mm bar.

Serviceability checks

Global analysis

Where semi-rigid joints are used to reduce deflection that is excessive using simple joints, both of the assumptions made in global analysis for ULS (halving $S_{j,ini}$ and assuming $\beta = 1$ for unequal loadings) will be conservative for checking deflections. This can be avoided by using the full stiffness $S_{j,ini}$ where $M_{j,Ed} \leq 2M_{j,Rd}/3$, and finding β iteratively from the results $M_{j,Ed}$. This will reduce the stiffness penalty for unequal loading, which is up to 60% here (Table 11.1). Here, the assumptions made in analysis for ULS will be used, and reviewed if deflections are high.

With unpropped construction, subsequent joint stiffness for serviceability checks depends on whether the construction loading for SLS causes yield within a joint before it becomes composite. In this Example, the weakest link is the beam end plates in the tension zone. They are found (below) not to yield. Otherwise, significant reductions would be needed in all three of the composite stiffnesses in Table 11.1.

Deflection for unpropped construction

Here, the four-member all-steel limited frame shown in Figure 11.13 will first be analysed for the maximum construction loading on both spans, 5.78 kN/m, using $S_j = S_{j,ini}/2 = 32.1$ kN m/mrad. Symbols ϕ_1 and ϕ_2 are the rotations at the joints. The bending moments are shown acting on the column. Elastic analysis of this frame by hand is not simple, as all three rotations are unknown. Here, $w_1 = w_2$ and the column rotation θ is zero from symmetry. Double integration of the curvatures as a function of x and the use of the boundary conditions for the members finds the end moment $M_{Ed,1}$ to be 67 kN m, with mid-span deflection of 13.5 mm. This moment is about 80% of the resistance, so the end plates do not yield.

Figure 11.13. Frame with semi-rigid joints and unequal beam loadings

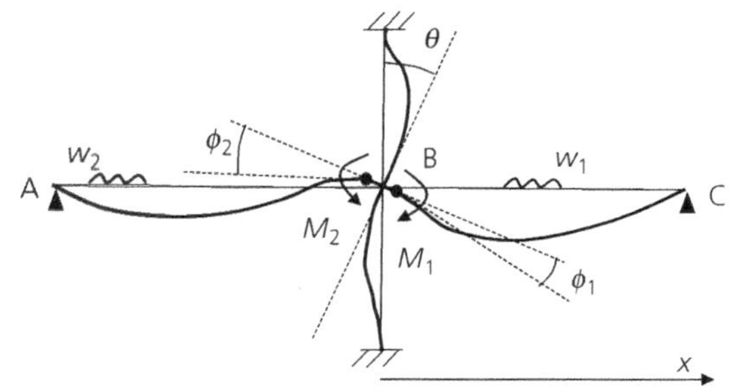

Figure 11.14. Analysis for composite beams, serviceability limit state

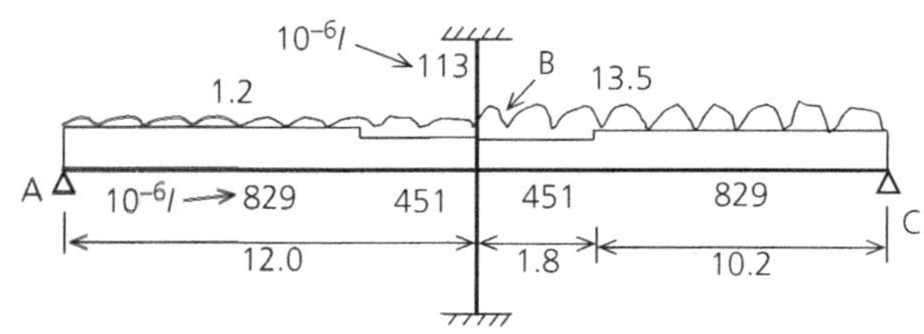

Maximum deflection occurs with maximum load on one span (BC here) and minimum load on the other. The composite phase uses the modular ratio $n = 20.6$. The floor finishes, 1.2 kN/m, act on both spans, and the frequent value of the imposed load, $0.7 \times 17.5 = 12.3$ kN/m, acts on span BC only. This is shown in Figure 11.14, where second moments of area are given in mm^4 units, and loads in kN/m.

Results from analysis of this frame are: end moment $M_{Ed,1} = 77$ kN m; deflection, 17.0 mm.

The total deflection of span BC for unpropped construction is, therefore, $13.5 + 17.0 =$ **30.5 mm**, plus the effect of shrinkage. This total deflection exceeds that for the fully-continuous beam (Table 9.3) by 6 mm. The additional deflection of the composite slabs, relative to that of the beams, is discussed in Example 9.1.

The effect of shrinkage of concrete

Accurate calculation is difficult where cracking occurs and semi-rigid joints are used, but an estimate can be made, as follows. The primary shrinkage deformations for the 24-m length of beam are shown in Figure 8.31(b). Calculations in Example 8.6 found that the radius of curvature is $R = 1167$ m. To allow for shrinkage in 85% of the span, the mean curvature over the whole span is assumed to be $0.85/1167 = 7.28 \times 10^{-4}\ m^{-1}$.

Calculations for span BC with no beam loading, and this sagging curvature added to both spans, give these results

- bending moment at B in span BC: **93 kN m**
- maximum deflection of span BC: **7.4 mm**.

Thus, shrinkage increases deflection to a total of **38 mm**, which is span/316. This result is not typical, as the shrinkage strain assumed in Example 8.7 is exceptionally high.

Cracking of concrete

It was found in Example 9.1 that with rigid joints, crack control to 0.3 mm was easily achieved using 12-mm longitudinal bars at 150 mm spacing. Here, the reinforcement is 16-mm bars at 250 mm spacing. This is also found to be satisfactory, even without allowing for the reduction in the hogging bending moment at support B caused by the flexibility of the semi-rigid joints.

Figure 11.9(d) shows that two bars adjacent to the column are 300 mm apart, which exceeds the limit for bar spacing from *Table 9.2* of EN 1994-1-1. However, the alternative condition in *Table 9.1* is satisfied by a wide margin, because the stress in the bars is well below 160 MPa. It can be concluded that the top reinforcement is unlikely to yield in service, and that wide cracks will not occur.

Review of the classification of the joint

The rules of clause 7.3.2.5(1) of EN 1993-1-8 for classification of beam-to-column joints in braced frames are illustrated in Figure 11.2(a). A joint is rigid if $S_{j,ini}\,L_b/EI_b > 8$ where L_b is the span between column centres, 12 m here. It is a nominal pin if designed using clause C.2 of EN 1993-1-8. If neither applies, it is semi-rigid, and no lower limit to $S_{j,ini}\,L_b/EI_b$ is defined for this class.

For the steel joint during construction, $10^{-6}\,I_b = 337.4\ \text{mm}^4$ from Example 8.6, and from Table 11.1, $S_{j,ini} = 64.3$ kN m/mrad. Hence,

$$S_{j,ini}L_b / EI_b = 64.3 \times 12 \times 1000 / (210 \times 337.4) = 10.9$$

so the joint is classified as rigid but was assumed here to be semi-rigid.

For the composite joint, cracking and creep may be neglected in finding the property I_b of the connected beam (*clause 11.2.3(3)*). Here, from Table 8.4 with $n = 10.3$,

$$10^{-6} I_b = 891\ \text{mm}^4$$

For both spans fully loaded, $S_{j,ini} = 135$ kN m/mrad, and

$$S_{j,ini}L_b / EI_b = 135 \times 12 \times 1000 / (210 \times 891) = 8.66$$

so the joint is just ‘rigid’. However, for imposed load on span BC only, $S_{j,ini,BA} = 53.5$ kN m/mrad, and the joint is obviously semi-rigid under the loading for maximum deflection, as assumed here. Results from an evaluation by this method should not be treated as exact, as many assumptions are made in the calculations for I_b and $S_{j,ini}$.

The calculations above show that the design of semi-rigid joints is complex. There is little evidence that they have been used in practice. However, the industry has become increasingly concerned with embodied carbon and the need to increase the efficiency of materials so their use may increase (BCSA, 2021; Dunant *et al.*, 2018). Design tools will be needed.

Roger P Johnson and Stephen J Hicks
ISBN 978-1-83662-921-4
https://doi.org/10.1108/978-1-83662-920-720253014

CHAPTER 12
Annex A (Informative). Stiffness of joint components in buildings

This chapter corresponds to *Annex A* in EN 1994-1-1, which has the following clauses.

- Use of this annex *Clause A.1*
- Scope and field of application *Clause A.2*
- Stiffness coefficients *Clause A.3*
- Deformation of the shear connection *Clause A.4*

Annex A is needed for the application of *clause 11*. It is 'Informative' because the 'component' approach to the design of steel and composite joints continues to be developed. The UK's National Annex is expected to permit it to be used. Its content is based on the best available research. It is informed by and generally consistent with two reports prepared on behalf of the steel industry (Couchman and Way, 1998; ECCS TC11, 1999).

A.2 Scope and field of application

This Annex supplements the provisions on stiffness of steel joints in clause 8 and Annexes A and B of EN 1993-1-8 (BSI, 2024). It covers conventional beam-to-column joints in regions where the longitudinal slab reinforcement is in tension, and the use of steel contact plates or end plates in compression. As in EN 1993-1-8, stiffness coefficients k_i are determined (***clause A.2(2)***), with dimensions of length, such that when multiplied by Young's elastic modulus for steel, the result is a conventional linear stiffness (force per unit extension or compression). *Clause A.2(2)*

The beam end moments $M_{Ed,1}$ and $M_{Ed,2}$, shown in *Figure A.1* and Figure 11.3, are assumed to have positive values when hogging, and as in EN 1993-1-8, $M_{Ed,1} \geq M_{Ed,2}$. Composite joints with $M_{Ed,2} < 0$ are outside the scope of EN 1994-1-1. Subscripts 1 and 2 are used here also for the properties of the connections on which these moments act.

Stiffness coefficients for all-steel joints are defined in Annex A of EN 1993-1-8. Those relevant to composite joints are listed in Section 11.2 of this guide, with the additional coefficients needed for a composite end-plate joint. There are also coefficients for concrete-encased webs in columns, *clause A.3(3)*.

A.3 Stiffness coefficients

Some background to this clause is available (Huber, 2000). Its scope is limited to those components not defined in EN 1993-1-8. The rules needed from EN 1993-1-8 are explained in Chapter 11 and used in Example 12.1. The coefficients are based on a combination of theory and test results.

During unpropped construction, tension in the end-plate joint in the Example is resisted by the top row of bolts (Figure 11.9), and EN 1993-1-8 applies. After construction, slab reinforcement becomes effective, and further checks use *clause 11* and *Annex A*.

The method of EN 1993-1-8 applies to an end-plate joint with r rows of bolts in tension, at distances $z_{r,1}$ and so on above the compression at bottom-flange level. Their stiffnesses (tension/extension) are $k_{eff,1}$, $k_{eff,2}$ and so on, and they act in parallel.

The joint used in the Examples is similar, with one row of bolts with stiffness k_{eff} and lever arm z_t, and one layer of reinforcement with stiffness $k_{s,red}$ and lever arm z_s, acting in parallel. Their equivalent stiffness coefficient is now derived.

Equivalent lever arm and stiffness

A bending moment M causes rotation θ, extensions $z_t\theta$ of the bolts and $z_s\theta$ of the reinforcement, and a total tensile force $(k_{eff} z_t + k_{s,red} z_s)\,\theta$, acting at an equivalent lever arm z_{eq} above the bottom flange.

Taking moments about bottom-flange level,

$M = (k_{eff} z_t^2 + k_{s,red} z_s^2)\,\theta = (k_{eff} z_t + k_{s,red} z_s)\,\theta\, z_{eq}$, so the equivalent lever arm is

$$z_{eq} = (k_{eff} z_t^2 + k_{s,red}\, z_s^2)/(k_{eff} z_t + k_{s,red} z_s) \qquad \text{(X12.1)}$$

The equivalent stiffness is $k_{eq} = M/\theta = (k_{eff} z_t + k_{s,red} z_s)\, z_{eq}$ (X12.2)

Longitudinal steel reinforcement in tension

Clause A.3.1.1(1) For this component, ***clause A.3.1.1(1)*** and *Table A.1* give formulae for the coefficient $k_{s,r}$ in terms of parameter β (explained under *clause 11.2* in Chapter 11). For reinforcement, a typical stiffness coefficient $k_{s,r}$ in *Table A.1* is

$$k_{s,r} = A_{s,r}/(hf(\beta)) \qquad \text{(X12.3)}$$

where h is the depth of the steel section of the column. For double-sided joints where $M_{Ed,1} > M_{Ed,2}$ with the simplified value $\beta = 1$, $k_{s,r} = A_{s,r}/(3.6h)$ for joint 1 and is infinite for joint 2. The denominator here is the length of reinforcement, the extension of which is assumed to contribute to the flexibility of the joint. When $M_2 = M_1$, this length is $0.5h$. Stiffness $k_{s,red}$ is $k_{s,r}$ modified to allow for slip, as explained in Chapter 11.

Bolted connection in tension

Symbol k_{eff} is used for the row of bolts because it is equivalent to four stiffnesses that act in series: those for the bolts, k_t, the end plate, $k_{t,ep}$, the column flange, $k_{t,fc}$, and the column web, $k_{t,wc}$, each with the same lever arm z_t. This is shown in Figure 11.3 for the conventional model for a beam-to-column joint of the type used in worked examples for this book. The concrete slab is assumed to be fully cracked. These four springs near the level of the steel top flange act in series, so their flexibilities are combined by addition. These are the reciprocals of the stiffnesses, as shown by Formula B.8 in EN 1993-1-8 for the stiffness of a single row of bolts in an end plate

$$k_{eff} = 1/(1/k_t + 1/k_{t,ep} + 1/k_{t,fc} + 1/k_{t,wc}) \qquad \text{(X12.4)}$$

Thus, the flexibility of four components in series exceeds that of any of them. A bolt row in tension for an end-plate joint is less stiff than the bolts. This is unlike the combination of a bolt row with the slab reinforcement in a composite joint, where the stiffnesses act in series.

Other stiffness coefficients

Clause A.3.1.2
Clause A.3.2.1(1)
Clause A.3.2.2(1)

The 'infinite' stiffness of a steel contact plate (***clause A.3.1.2***) simply means that one term in an equation that combines flexibilities is zero. In ***clause A.3.2.1(1)***, the stiffness of a column web panel in shear is reduced below the value in EN 1993-1-8 because the force applied by a contact plate may be more concentrated than would occur with other types of end-plate connection. Similarly, the stiffness for a web in transverse compression has been reduced in ***clause A.3.2.2(1)***, where the value 0.2 replaces 0.7 in EN 1993-1-8.

Concrete encasement

Encasement in concrete increases the stiffness of the column web in shear, which is given for an uncased web in clause A.4.2 of EN 1993-1-8 as

$$k_{wp} = 0.38\, A_{wp}/(\beta z)$$

Clause A.3.3.1(1) The addition to k_{wp} given in ***clause A.3.3.1(1)*** has a similar form

$$k_{wp,c} = 0.06\,(E_{cm}/E_a) b_c h_c/(\beta z)$$

where (E_{cm}/E_a) is the modular ratio and $b_c h_c$ is the area of the column cross-section, as shown in *Figure 8.24(a)*.

For an encased column web in compression (***clause A.3.3.2***), the relationship with the stiffness of the steel web is similar to that for the web in shear. The coefficient in the additional stiffness $k_{c,wc,c}$ is 0.5 for an end plate, but reduces to 0.13 for a contact plate because of the more concentrated force. *Clause A.3.3.2*

A.4 Deformation of the shear connection

The background to the rather complex provisions of ***clause A.4*** first appeared in a report of the European Convention for Constructional Steelwork in 1997. They are based on linear partial-interaction theory for the shear connection. *Formulae A.6* to *A.8* are derived as equations 7.26 and 7.28 in Aribert (1999). *Clause A.4*

Formula A.5 in ***clause A.4(2)*** can be rearranged as follows: $1/(k_{slip}k_{s,r}) = E_s/K_{sc} + 1/k_{s,r}$ *Clause A.4(2)*

This format shows that the flexibility E_s/K_{sc} has been added to that of the reinforcement, $1/k_{s,r}$, to give the combined stiffness, $k_{slip}k_{s,r}$. It also shows that, unlike the k_i's, K_{sc} is a conventional stiffness, force per unit extension. Division by E_s gives its equivalent 'k' value.

The values for the stiffness of a shear connector k_{sc} in ***clause A.4(3)*** assume that the mean load per connector will be around 90% of the design strength and 0.9 $P_{Rk}/\gamma_{VS} \approx 0.7\ P_{Rk}$. This stiffness is, therefore, for ultimate limit state considerations, which is why the elastic slip δ_e derived from this property is used within the ductility categories in *clause 5.4.2.1*. When calculating deformations such as in *clause 9.3.1(5)*, a lower load level within the shear connector is appropriate, and a value of around 0.5 P_{Rk} has been recommended by Hanswille *et al.* (2024). The use of k_{sc} is limited to slabs in which the reduction factor k_t (*clause 8.6.9.2*) is unity. It may not apply, therefore, to pairs of studs in each trough, as used in Example 8.7. *Clause A.4(3)*

In fact, this stiffness, the *connector modulus*, can vary widely. Johnson and Buckby (1986) refer to a range from 60 kN/mm for 16-mm studs to 700 kN/mm for 25-mm square bar connectors. The value 100 kN/mm is of the correct magnitude, but designs that are sensitive to its accuracy should be avoided.

Comments on stiffness of joints

The following example shows that calculation of the stiffness of a joint can be complex. It is unlikely to become a routine process. For the most widely used types of joint there is a need for simplified formulae based on parametric studies, for tables of results, and for research on the extent to which highly approximate stiffnesses would be sufficient for use in practice.

Example 12.1: elastic stiffness of an end-plate joint

The rotational stiffness $S_{j,ini}$, modified to give S_j, is needed for elastic or elastic-plastic global analysis, both for finding the required rotation of the joint, and for checking deflection of the beams. The stiffnesses $S_{j,ini}$ found here are repeated in Table 11.1 for use in Example 11.1, where further comment is given.

Sequence followed and summary of results

This list of steps may be useful for those who do not wish to follow the details.

1. The effective stiffness from the bolts in tension, k_{eff}, is found from its four components.
2. The contribution from the reinforcement depends on the transformation factor β. Its value depends on the end moments at the joint, $M_{Ed,1}$ and $M_{Ed,2}$, which depend on the stiffness of the joints, leading to iteration. Usually the simplified values ($\beta = 0$ or 1) can be used. Here, results for $\beta = 0$ (equal end moments) and 1 (unequal end moments) are found.
3. The stiffnesses of the reinforcement are found: a value $k_{s,r}$ for both joints with $\beta = 0$, and for unequal end moments, values $k_{s,r,1}$ and $k_{s,r,2}$ for the two joints that are below and above $k_{s,r}$. For $M_{Ed,1} = M_{Ed,2}$, with $\beta = 0$, $\mathbf{k_{s,r} = 10.05}$ **mm**.
4. The stiffness of the shear connection, K_{sc}/E_s is found, and from it, the reduction factors k_{slip} that modify $k_{s,r}$ to allow for slip of the shear connection. The result for $\beta = 0$ is $\mathbf{k_{s,red} = 1.75}$ **mm**.
5. The tensile stiffnesses from the bolts and the reinforcement act at different distances from the centre of compression. An equivalent distance z_{eq} is calculated, so that the stiffnesses k_{eff} and $k_{s,red}$ can be added. The result for $\beta = 0$ is $k_{eq} = 4.22$ mm with lever arm $z_{eq} = 0.461$ m.

6. The only stiffness in compression small enough to need inclusion comes from the web of the column. It is found to be $k_{c,wc} = 10.6$ mm. It acts in parallel with the stiffness in tension to give the initial stiffness of the joint, $k_j = 3.02$ mm.
7. Because of the way in which the k values are defined, the initial stiffness of the joint is found in moment/rotation terms from $\boldsymbol{S_{j,ini}} = E_a z_{eq}^2 k_j = \mathbf{135\ kN\ m/mrad}$.
8. Similarly, the stiffnesses for $\beta = 1$ (unequal end moments) are found

$$\boldsymbol{S_{j,ini,1}} = \mathbf{53.5\ kN\ m/mrad};\ \boldsymbol{S_{j,ini,2}} = \mathbf{67.4\ kN\ m/mrad}.$$

9. Further calculations give the stiffness of joints during construction and the effect of end-plate yield.

Calculations

Effective stiffness of bolts in tension

The stiffnesses k_i shown in Figure 11.3 are calculated in turn. Formulae for the steel components are taken from Annex A of EN 1993-1-8. Dimensions are shown in Figures 11.8 to 11.10.

Column web in transverse tension, unstiffened

From Formula A.32 of EN 1993-1-8, $k_{t,wc} = 0.7\ b_{eff,t,wc}\ t_{wc}/d_{wc}$

where d_{wc} is the 'straight portion of the column web'. (It is incorrectly shown as d_c in Formula A.32.) Symbol $b_{eff,t,wc}$ is the smaller of the effective lengths ℓ_{eff} from Table A.2 of EN 1993-1-8, found for the column T-stub. It was found in Example 11.1 to be $2\pi m$, with $m = 23.2$ mm (Figure 11.10), so $b_{eff,t,wc} = 146$ mm. Hence,

$$\boldsymbol{k_{t,wc}} = 0.7 \times 146 \times 10/164 = \mathbf{6.22\ mm}$$

Column flange in bending

From Formula A.35 of EN 1993-1-8, $k_{t,fc} = 0.9\ \ell_{eff}\ (t_{fc}/m)^3$

where $\ell_{eff} = 2\pi m$ as above, and $t_{fc} = 17$ mm (Figure 11.9), so

$$\boldsymbol{k_{t,fc}} = 0.9 \times 146\ (17/23.2)^3 = \mathbf{51.7\ mm}$$

End plate in bending

From Formula A.36 of EN 1993-1-8, $k_{t,ep} = 0.9\ \ell_{eff}\ (t_p/m)^3$

The smallest ℓ_{eff} for the end plate was 213 mm (Example 11.1), and from Figure 11.10, $m = 31.2$ mm. Hence,

$$\boldsymbol{k_{t,ep}} = 0.9 \times 213\ (12/31.2)^3 = \mathbf{10.9\ mm}$$

Bolts in tension

From Formula A.40 of EN 1993-1-8, $k_t = 1.6\ A_s/L_b$, where L_b is the bolt elongation length from Table 8.2 of EN 1993-1-8. This is the grip length plus an allowance for the bolt head and nut, and is 44 mm for these M20 bolts. For a row of two bolts, elastic theory finds the coefficient 1.6 to be 2. The reduction to 1.6 is probably based on calibration from tests. The tensile stress area is 245 mm^2 per bolt. Hence,

$$\boldsymbol{k_t} = 1.6 \times 245/44 = \mathbf{8.91\ mm}$$

Stiffness in tension

Stiffnesses $k_{t,wc}$, $k_{t,fc}$, $k_{t,ep}$ and k_t (bolts and end plate) are in parallel, Figure 11.3. From the results above, for a single row of two bolts,

$$1/k_{eff} = 1/6.22 + 1/51.7 + 1/10.9 + 1/8.91 = 0.384\ \text{mm}^{-1} \tag{X12.5}$$

Hence, $\boldsymbol{k_{eff}} = \mathbf{2.60\ mm}$. This symbol is from EN 1993-1-8. It is not used in EN 1994-1-1.

It is notable that only one of the four terms (for the column flange) is negligible. This stiffness does not depend on β. With lever arm $z_t = 0.383$ m, it acts in parallel with the stiffness $k_{s,red}$ of the reinforcement with the much higher lever arm, $z_s = 0.543$ m.

Slab reinforcement in tension

Stiffness $k_{s,r}$, given in *Table A.1*, is discussed above (Equation X12.3). The area of reinforcement is $A_{s,r} = 1206$ mm^2 and the column width is $h = 240$ mm. There are three cases:

(*a*) for $M_{Ed,1} = M_{Ed,2}$ with $\beta = 0$, then for both joints,

$$\boldsymbol{k_{s,r}} = A_{s,r}/(h/2) = 1206/(0.5 \times 240) = \mathbf{10.05\ mm}$$

(*b*) for $M_{Ed,1} > M_{Ed,2}$ with $\beta = 1$, $k_{s,r} = A_{s,r}/(3.6\ h)$ for the joint in span BC whence,

$$\boldsymbol{k_{s,r,1}} = 1206/(3.6 \times 240) = \mathbf{1.40\ mm}$$

(*c*) for the joint in span AB, $k_{s,r} = A_{s,r}/[h\{(1 - \beta)/2\}]$, so $\boldsymbol{k_{s,r,2}} \rightarrow \infty$.

These results for $k_{s,r}$ will be corrected (Equation X12.6) for the effects of slip.

Deformation of the shear connection and reduced stiffness of reinforcement

Reference should be made to *clause A.3* for K_{sc}/E_s, the stiffness of the shear connection. In its notation

- hogging length of beam, taken as 15% of the span, $L = 0.15 \times 12 = 1.80$ m
- distance of bars above centre of compression, $h_{st} = 543$ mm from Figure 11.9
- height of bars above centroid of steel beam, $d_{sb} = 225 + 100 = 325$ mm from Figure 11.9
- second moment of area of steel beam, $10^{-6}\ I_a = 337$ mm^4
- stiffness of 19-mm headed stud, $k_{sc} = 100$ kN/mm.

From Example 8.7, the number of stud connectors in length L (1.8 m) is $n_{Ls} = 9$

From *Formula A.8*, $\xi = E_a I_a/(d_{sb}{}^2\ E_s\ A_s) = 210 \times 337/(0.325^2 \times 200 \times 1206) = 2.78$

From *Formula A.7*, $\nu = [(1 + \xi)\ n_{Ls}\ k_{sc}\ L\ d_{sh}{}^2/E_a I_a]^{1/2}$

$$= [3.78 \times 9 \times 0.100 \times 1.8 \times 325^2/(210 \times 337)]^{1/2} = 3.02$$

From *Formula A.6*, $K_{sc} = n_{Ls}\ k_{sc}/[\nu - (\nu - 1)\ (h_{st}/d_{sb})/(1 + \xi)]$

$$= 9 \times 100/[3.02 - 2.02(543/325)/3.78] = 423\ \text{kN/mm}$$

Hence, $K_{sc}/E_s = 423/200 = 2.11$ mm. From *Formula A.5*, the reduction factor to be applied to $k_{s,r}$, the stiffness coefficient for the reinforcement, is $k_{slip} = 1/[1 + k_{s,r}/2.11]$

Its values are $\boldsymbol{k_{slip}} = \mathbf{0.174}$, $\boldsymbol{k_{slip,1}} = \mathbf{0.602}$, $\boldsymbol{k_{slip,2}} = \mathbf{0}$.

The reduced value is $k_{s,red} = k_{slip}\ k_{s,r}$. For unequal loading, for span AB, $k_{s,r} \rightarrow \infty$ (above) and $k_{slip,2}$ is zero, so $k_{s,red}$ appears to be indeterminate ($= 0 \times \infty$). However, in the algebraic expression for $k_{slip}\ k_{s,r}$, the product simplifies to $k_{s,red} = K_{s,c}/E_s$, which is consistent with values found for β close to 1.0. The results are

$$\boldsymbol{k_{s,red}} = \mathbf{1.75\ mm},\ \boldsymbol{k_{s,red,1}} = \mathbf{0.84\ mm},\ \boldsymbol{k_{s,red,2}} = \mathbf{2.11\ mm} \qquad \text{(X12.6)}$$

Thus, in this method, allowance for flexibility of the shear connection can severely reduce the contribution of reinforcement to the stiffness of the joint.

Stiffness of joints, for $M_1 = M_2$ ($\beta = 0$) and $M_1 > M_2$ ($\beta = 1$)

From Formula B.9 of EN 1993-1-8 the equivalent (effective) lever arm for the combined stiffness is

$$z_{eq} = (k_{s,red} z_s^2 + k_{eff} z_t^2)/(k_{s,red} z_s + k_{eff} z_t) \qquad \text{(X12.7)}$$

and the combined stiffness (Equation X12.2) is $k_{eq} = (k_{s,red} z_s + k_{eff}\ z_t)/z_{eq}$ (X12.8)

With $z_s = 0.543$ m, $z_t = 0.383$ m, and $k_{s,red}$ from Equation X12.6, the results are

$$z_{eq} = 0.461 \text{ m}, \; z_{eq,1} = 0.433 \text{ m}, \; z_{eq,2} = 0.469 \text{ m} \tag{X12.9}$$

$$\boldsymbol{k_{eq} = 4.22 \text{ mm}, \; k_{eq,1} = 3.35 \text{ mm}, \; k_{eq,2} = 4.57 \text{ mm}} \tag{X12.10}$$

The reinforcing bars increase the non-composite stiffness in tension, k_{eff}, by 62% for $M_{Ed,1} = M_{Ed,2}$. For unequal end moments, the increases are 29% for span 1 and 76% for span 2. Hence, use of the approximate values for β gives unreliable results where the end moments are unequal but similar.

Stiffness in compression

Unlike in tension, the location of the compressive force at the joint is not concentrated at the bolts, but at bottom-flange level. The stiffness is that of the column web, k_{wc}. From Example 11.1, the width of the web, $b_{eff,c,wc}$, is 248 mm, $t_{wc} = 10$ mm, and the effective compressed length, d_{wc}, is 164 mm. From Formula A.26 of EN 1993-1-8,

$$\boldsymbol{k_{c,wc}} = 0.7\, b_{eff,c,wc}\, t_{wc}/d_{wc} = 0.7 \times 248 \times 10/164 = \mathbf{10.6 \text{ mm}}$$

Stiffness of joints

From clause B.4.1(4) of EN 1993-1-8, the initial flexibility for both composite joints at B is given by flexibilities $1/k_{eq}$ and $1/k_{c,wc}$ acting in series (i.e. to be added), with lever arm z_{eq},

$$\boldsymbol{S_{j,ini}} = E_a z_{eq}^2/(1/k_{eq} + 1/k_{c,wc}) = 210 \times 0.461^2/(1/4.22 + 1/10.6) = \mathbf{135 \text{ kN m/mrad}} \tag{X12.11}$$

This is applicable only where $M_{Ed,1} = M_{Ed,2}$. Results for unequal end moments ($\beta = 1$) now follow.

Column web panel in shear, for unequal end moments ($M_1 > M_2$)

The flexibility of the column web panel in shear, $1/k_{wp}$, has the same lever arm, z_{eq}, and is now added to the flexibilities $1/k_{eq}$ and $1/k_{c,wc}$. From Formula A.13 of EN 1993-1-8,

$$k_{wp} = 0.38\, A_{vc}/(\beta z) \tag{X12.12}$$

Stiffness k_{wp} is relevant only for unequal beam loading, for which the approximate transformation parameter β is 1.0. The shear area of the column web is 3324 mm^2 (see Example 11.1). For span BC, $z_{eq,1} = 0.433$ m, so from Equation X12.12,

$$\boldsymbol{k_{wp,1}} = 0.38 \times 3324/433 = \mathbf{2.92 \text{ mm}}$$

From calculation similar to Equation X.12.11,

$$\boldsymbol{S_{j,ini,1}} = E_a z_{eq,1}^2/(1/k_{eq,1} + 1/k_{wp,1} + 1/k_{c,wc}) = 210 \times 0.433^2/(1/3.35 + 1/2.92 + 1/10.6)$$
$$= \mathbf{53.5 \text{ kN m/mrad}}$$

Similarly, $k_{wp,2} = 2.69$ mm, and $\boldsymbol{S_{j,ini,2}} = \mathbf{67.4 \text{ kN m/mrad}}$

Stiffness of joints during unpropped construction

Reinforcement is ineffective, so the initial stiffness of each joint at the construction stage is found with $k_{s,red} = 0$, $z = z_t$, and $k_{eq} = k_{eff} = 2.60$ mm (above). From Formula B.6 in EN 1993-1-8, for equal end moments

$$\boldsymbol{S_{j,ini,steel}} = 210 \times 0.383^2/(1/2.60 + 1/10.6) = \mathbf{64.3 \text{ kN m/mrad}} \tag{X12.13}$$

For unequal end moments, flexibility $1/k_{wp,1}$ must be added. Then,

$$\boldsymbol{S_{j,ini,steel,1}} = \mathbf{37.5 \text{ kN m/mrad}}$$

These results apply only for loading until the resistance of the joint is reached. For further loading, if any, each beam is effectively simply-supported, which increases deflections.

Effect of end-plate yield during unpropped construction

Stiffnesses for serviceability verifications depend on whether SLS loading on the steel beam has caused the end plates to yield during construction. Then, the tensile stiffness represented by k_{eff} becomes ineffective for actions on the composite member, leaving only the top reinforcement. For the stiffness of the composite joint,

$$k_{eff} = 0,\ z_{eq} = z_s,\ k_{eq} = k_{s,red} = 1.75\ \text{mm}$$

and Formula B.6 gives reduced values for $S_{j,ini}$ for use with yielded end plates. For example, for $M_1 = M_2$, where $S_{j,ini,ep}$ was 135 kN m/mrad,

$$\boldsymbol{S_{j,ini,ep}} = E_a\ z_s^2/(1/k_{s,red} + 1/k_{c,wc}) = 210 \times 0.543^2/(1/1.75 + 1/10.6) = \mathbf{93\ kN\ m/mrad} \quad \text{(X12.14)}$$

Comment on the influence of unequal end moments on joint stiffness

Inequality of the bending moment on the two sides of the column increases β and affects both $k_{s,red}$ (for reinforcement and shear connection) and k_{wp} (for web panel shear) and hence the joint stiffness $S_{j,ini}$. Results for a typical example are shown in Figure 12.1.

In this Eurocode method, moment inequality is being used to represent inequality of vertical shear, which means different amounts of slip in the connectors above the joints; more at joint 1, with the larger bending moment, and less at joint 2; hence the different slopes of the curves for $k_{s,red}$. Shear deformation of the column web also reduces the stiffness of the joints on both sides of the column.

Figure 12.1. The effect of unequal end moments on stiffnesses of end-plate joints

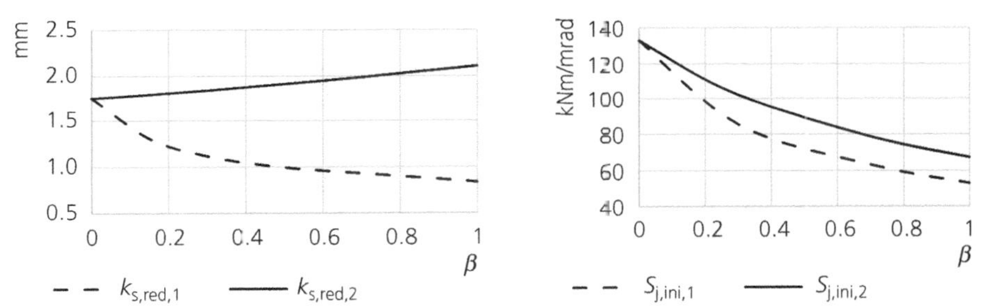

Roger P Johnson and Stephen J Hicks
ISBN 978-1-83662-921-4
https://doi.org/10.1108/978-1-83662-920-720253015

CHAPTER 13
Annex B (Normative). Standard tests

This chapter corresponds to *Annex B* in EN 1994-1-1, which has the following clauses.

B.2 Scope and field of application

Annex B presents standard tests for shear connectors and composite floor slabs. These tests were introduced in the original 1985 edition of Eurocode 4, have been modified in subsequent ENV and EN versions, and were included in the absence of guidelines for European Technical Assessments (ETAs). As the tests in *Annex B* have now been widely adopted, and ETAs can rely on the information presented, this annex, previously Informative, is now Normative.

An objective of standard tests is to provide guidance to designers in situations where calculation models are not sufficient. This commonly occurs for two components of composite structures: shear connectors and profiled steel sheeting. Their existing design rules are based mainly on test data obtained over many decades, including the use of various procedures and types of test specimen for which there was no international standard. The tests determine the shearing deformation caused by longitudinal shear force applied across an interface between the steel and concrete components of a composite member.

There has been some consensus on details of the *m-k* test for resistance of composite slabs to longitudinal shear. However, evidence has accumulated that both this test and the UK's version of the push test for shear connectors (first standardised in 1965) have significant weaknesses. These restrict the development of new products, typically by giving results that are misleading (*m-k* test, now omitted from Eurocode 4), or over-conservative (push test).

The push test

When the specification for an existing test is changed, past practice should, in principle, be re-evaluated. This is one reason why the UK's push test survived so long. The new push test, *clause B.3*, was in drafts of Eurocode 4 in the 1980s, and was based on research work in the preceding decade. Almost all non-commercial push tests since that time have used slabs wider than the 300 mm of the 1965 test, and most tests have followed *clause B.3*.

The new test generally gives higher results and so does not raise questions about past practice. It costs more, but the results are more consistent and relevant to the behaviour of connectors in composite beams and columns. Its purpose is to compare performance of different connectors and materials. It is not possible for it to reproduce conditions in beams, because these vary so widely.

In a beam, there is normally a compressive force between the concrete slab and the steel beam that is little less than the total load, and can be up to 10% of the longitudinal shear. Non-standard push tests that reproduce this give slightly higher shear resistances than standard tests (Hicks *et al.*, 2020). The new tests include the optional use of transverse compression, in *clause B.3.2.3(3)*. Its use is a requirement in some countries.

Tests for shear resistance of composite slabs

The empirical *m-k* test does not distinguish sufficiently between profiles that fail in a ductile manner and those that fail suddenly, and does not exploit the use of end anchorage or the ability of many modern profiles to provide partial shear connection. However, its use was, for several decades, the principal basis for the design of composite slabs for longitudinal shear. Re-testing of the scores of types of composite slabs used in existing structures is impracticable. A method for conversion of existing *m-k* data is available (Johnson, 2004).

Clause B.4, therefore, provides a method of testing that essentially follows the *m-k* procedure, though with a tighter specification relating to the mode of failure and other changes based on recent experience of its use. The test results are used with the partial-interaction design method for composite slabs.

Properties of materials

Ideally, the strengths of the materials in a test specimen should equal the characteristic values specified for the application concerned. This is rarely possible in composite test specimens that include three different materials. It is, therefore, necessary to adjust resistances found by testing, or to limit the range of acceptable measured strengths of a material so that any adjustment would be negligible. Relevant provisions are given within *clauses B.3* and *B.4.*

The influence of cracking of concrete may be assumed to be allowed for in the test procedures. Tests that fully reproduce the effects of shrinkage and creep of concrete are rarely practicable; but these effects can normally be predicted once the behaviour in a short-term test has been established, and have little influence on ultimate strength except in slender composite columns.

B.3 Tests on shear connectors

B.3.1 General

The property of a shear connector that is needed for design is a curve of the type shown in *Figure B.3*, that relates longitudinal slip, δ, to shear force per connector, P. It is difficult to deduce such curves from results of tests on composite beams, mainly because bending resistance is insensitive to the degree of shear connection, as shown by curve CH in Figure 8.4(a).

Clause B.3.1
Clause B.3.1(2)

Testing to ***clause B.3.1*** can be a costly alternative to the use of the design rules in *clause 8* because of the many variables to be studied (***clause B.3.1(2)***). *Annex B* defines test specimens for both static and fatigue loading, but gives test procedures for static loading only.

B.3.2 Testing arrangements

Clause B.3.2.1(1)
Clause B.3.2.1(2)

Clause B.3.2.1(1) limits the scope of 'standard' tests, with composite slabs given as an example where 'specific' tests are applicable. The requirement of ***clause B.3.2.1(2)*** for at least three '*nominally identical*' tests to be done '*for each variable to be investigated*' can be unrealistic, and is discussed below.

Clause B.3.2.2(1)
Clause B.3.2.3

A 'standard' specimen is closely defined in ***clause B.3.2.2(1)***, *Figure B.1* and Figure 13.1. A specimen for 'specific' push tests is defined in more general terms in ***clause B.3.2.3***. It can be inferred from *clause B.3.2.1(1)* that separate 'specific' push tests should be done to find the properties of connectors in columns and in L-beams, which commonly occur at external walls of buildings and adjacent to large internal openings in floors. This is rarely done, although a connector very close to a free edge of a slab is likely to be weaker and have less slip capacity than one in a T-beam. This problem can be avoided by appropriate detailing, and is the reason for the requirements of *clauses 8.6.10.3* on longitudinal splitting and *8.6.10.4* on dimensions of haunches.

The standard push-test specimen is mainly suitable for solid slabs and stud shear connectors small enough for shank failure (the strongest mode) to occur. Its reinforcement might not be stiff enough to ensure this – for example, if 25-mm diameter studs were used.

Clause B.3.2.3(2)

Clause B.3.2.3(2) on specific push tests gives guidance on specimens for other situations, but is not comprehensive, especially where profiled sheeting is present. The rules *a)* to *j)* in *paragraph (2)* provide guidance on the applicability of test results in situations that differ from those in the tests.

Clause B.3.2.3(4)

Clause B.3.2.3(4) refers to the clause that defines the maximum amount of transverse tension below which its adverse effect on P_{Rd} can be ignored. Hence, tests with transverse compressive loading are not applicable where allowance has to be made in design for effects of transverse tension.

Figure 13.1. Test specimen for the standard push test (dimensions in mm) (from BS EN 1994-1-1, by permission of the British Standards Institution)

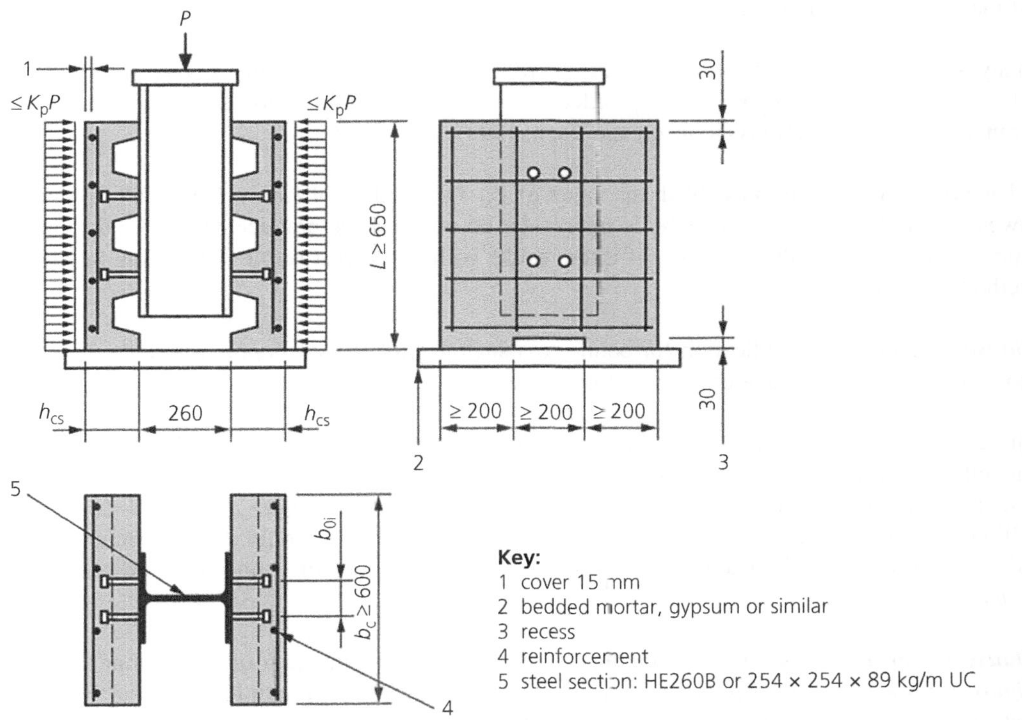

B.3.3 Preparation of specimens

It is more convenient to cast the concrete for push-test specimens with the slabs vertical. The requirement in *paragraph (1)* for horizontal casting (as would be used in practice) leads to more uniform compaction of the concrete around the connectors and reduces the scatter of test results.

B.3.4 Testing procedure

The cyclic loading in *paragraph (1)* ensures that if the connector tested is susceptible to progressive slip, this will become evident. *Paragraph (6)* on relaxation of loading is new. This was usually allowed for in earlier testing, but the method used was often not recorded. The new rule for 5-min delays so lengthens testing that it may be revised.

B.3.5 Test evaluation

The former approximate method for evaluation of test results is now limited to headed studs, ***clause B.3.5(1)***. The reason is that for other shear connectors, the coefficients of variation may not be known from a sufficient number of previous tests to permit the use of 'additional prior knowledge' according to EN 1990. Normally, three tests are conducted on nominally identical specimens to determine the characteristic resistance P_{Rk} for concrete and connector material of specified strengths f_{ck} and f_u, respectively. Let P_m be the mean and P_{min} the lowest of the three measured resistances per connector. If all three results are within 10% of P_m, $P_{Rk} = 0.9\ P_{min}$. Otherwise, more tests are required, with evaluation to Annex D of EN 1990. *Clause B.3.5(1)*

The 'resistance' is usually assumed to be the highest point on the load-slip curve for a test. It is sometimes found that after initial failure there is a higher peak at a slip that can be 15 mm or above. This may not represent behaviour in a beam, so its value should not be used.

Formula B.1 for P_{Rd} in *clause B.3.5(1)* was in the previous EN 1994-1-1. It allows for variation in the strength of the material of the studs. *Formula B.2* for variation of the strength of the concrete is new. It refers to the mean value of the concrete cylinder strength obtained from EN 1992-1-1. This presumably means the value f_{cm} that corresponds to the *specified* characteristic cylinder strength f_{ck}. Its effect is to reduce test results where the concrete is stronger than specified.

Clause B.3.5(3) For all other types of shear connector, ***clause B.3.5(3)*** requires evaluation to Annex D of EN 1990. The use of clause D.7 of EN 1990 for finding a characteristic value from a small number of test results would severely penalise a three-test series that took no account of prior knowledge. This requires an estimate of the expected coefficient of variation, V_x, of mean values from similar test series.

Many of the reported results are from research programmes, with many different types of test specimen – for example, the results for studs in profiled sheeting have been found to be samples from seven different statistical populations (Johnson and Yuan, 1998).

It has not been possible to establish the value of V_x. The method of *clause B.3.5(1)*, of reducing the lowest of the three results by 10%, is mainly based on previous practice. It can be deduced from Annex D of EN 1990 that for a set of three results with the lowest result 10% below the mean, the method implies that $V_x = 11\%$.

Clause B.3.5(4) ***Clause B.3.5(4)*** finds application for connectors such as blocks with hoops, where the block resists most of the shear, and the hoop resists most of the uplift.

Clause B.3.5(5) The classification of a connector as ductile (*clause 8.6.1(5)*), depends partly on its characteristic slip capacity, which is defined in ***clause B.3.5(5)***. From the definition of P_{Rk} (*clause B.3.5(1)*), all three test specimens will have reached a higher load, so the slips δ_u in *Figure B.3* are all taken from the falling branches of the load/slip curves. It follows that a push test should not be terminated as soon as the maximum load is reached. The determination of ductility of connectors is discussed under *clause 5.4.2.1*.

Clause B.3.5(6) ***Clause B.3.5(6)*** defines further properties needed for the classification of connectors for ductility. *Figure B.3* shows a load-slip curve from a particular test, with maximum load P_e. For three or more tests, there will be other curves. The characteristic resistance P_{Rk} is found from them and is the same for each curve. However, the slip s_e at 0.7 P_{Rk} will be slightly different for each curve. The mean of the values s_e should be used to determine stiffness k_{sc} and slip δ_{uk}.

B.4 Testing of composite floor slabs

B.4.1 General

A likely mode of failure of a composite slab is by longitudinal shear, loss of interlock occurring at the steel-concrete interface. Resistance to longitudinal shear is difficult to predict theoretically. The pattern and height of indentations or embossments and the shape of the sheeting profile all have significant effect. There is no established method to calculate this resistance, so the methods of EN 1994-1-1 rely on testing.

Clause B.4.1(1) Tests are needed for each new shape of profiled sheet. They are normally done by or for the manufacturer, who will naturally be concerned to minimise their cost. Their purpose (***clause B.4.1(1)***) is to determine the design longitudinal shear strength $\tau_{u,Rd}$ required for the partial shear connection method. The procedures for verifying resistance to longitudinal shear are given in *clauses 10.7.3* and *10.7.4*. Comments on them are relevant here.

Clause B.4.1(2) The tests also determine whether the shear connection is brittle or ductile (***clause B.4.1(2)***). Sheetings with brittle behaviour are outside the scope of EN 1994-1-1 (*clause 10.7.2(2)*). In view of the purpose of the tests, failure must be in longitudinal shear (*clause B.4.2(6)*).

Number of tests

Clause B.4.1(3)
Clause B.4.1(4) The list of relevant variables in ***clause B.4.1(3)*** and the concessions in ***clause B.4.1(4)*** define the number of tests required and the applicability of their results. One of the variables, slab thickness, not referred to again in *clause B.4*, is discussed below.

The rules of *clause B.4.1(4)* are based on the principle that results may be unconservative if used in design of slabs with thinner sheeting or weaker materials than those in the tests. Strengths of materials measured in practice are almost always higher than specified characteristic values, which is why test results may be used for concretes with specified f_{ck} not less than 80% of the mean strength in the tests.

For the sheeting there are two conditions. The nominal yield strength to be used in practice should not be greater than the *nominal* strength of the sheets in the tests; and the *measured* mean yield strength in the tests ($f_{ypm,t}$) should not exceed 1.25 f_{yp}, where f_{yp} is the nominal yield strength to be used in practice. This is consistent with an entry in Table E.1 of EN 1993-1-1, 'Assumed variability of material properties', for steel grades S235 and S275, which assumes mean measured strengths to be 125% of nominal strengths. For background, see Hicks *et al.* (2014).

Manufacturers are unlikely to provide a value for $f_{ypm,t}$ in their technical literature. There have been cases where their test and evaluation report has been made available to a designer who sought evidence to support the characteristic values.

Nominal yield strengths of sheeting, f_{yp}, are used in design as characteristic values, to be divided by the relevant γ_M (often 1.0) to give the design strength $f_{yp,d}$ used in *clause 10.7.2.*

To illustrate the practical problem that arises from *clause B.4.1*, let us assume that a manufacturer seeks to determine shear resistance for a new profile, as the basis for design data for a range of sheet thicknesses, slab thicknesses and spans, and concrete strengths, with both lightweight and normal-weight concrete. The nominal strength of the new sheeting will not be varied. How many tests are required? It is assumed that the new profile is found to satisfy *clause 10.7.2(3)* on ductility.

From *clause B.4.2(6)*, tests are done in groups of at least four. The variables are now considered in turn, to find the minimum number of tests needed for each one, and hence, the number of tests for a complete investigation.

(*a*) Thickness of sheeting: test the thinnest sheeting. As shear resistance is influenced by local bending of individual plate elements in the sheeting profile, results may not be applied to thinner or significantly weaker sheets, which would be more flexible. (One set of four tests.)
(*b*) Type of sheeting, meaning the profile, including any overlap details, and the specification of embossments and their tolerances. Ensure that the embossments on sheets tested satisfy *clause B.4.3(2)* and standardise the other details. (One test.)
(*c*) Steel grade: test sheeting with the same or a lower *nominal* strength than that of the new profile. Check that its *measured* strength satisfies the conditions above. (One test.)
(*d*) Coating: this should be standardised, if possible, unless it can be shown that a different coating has no effect on the structural properties. (One test.)
(*e*) Density of concrete: test the lowest and highest densities to be used in practice. (Two tests.)
(*f*) Grade of concrete: choose a concrete mix for the tests that is expected to have a mean strength not exceeding 1.25 times the lowest value of f_{ck} to be specified. Test results will be slightly conservative for stronger concretes. (One test.)
(*g*) Slab thickness: test the thinnest and thickest slabs. (Two tests.)
(*h*) Shear span: account is taken of this in the provisions for use of the test results.

This gives a total of $4 \times (1)^5 \times (2)^2 = 16$ tests. It might be possible to limit the scope of the investigation to profiles with normal-density concrete. This reduces the number of tests to 8, which is more realistic. If from other evidence it were believed that the results for parameters (*a*) and (*f*) would be over-conservative for thick sheets and strong concrete, respectively, more tests would be needed.

In practice, manufacturers may prefer to test only the thinnest slabs. In Appendix B of the second edition of this guide, it was shown that a lower longitudinal shear strength is found for thinner slabs in comparison with companion tests on thicker slabs.

B.4.2 Testing arrangement

Loading is applied symmetrically to a simply-supported slab of span L, at points distant $L/4$ from each support (***clause B.4.2(3)***). The failure loading is much heavier than the slabs, so the shear spans L_s (*Figure B.4*) are subjected to almost constant vertical shear. *Clause B.4.2(3)*

The spans of the test specimens should follow ***clause B.4.2(6)***. The choice of the longest span to be tested requires prior evaluation of the bending resistance of the composite section, because a shear strength is too low if determined from a test in which flexural failure occurs. The 'as long as possible' *Clause B.4.2(6)* requirement is to allow for possible reduction of shear resistance with increase of shear span. There may be some arching action in the short-span test, which could increase shear resistance. Its result 'should only be used' for checking ductility, so its result for τ_u is not used when finding the characteristic shear strength, following *clause B.4.5(3)*. Spans are considered in Example 13.1. If possible, the tests should encompass the range of spans required for use in practice.

B.4.3 Preparation of specimens

Clause B.4.3(1)
Clause B.4.3(2)

Clauses B.4.3(1) and ***B.4.3(2)*** are intended to minimise the differences between the profiled sheeting used in the tests and that used in practice. The depth of embossments has been found to have a significant effect on the resistance.

Clause B.4.3(6)

Propped construction increases the longitudinal shear. It is required for the test specimens to enable the results to be used with or without propping (***clause B.4.3(6)***). It also ensures that there is no ponding during concreting, so a constant thickness h_{cs} can be assumed when determining the design values.

Clause B.4.3(8)
Clause B.4.3(9)

From ***clause B.4.3(8)***, the specimens for the determination of concrete strength, defined in ***clause B.4.3(9)***, should be cured under the same conditions as the test slabs. This cannot of course be the curing under water normally used for test cubes and cylinders. When deviation of strength from the mean is significant and the concrete strength is taken as the maximum value (*clause B.4.3(9)*), the applicability of the test results is reduced (*clause B.4.1(4)*).

Clause B.4.3(10)

The test procedure for the strength of the profiled sheeting (***clause B.4.3(10)***) is given elsewhere (BSI, 2019).

B.4.4 Test loading procedure

Clause B.4.4(1)
Clause B.4.4(3)

The initial cyclic loading in a test (***clauses B.4.4(1)*** to ***(3)***) is intended to destroy any chemical bond between the sheeting and the concrete, so that the subsequent test to failure gives a true indication of the long-term resistance to variations of longitudinal shear. The number of cycles required has been reduced from the previous 5000 to 25, which is now considered to be sufficient, and reduces the cost of testing.

B.4.5 Determination of the design values for $\tau_{u,Rd}$

Clause B.4.5(1)

Clause B.4.5(1) refers to the use of 'measured dimensions' in the method for finding $\tau_{u,Rk}$ from tests, but it is not clear whether the total (A_p) or the effective (A_{pe}) cross-sectional area of the sheeting should be used. In principle, the same cross-sectional area should be used in determination of shear resistance and in design calculations using that resistance. However, any error from using the wrong area will be small in comparison with the scatter found in tests for shear resistance, which is allowed for.

This clause refers to the partial-interaction curve shown in *Figure B.5*. It assumes that an undefined slip can occur at the interface between the sheeting and the concrete, and relies on ductile behaviour. This figure is for sagging bending and is constructed for a group of tests on specimens with nominally identical cross-sections as follows.

Clause B.4.5(2)

The measured values of the required dimensions and strengths of materials are determined, and used to calculate the full-interaction plastic moment of resistance of a test specimen, $M_{pl,Rm}$ and the corresponding compressive force in the concrete slab, $N_{c,fm}$. The process given in ***clause B.4.5(2)*** then provides co-ordinates for the partial-interaction curve.

Clause B.4.5(3)

Clause B.4.5(3) gives the method for finding for each test a value of the ultimate shear strength, τ_u. The use of breadth b_c in *Formula B.8* shows that this is an equivalent stress on a horizontal plane, not a shear stress at the surface of the sheeting.

Clause B.4.5(4)

Figure B.5 shows an overhang L_o beyond the shear span L_s, along which slip will occur. This is allowed for in *Formula B.8*, which assumes that the shear strength is uniform along the total length ($L_s + L_o$). In reality, the strength includes a contribution from friction at the interface between the sheeting and the concrete, arising from the transmission of the vertical load across the interface to the support. This effect is assumed to be included in *Formula B.8* for τ_u. A more accurate formula for τ_u is given in ***clause B.4.5(4)***, for use with the alternative method of *clause 10.7.3(2)*, where relevant comment is given.

The rules for calculating the lower 5% fractile value, $\tau_{u,Rk}$ (*clause B.4.5(5)*), are similar to those for the characteristic shear strength of a stud connector, *clause B.3.5(1)*, except that its corrections for strengths of materials are replaced by rules for the applicability of results, *clause B.4.1(4)*.

Example 13.1: tests on composite floor slabs and use of the partial-interaction method

In this example, a value for $\tau_{u,Rd}$ is determined from a set of tests that met the '*basic requirements of the partial connection method*' (*clause 10.7.2(8)*), and were done to the relevant Netherlands standard, RSBV 1990. It will often be found that existing tests, like these, were not fully in accordance with *clause B.3.2(6)*. This Example illustrates potential problems in using test data of this type for the method defined in this Normative annex.

This set of tests on eight simply supported composite slabs has been reported (van Hove, 1991). Figure 10.5 shows the cross-section of the profiled sheeting used. The details of the slabs (given below) were not all the same. The value of $\tau_{u,Rd}$ determined here is used in Example 10.1, Two-span continuous composite slab.

Test specimens and procedure

All of the composite slabs had the same width, 915 mm, which satisfies *clause B.4.3(5)*, and were cast using the same mix of lightweight-aggregate concrete, with propped construction (*clause B.4.3(6)*). (Where a clause number is given without comment, as here, it means that the clause was complied with.) The tests to failure were done at ages between 27 and 43 days, when the strengths of test specimens stored with the slabs (*clause B.4.3(8)*) were as given in Table 13.1. Steel mesh with 6-mm bars at 200 mm spacing was provided in each slab, equivalent to A142 square mesh (*clause B.4.3(7)*).

For specimens 1 to 4, the overall thickness and span were $h_{cs} = 170$ mm and $L = 4500$ mm, where the notation is as in *Figure B.4*. For specimens 5 to 8, $h_{cs} = 120$ mm and $L = 2000$ mm. The distance between the centre-line of each support and the adjacent end of each slab was 100 mm (*clause B.4.2(4)*). The surfaces of the sheeting were not degreased (*clause B.4.3(1)*).

The tensile strength and the yield strength of the sheeting were found from coupons cut from its top and bottom flanges (*clause B.4.3(10)*). The mean values were $f_u = 417$ N/mm^2 and $f_{y,0.2} = 376$ N/mm^2. The stress 376 N/mm^2, measured at the 0.2% proof strain, is significantly higher than the yield strength found in another series of tests (Elliott and Nethercot, 1991), which was 320 N/mm^2. The nominal yield strength for this sheeting was changed from 280 N/mm^2 to 350 N/mm^2 at about this time. These tests comply with the condition of *clause B.4.1(4)* on yield strength only if the nominal yield strength was 350 N/mm^2.

Further properties of the sheeting (van Hove, 1991) in the notation of *Annex B* are $e = 30$ mm and $e_p = 33$ mm, $A_p = 1178$ mm^2 per metre width. Plastic resistance moments of sheeting are often given per metre width (here, 5.65 kN m/m). Calculations to *Annex B* use properties of a test specimen, with $b_c = 0.915$ m, so $M_{pa,m} = 0.915 \times 5.65 = 5.17$ kN m, a correction easily forgotten.

The slabs were tested under four-point loading as shown in Figure 13.2, not in accordance with *clause B.4.2(3)*, which specifies two-point loading. A shear-force diagram for two-point loading that has the same area as the actual shear-force diagram and the same maximum vertical shear gives $L_s = L/4$, as it is in *Annex B*. Crack inducers were provided at distances $L/8$ each side of mid-span, rather than the $L/4$ now specified.

Cyclic loading

Clause B.4.4 specifies upper and lower limits for the cyclic loading. For the long-span slabs the low and high cyclic loads were about 50% below these limits. They were about 20% below for the short-span slabs. Each loading was for 10 000 cycles, far above the 25 cycles now in *Annex B*.

It is concluded that the methods for these tests are close enough to those of *Annex B* for the results to be applicable in practice.

Results of tests to failure

To satisfy *clause 10.7.2(3)* on ductility, the failure load on the specimen should exceed the load at an end slip of 0.1 mm by more than 10%. The results in Table 13.1 show that the lowest excess was 13%. The failure load is taken as less than the maximum load where the latter is reached at a deflection exceeding $L/50$. Here, Table 13.1 shows that the correction would be negligible. All failures are therefore 'ductile'.

Figure 13.2. Loading used in the tests on composite slabs

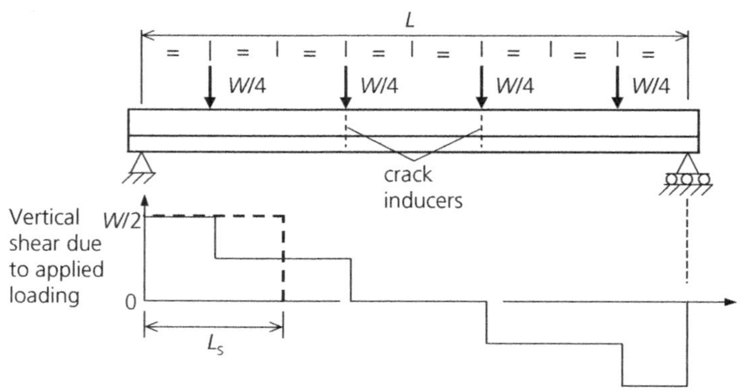

Table 13.1. Results of tests on composite floor slabs

Test No.	f_{cu}	Load at 0.1 mm slip	Load at $\delta = L/50$	Max. load	M_{test}	$M_{test}/M_{pl,Rm}$	η	τ_u
	N/mm²	kN	kN	kN	kN m			N/mm²
1	31.4	63.3	75.5	75.5	44.8	0.840	0.821	0.203
2	30.4	65.3	75.5	75.5	44.8	0.840	0.821	0.203
3	32.1	64.3	73.9	73.9	43.9	0.823	0.803	0.199
4	34.2	66.8	75.4	75.4	44.7	0.838	0.819	0.203
5	35.3	52.2	94.0	94.2	23.8	0.712	0.653	0.340
6	37.4	54.2	90.9	94.0	23.7	0.708	0.648	0.338
7	37.2	52.2	91.7	93.9	23.7	0.708	0.648	0.338
8	36.9	56.2	94.7	95.4	24.1	0.720	0.663	0.345

Clause B.4.2(6) requires a minimum of four tests on specimens of the same overall depth h_{cs}: three with a long shear span, to determine τ_u, and one with a short shear span, but not less than $3h_{cs}$. The tests available satisfy the $3h_{cs}$ condition, and $\tau_{u,Rd}$ can be found from the long-span tests 1 to 4.

The maximum recorded end slips in those tests, only 0.3 mm, occurred at maximum load. The tests were then discontinued, at deflection span/50. It is unlikely that the maximum longitudinal shear was reached. The specimens had the high span/depth ratio of 26.5, and probably failed in flexure. Then, the condition of *clause B.4.2(6)* for a shear span to be 'as long as possible while still providing failure in longitudinal shear' would not be satisfied, and the value $\tau_{u,Rd}$ found would be lower than if a shorter span had been used for tests 1 to 4. This is confirmed below by finding $\tau_{u,Rd}$ also from the short-span tests 5 to 8.

The partial-interaction diagram

Measured cube strengths and maximum loads for the eight tests are given in Table 13.1. For full shear connection, the plastic neutral axes are above the sheeting, so *clause 10.7.2(7)* on the plastic neutral axis applies. Calculations are given for tests 1 to 4. The mean cube strength is 32.0 N/mm². The cylinder strength (using the strength classes in Table 3.5 of EN 1992-1-1) is $f_{cm,t} = 26.4$ N/mm².

The longitudinal force for full interaction in a slab 0.915 m wide is

$$N_{c,fm} = A_p f_{yp} = 1178 \times 0.915 \times 0.376 = 405 \text{ kN} \quad \text{(X13.1)}$$

The stress blocks for full and partial interaction are shown in Figures 13.3(b) and 13.3(c), respectively. For partial interaction, the tensile force in the sheeting is shown as N_c, equal to the compression in the slab, plus $N_{p,b}$, equal to the flexural tension in the sheeting. The two forces $N_{p,b}$ provide the bending resistance $M_{pr,m}$ given in *Formula B.6*.

Figure 13.3. Bending resistance of composite slab for tests 1 to 4: (a) slab and sheeting; (b) stress blocks for full interaction; (c) stress blocks for degree of shear connection η (units: mm and kN)

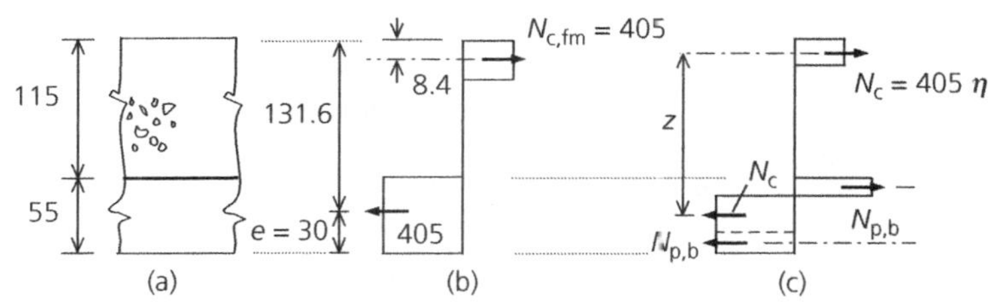

From *Formula B.4* the depth of the concrete stress block is

$$z_{pl} = 405/(0.915 \times 26.4) = 16.8 \text{ mm} \quad \text{(X13.2)}$$

Calculations are done for a range of chosen values of the degree of shear connection; $\eta = 0.6$ is now used here as an example. From *Formula B.5*, the lever arm is

$$z = 170 - 16.8/2 - 33 + 3 \times 0.6 = 130.4 \text{ mm}$$

From *Formula B.6*, $M_{pr,m} = 1.25 \times 5.17 \times 0.4 = 2.59$ kN m

From *Formula B.7*, $M = 405 \times 0.6 \times 0.130 + 2.59 = 34.3$ kN m

Calculation with $\eta = 1$ gives $M = M_{pl,Rm} = 53.3$ kN m

For $\eta = 0.6$, $M/M_{pl,Rm} = 34.3/53.3 = 0.643$, and (0.6, 0.643) is a point on the partial-interaction curve.

The curve found in this way, shown as FG in Figure 13.4, is almost linear. Similar calculations for tests 5 to 8 give curve DE in the figure.

In *clause B.4.5(3)*, the bending moment M_{test} is defined as being '*at the cross-section under the point load*', on the assumption that two-point loading is used in the tests. Here, four-point loading was used. At failure, there was significant slip throughout the length of 3*L*/8 between each inner point load and the nearer support, so for these tests, M_{test} was determined at an inner point load, and L_s was taken as 3*L*/8.

Maximum loads, M_{test}, η and τ_u for tests 1 to 8 are given in Table 13.1. The route ABC shown in Figure 13.4 gives η for test 1 as 0.821. From *Formula B.8*

$$\tau_u = 0.821 \times 405/[0.915(3 \times 4.5/8 + 0.10)]/1000 = 0.203 \ N/\text{mm}^2$$

The values from tests 2 to 4 show no deviation exceeding 10%, so from *clause B.4.5(5)*, τ_u is taken as 10% below the lowest value, and with $\gamma_{VS} = 1.25$,

$$\boldsymbol{\tau_{u,Rd} = 0.9 \times 0.199/1.25 = 0.144 \text{ N/mm}^2}$$

The interaction curve DE for specimens 5 to 8 is slightly higher than FG in Figure 13.4. Its value at $\eta = 0$, $M_{p,a}/M_{pl,Rm}$, is higher because for these thinner slabs (h_{cs} = 120 mm), $M_{pl,Rm}$ is lower, at 33.5 kN m/m. Using the preceding method for these results gives $\tau_{u,Rd}$ = **0.24 N/mm²**. This much higher result confirms the suspicion, noted above, that longitudinal shear failure was not reached in specimens 1–4.

Figure 13.4. Partial-interaction diagram from tests

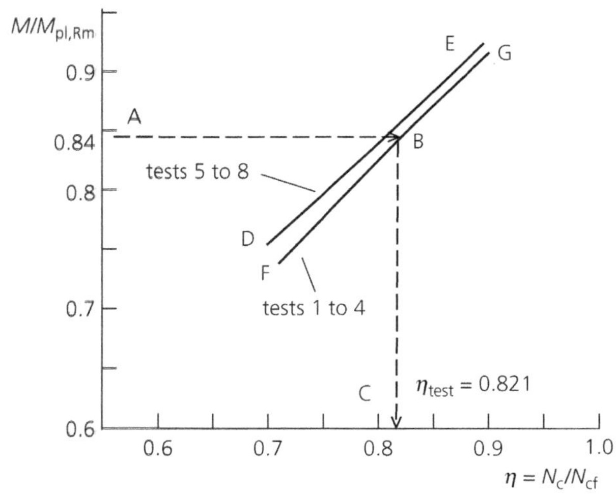

Comments

Where the test data are in accordance with the specification in *Annex B*, determination of $\tau_{u,Rd}$ is straightforward, as values of η can be found by replacing the graphical method (used here for illustration) by direct calculation. However, where tests are being planned, or other data are being used, as here, the work requires understanding of the basis of the provisions of *Annex B*. It is important, and can be difficult, to ensure that longitudinal shear failures occur in the tests.

Roger P Johnson and Stephen J Hicks
ISBN 978-1-83662-921-4
https://doi.org/10.1108/978-1-83662-920-720253016

CHAPTER 14
Annex D (Normative). Composite beams with web openings

Openings in steel webs are the subject of both *Annexes D* and *E*, as explained under 'Scope', below. This chapter corresponds to *Annex D* of EN 1994-1-1, which has the following clauses.

- Use of this annex — *Clause D.1*
- Scope and field of application — *Clause D.2*
- Method of design — *Clause D.3*
- Analysis — *Clause D.4*
- Ultimate limit states — *Clause D.5*
- Serviceability limit states — *Clause D.6*

Openings were being provided in the webs of steel I-beams long before the Eurocodes appeared, but no design rules for them were included in either Eurocode 3 or 4. Beams have two main types of opening: individual holes for the passage of service pipes and ducts, and regularly spaced openings continuous along the span. Those known as 'castellated beams' have a regular repeated pattern of hexagonal openings, and a depth about 50% greater than that of the original I-beam. Their bending resistance and flexural stiffness are greater than those of the original beam. They have been widely used for long lightly-loaded spans, whether service runs are needed or not. Castellated beams were followed in the 1990s by 'cellular' beams, which typically have regularly-spaced circular web openings over most or all of the span (Figure 14.1).

Figure 14.1. Beams with web openings and a composite floor slab (courtesy of Composite Profiles)

Design methods for these beams, that depended on the shape of the openings, were provided by their producers. Both types of beam are within the scope of *Annexes D* and *E*, which give the first standardised set of methods. Background information for *Annex D* is available: Lawson and Hicks (2011).

Symbols in *Annexes D* and *E*

Many symbols that are not in the main lists in *clauses 3.2.1* to *3.2.4* are given separately in *clause 3.5*, so users may have to search in both places. Some symbols are complex, with several subscripts. The following lists provide an alternative presentation of the symbols.

Widely used upper case main symbols

A	area of cross-section
E	modulus of elasticity
F	force (tension or compression)
I	second moment of area
M	bending moment
N	axial force
P	shear force at a connector
V	vertical shear force

Lower case main symbols

a	length or distance
b	width
d	distance (horizontal or vertical in the plane of the cross-section of the member)
e	eccentricity
f	strength of a material
h	height or depth
k	constant
n	number
r	factor
s	spacing
t	thickness
w	deflection; width
z	vertical distance from the centroid of a T-section to the surface of its flange
ρ	reinforcement ratio
ϕ	diameter of reinforcing bar

Subscripts in matched pairs

A, B	at the low moment or high moment side of an opening, respectively, referring to bending moment in the concrete slab
bT, tT	bottom T-section or top T-section, respectively, at an opening
comp, ten	vertical compressive or tensile force, respectively, at a slab-T interface
Ed, Rd	design (factored) action effect or resistance, respectively, of a cross-section of a member
el, pl	elastic or plastic resistance, respectively, of a cross-section
net, gross	composite cross-section at, or not at, an opening

Other subscripts

a	structural steel component of a cross-section of a member
add	additional
av	average of values at centres of adjacent openings
b	bottom, referring to reinforcement in a slab or to a T-section
b	referring to a solid web composite beam, or to local bending
bo	openings along a length of a beam
c	concrete floor or deck slab; centroid

cs	compression in slab developed by shear connectors
eff	effective, for length of an opening for buckling and deflection calculations
eff,0	related to the effective width of a slab for local bending
eff,v	related to the effective width of a slab for local vertical shear
elN	elastic analysis taking account of axial force
f	flange
hm	at the high-moment end of an opening
lim	limit
min	minimum
N	account taken of axial compression in the bending resistance of a member
o	at an opening or the centre of an opening in a steel web
oa	related to the steel component at an opening
oc	in or for the concrete slab above an opening
p	related to the width of a web post between openings
red	reduced
s	reinforcement
ser	serviceability limit state
sl	shear connectors between the centre-line of an opening and its nearer support
slab	related to the vertical shear in a stiff concrete slab above an opening
st	stiff slab
t	tension; transverse
tr	horizontal direction transverse to the axis of the member considered
V	related to the vertical shear at a cross-section
v	related to additional deflection caused by Vierendeel bending at an opening
v	related to shear area of a steel T
vc	Vierendeel bending resistance due to composite action
Vier	Vierendeel bending
w	steel web between openings
wp	web post
y	dimension horizontally transverse to the axis of the member considered

D.2 Scope and field of application

D.2.1 General

Many cellular beams, widely used for roof structures, are not composite. They are designed to EN 1993, 'Design of Steel Structures, Part 1-13: Beams with large web openings' (BS EN 1993-1-13:2024). *Annexes D* and *E* of EN 1994-1-1 are its supplement for composite beams, and give calculation models that are extensions of those for non-composite beams. Their explanation here needs much reference to EN 1993-1-13.

Clause D.2.1(3)

Clause D.2.1(3). To reflect current practice, EN 1993-1-13 and *Annex D* include openings of several shapes: rectangular, circular or similar, hexagonal and sinusoidal. They are characterised by their maximum length a_o, height h_o and eccentricity e_o from mid-depth of the web. For local bending, openings are modelled as 'equivalent rectangles' with length a_{eq} and height h_{eq}. The values for rectangles are their actual dimensions. For circles, $a_{eq} = 0.45a_o$ and $h_{eq} = 0.9h_o$, from EN 1993-1-13, Table 8.3. The rules include the use of horizontal or vertical steel stiffeners at the edges of openings.

To simplify explanations, the scope of this guide is limited to unstiffened rectangular openings with the centre at the same level as the centre-line of the beam (i.e. $e_o = 0$) unless stated otherwise. The steel component is assumed to be a rolled I-section, so that the T-sections above and below an opening have the same cross-section.

Clause D.2.1(4)
Clause D.2.1(5)

Clause D.2.1(4). The scope of *Annex D* is limited to simply supported beams in buildings with the concrete slab above an I-section rolled steel beam or plate girder. From ***clause D.2.1(5)***, the shear connectors must be in ductility category D2 or D3 as defined in EN 1994-1-1. Although *Annex D* is not limited to studs, they are likely to be used because connectors at openings need to have resistance to vertical tension. This resistance, given in *Annex H*, is provided only for studs.

Clause D.2.1(6)

A further limit to the scope of *Annex D*, ***clause D.2.1(6)*** is that the slab above the opening must be 'flexible' rather than 'stiff'. The conditions for the slab to be 'stiff' are given in *clause E.2(3)* – that is, the thickness of the slab must at least equal the depth of the steel T-section below it and there must be more than 0.3% of longitudinal reinforcement. Design, to *Annex E*, is then more complicated. In practice, it may be possible to avoid it – for example, two layers of A192 mesh (7-mm bars) provide less than 0.3% of reinforcement for solid slabs more than 128 mm thick.

D.2.2 Dimensional limits of openings

The limits arise from the range of validity of the ultimate-strength models used in the design methods. These assume plastic behaviour of steel, resistances of reinforced concrete taken from EN 1992-1-1, and shear resistance of connectors based on the standard tests in EN 1994. Local buckling of the web at free edges of openings is allowed for by using equivalent cross-sections from EN 1993. Many of the limits are given in EN 1993-1-13. They are mostly ratios of lengths.

Clause D.2.2(3)

Many design checks can be omitted for openings that are defined by the rules of ***clause D.2.2(3)***, and referred to in this guide as 'small'. They also have to be 'widely spaced'. These are defined in clause 3.1.7 of EN 1993-1-13 as openings 'where web buckling next to the openings is not influenced by adjacent openings'. Widely spaced means not 'closely spaced' (*clause D.2.2(7)*), which is defined in ***clauses D.2.2(4)*** and ***(5)***.

Clause D.2.2(4)
Clause D.2.2(5)

Additional design checks are required for web posts between closely spaced openings in *clause D.4.4* on analysis and in *clause D.5.2*. *Clause E.2(5)* excludes stiff slabs with $s_o/a_{eq} < 2$ from the scope of *Annex E*, and hence from Eurocode 4, for distances s_o between the centres of adjacent openings.

Figure 14.2 provides for rectangular openings an overview of the limits to dimensions and scope, and shows their interaction with the distinctions between stiff and flexible slabs and between widely spaced and closely spaced openings. These depend mainly on ratios between three dimensions of a pair of openings (Figure 14.3): their length a_o, height h_o and the width w_p of the web post that separates them

Figure 14.2. Relationships between definitions and limits to dimensions for unstiffened rectangular openings with $e_o = 0$

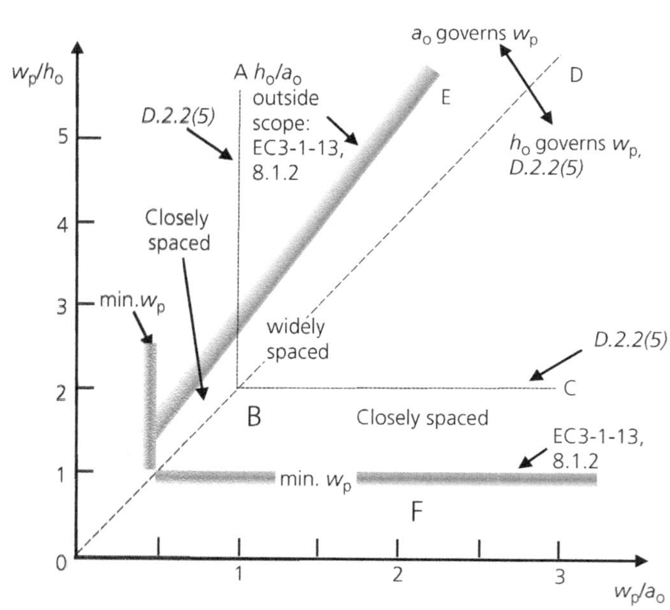

(this is symbol s_o in EN 1993-1-13). The axes of the chart are $x = w_p/a_o$ and $y = w_p/h_o$, so a point (x,y) represents an opening with the proportions a_o/h_o as this equals y/x.

Rules for w_p depend on which of a_o and $2h_o$ is the greater. In Figure 14.2, a_o exceeds $2h_o$ above line OBD. The rule for minimum width of a web post (EN 1993-1-13, Table 8.1) appears in the Figure as the two lines labelled 'min. w_p'. *Clause D.2.2(5)* gives lines AB and BC in Figure 14.2 as the boundary between close and wide spacing. Table 8.1 of EN 1993-1-13 requires a_o not to exceed $2.5h_o$, so openings that plot above line OE are outside the scope of *Annex D*.

D.3 Method of design

D.3.1 General

Clause D.3.1(1) refers to *Vierendeel* bending. A Vierendeel frame or truss is one where the vertical shear is resisted mainly by the bending strength of the joints, unlike a pin-jointed truss, where the joints are assumed to have no resistance to bending.

Figure D.1 shows all of the failure modes treated in *Annex D*. Figure 14.3 shows regions liable to crush, crack, yield or buckle. At the centre of an opening, the cross-section is analysed as three separate members: the upper and lower steel T-sections and the concrete slab, Figure 14.4(a). The web post is another 'member', assumed to be connected by rigid joints to the two T's, Figure 14.4(b). The diagrams here all assume that the nearer support is to the left and that the global shear acts clockwise on the length of beam shown, and show a concrete slab that is not composite.

Figure 14.3. Failure modes adjacent to web openings

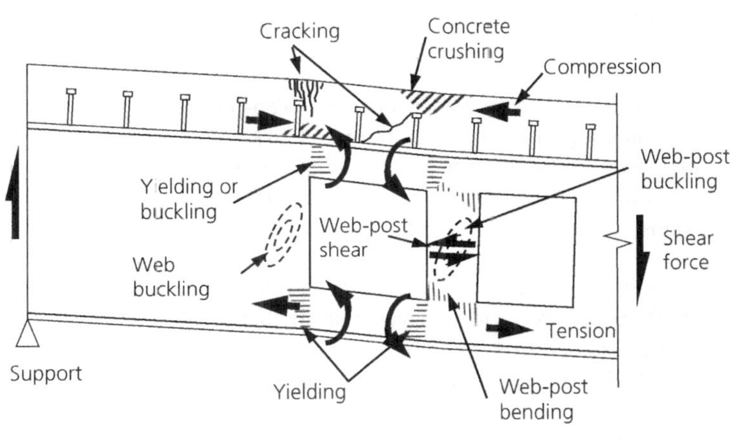

Figure 14.4. Assumed true hinges and plastic hinges (a) at an opening and (b) adjacent to openings

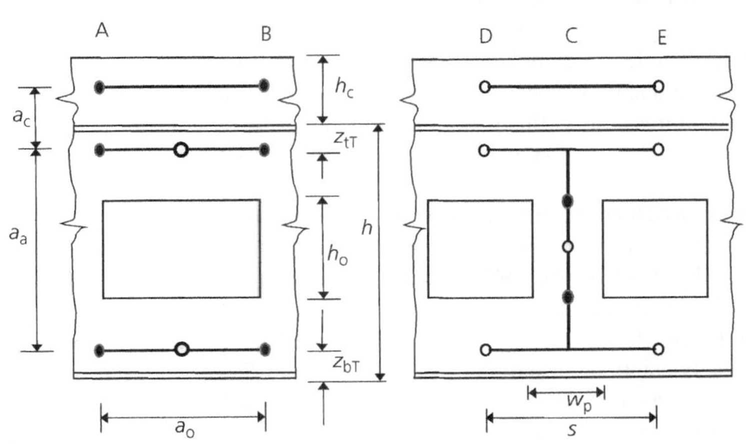

The local action effects in these members of a given beam will, for a given pattern of loading, have different ratios to their resistances. The basis of design to *Annex D* is that as load increases, failure of a cross-section of the beam will not occur until either

(*a*) there are sufficient plastic hinges in the members to create a mechanism, or
(*b*) sufficient resistances have been reached, after permitted redistribution of action effects between the members, for the cross-section to fail.

Simplified models are used for the possible types of local failure. Most are shown as plastic hinges in Figure 14.4. They are assumed to occur in horizontal members at the two ends of an opening, and at cross-sections at the ends of the web post. Each member resists axial force N_{Ed}, bending moments M_{Ed} and transverse shear forces V_{Ed}. Resistance to longitudinal shear between the slab and the adjacent T-section must also be verified.

The members are represented in Figure 14.4 by the lines of action of their axial forces. These are at the centres of area of the T-sections and at a level in the slab that is discussed later.

The global bending moment and vertical shear both vary along the lengths a_o and s shown. It is assumed that this variation is small enough for the action effects at vertical cross-sections A and B both to correspond to the global action effects half-way between them, and at section C to the mean of the global effects assumed for the adjacent openings.

Then, the member cross-sections shown by the open circles in Figure 14.4 can be assumed to resist no global bending moment ('true hinges'), and local analyses become possible. In effect, the local loading on the relevant length of beam is being ignored. It follows that where a significant point load (e.g. from a secondary beam to a primary beam) acts on a beam with regular openings, the local opening may need a steel infill plate.

Clause D.3.1(2)

An opening obviously weakens the beam where it occurs. Web-post regions, Figure 14.4(b), are obviously less critical, with buckling at a free edge the main risk (*clause D.4.2.4*). The resistance of a web post to longitudinal shear in the beam has to compensate for the loss caused by the opening, and narrow web posts can fail in this way or in bending. ***Clause D.3.1(2)*** refers to *Annex E* for additional failure modes in stiff slabs.

The plastic hinge locations shown in Figure 14.4 are sections of maximum bending moment ('Vierendeel bending'), combined with axial force and transverse shear that is constant along the member in the bottom T. The level of the assumed true hinge within the web post follows from the value for $M_{wp,Ed}$ (shown in Figure 14.5), which can be chosen by the designer.

Clause D.3.1(4)

The resistances of the steel cross-sections are found from the rules for combined *M-N-V* action effects in EN 1993-1-13. The alternative methods referred to in ***clause D.3.1(4)*** are for openings of circular and sinusoidal shape. *Annex D* gives no guidance on how to apply them where there is interaction with a concrete slab.

D.3.2 Equivalent length of openings

Clause D.3.2(2)

There is comment on equivalent lengths a_{eq} in comment on *clause D.2.1(3)*. Effective lengths a_{eff} (***clause D.3.2(2)***) are needed in calculations for buckling and deflections, and affect the slenderness class of each steel component. A T-section can be in Class 3 for global compression and in Class 2 for Vierendeel (local) compression, as this is not uniform along the length of the edge above an opening. For rectangles, $a_{eff} = a_{eq} = a_o$ and symbol a_o is used in this book. For other shapes, $a_{eff} > a_{eq}$.

D.3.3 Effective width of concrete slab at an opening

Clause D.3.3(1)

The four effective widths of a concrete slab at an opening can differ widely, and should not be confused. For global bending of the beam, ***clause D.3.3(1)*** refers to b_{eff} as given in *clause 7.4.1.2*. It is not affected by the opening and is usually much greater than the others. From *clause D.5.1.2(7)* it may also be used for Vierendeel (local) bending, but determined 'at the high-moment end of the opening'. This means that it depends on the location of the opening, but use of the variation shown in *Figure 7.1* is optional. Here, the width is taken at the centre of the opening nearest to a support, which is conservative for other openings. The resistance of the slab is in fact governed by the shear connection. Changes of b_{eff} have a tiny effect on resistances.

Clause D.3.3(2) defines the effective width for vertical shear in a 'flexible' slab, $b_{eff,V}$, in terms of the thickness of the slab and the flange width of the adjacent T-section. *Clause D.3.3(2)*

The other effective widths are for a stiff slab. The symbols used are $b_{eff,b}$ for local bending and $b_{eff,V}$ for vertical shear. They are defined in *clauses E.4(1)* and *(2)*, where comment is given.

D.4 Analysis

D.4.1 General

The 'internal forces' are the action effects in the 'members' explained above, which in total are in equilibrium with the external action effects of global bending moment and vertical shear. From EN 1993-1-13, clause 8.1.1(3), any axial force in the steel section must not exceed $N_{a,pl,Rd}/50$ at the centre of the largest opening and should be shared between the two T-sections.

The use of 'may' in ***clause D.4.1(2)*** shows that the use of *clauses D.4.2* to *D.4.4* is optional. This choice is given so that redistribution of action effects between the members can be assumed in a way that makes best use of their resistances. *Clause D.4.1(2)*

D.4.2 Axial forces at the mid-length of openings

The vertical spacings of the axes of the members, the 'lever arms', are shown in Figure 14.5 as a_c and a_a (symbols not used in Eurocode 4). From *Figure D.2*, they are

$$a_c = d_c + z_{tT}$$

where z_{tT} is the distance of the centre of area of the top T below the top of the T. The corresponding distance for the bottom T is z_{bT} so that

$$a_a = h_a - z_{tT} - z_{bT}$$

The axial forces resist the global bending moment $M_{a,Ed}$ at the centre of the opening. The tension in the bottom T is lowest where the resistance of the slab, $N_{oc,Rd}$, is not less than $M_{a,Ed}/(a_c + a_a)$, because then, compression in the top T can be assumed to be zero, and the forces are as in *Formula D.2* of ***clause D.4.2(2)***. Resistance $N_{oc,Rd}$ is defined in *clause D.5.1.2(6)*. Where the slab is stiff, *Annex E* could give a lower value. *Clause D.4.2(2)*

Where the slab needs help from the top T, it is assumed that $N_{oc,Ed} = N_{oc,Rd}$. *Formula D.6* then gives the tension in the bottom T higher than the value from *Formula D.2*, and *Formula D.9* gives the compression in the top T. The resistance of the bottom T to tension may be reduced by coexisting bending moment and conversely. In *Annex D*, axial force takes precedence and may lead to reduction in the bending resistance. If the bottom T tension exceeds its resistance, a new design is needed.

Calculation of these forces begins with three interrelated unknown values: $V_{oc,Ed}$, d_c and z_c. A method is given in Example 14.1.

D.4.3 Distribution of shear forces

Freedom is given to share the global vertical shear at the centre of an opening, $V_{o,Ed}$, between the slab and the two T's, subject to their shear resistances, by the use of 'may' in this clause. It can be difficult to find the resistances of the T's before their shares of $V_{o,Ed}$ are known, so ***clause D.4.3(3)*** suggests that the slab is assumed to resist as much shear as possible. Then, $V_{cc,Ed} = V_{oc,Rd}$, the shear resistance of the slab, which is defined in *clause D.5.1.3*, and discussed with that clause. *Note 2* of *clause D.4.3(3)* also allows the shear $V_{c,Ed}$ to be taken as zero. This is a useful simplification for mid-span regions where vertical shear is low. *Clause D.4.3(3)*

Clause D.4.3(4) says that the remaining vertical shear, $V_{oa,Ed} = V_{o,Ed} - V_{oc,Rd}$, may be shared between the T's according to their plastic shear resistances from EN 1993-1-1. Where the web post is expected to be the weakest local member, a better strategy, ***clause D.4.3(5)***, may be to choose the shear resisted by the bottom T, $V_{bt,Ed}$, to be such that the web-post bending moment $M_{wp,Ed}$ is minimised. Further guidance is provided in Section D.4.4. *Clause D.4.3(4)* *Clause D.4.3(5)*

Figure 14.5. Action effects for analysis of a web post

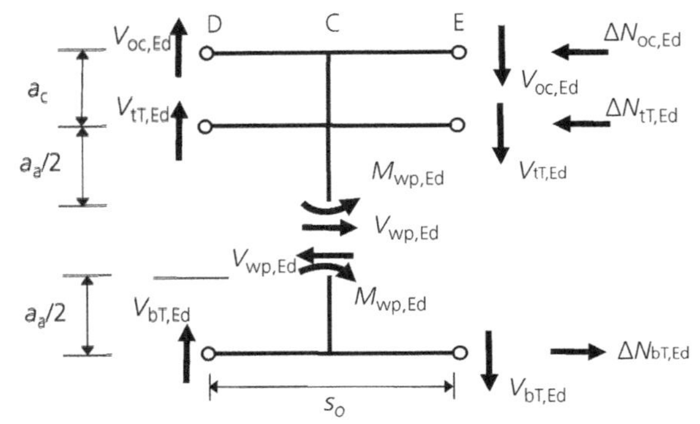

D.4.4 Forces and moments in web posts between closely spaced openings

Figure 14.5 shows the action effects at the assumed hinge positions and at the mid-depth of the web post, and the assumed shares $V_{oc,Ed}$, $V_{tT,Ed}$ and $V_{bT,Ed}$ of the design vertical shear at cross-section C, $V_{av,Ed}$. These are slightly different from the actual values at cross-section C, because their sum is $V_{av,Ed}$, not $V_{o,Ed}$.

The bending moment at mid-depth of the web post, $M_{wp,Ed}$, is the only unknown value at this stage. The global compression in the slab is lower at section D than at E. This change, caused by the shear connection, is shown as the force $\Delta N_{oc,Ed}$. The part of the force $N_{oc,Ed}$ that remains constant along length DE is not shown. Similarly, the changes in the other members are shown as $\Delta N_{tT,Ed}$ and $\Delta N_{bT,Ed}$. For equilibrium of the bottom T, the shear force in the web post, $V_{wp,Ed}$, equals $\Delta N_{bT,Ed}$, and *Annex D* uses both symbols.

If the change $\Delta N_{tT,Ed}$ in the top T has been chosen to be zero (as explained above), the equation of moments for the whole diagram gives

$$V_{wp,Ed} = \Delta N_{bT,Ed} = \Delta N_{oc,Ed} = V_{av,Ed}\ s_o/(a_a + a_c) \qquad (D.14)$$

If $\Delta N_{tT,Ed} > 0$, taking moments about the level of the top T gives

$$V_{wp,Ed} a_a = V_{av,Ed} s_o - \Delta N_{oc,Ed}\ a_c \qquad (D.15)$$

Clause D.4.4(1) The second equation gives a higher value for $V_{wp,Ed}$ than the first, except when $\Delta N_{tT,Ed} = 0$. Then, they give the same value, so the advice in ***clause D.4.4(1)*** to use the larger value is always correct.

Clause D.4.4(2) ***Clause D.4.4(2)*** gives the bending moment in the web post at mid-depth of the opening. In this simplified explanation, $e_o = 0$ and $z_t = z_b$, so *Formula D.16* becomes

$$M_{wp,Ed} = (V_{av,Ed} - 2V_{bT,Ed})s_o/2 - \Delta N_{oc,Ed}\ a_c/2 \qquad (X14.1)$$

It is derived from the equations of rotational equilibrium of the separated parts of the structure in Figure 14.5, assuming that the change in M_{Ed} along the length s_o is $V_{av,Ed}\ s_o$.

By changing $V_{bT,Ed}$, the equation can be used to reduce $|M_{wp,Ed}|$, provided that the T's have sufficient shear resistance. EN 1993-1-13 requires that $M_{wp,Ed}$ should not exceed the *elastic* resistance of this cross-section, which is $M_{wp,el,Rd} = t_w w_p^2 f_y/6$.

Symbol $\Delta N_{oc,Ed}$ appears only in *clause D.4.4* and *Figure D.4*. Its value depends on whether $V_{wp,Ed}$ is found from *Formula D.14* or *Formula D.15*. If *D.14* governs, then $\Delta N_{tT,Ed} = 0$, and then $\Delta N_{oc,Ed} = V_{wp,Ed}$. If the moment $M_{wp,Ed}$ is high enough to need resistance to compression from the top T as well as from the slab, then *D.15* governs, and $\Delta N_{oc,Ed} = \Delta N_{oc,Rd}$.

A note in *clause D.4.4(2)* that refers to *clauses D.2.2(4)* and *(5)* is a reminder that *clause D.3.4* is applicable only to closely spaced openings.

D.4.5 Classification for global bending

The steel webs that edge unstiffened openings are relatively thin. Those above an opening are subjected to almost uniform global compression, greatest in regions of low vertical shear. Design should avoid a free edge of a web in Class 4, because elastic analysis is not consistent with the ultimate-strength basis for *Annex D*. Class 3 webs are discussed below.

Clause D.4.5(1) refers to the classification rules in *clause 7.5.2*, which in turn refer to the well-known rules in EN 1993-1-1. ***Clause D.4.5(2)*** refers to EN 1993-1-13. Its clause 7.4 is relevant. The web at an opening should be treated as an outstand flange in uniform compression. The method depends on the slenderness of the web outstand, $d_t/t_w\varepsilon$, where d_t is the depth of the outstand, t_w its thickness, and ε is the usual Eurocode 3 factor that allows for the yield strength of the steel. If for the T, $10 < d_t/t_w\varepsilon \leq 14$, it is in Class 3. Where the flanges are in Class 1 or 2, which is usual, the web can to be treated as in Class 2, by replacing it by an equivalent web with depth reduced to the Class 2 limit, $d_t/t_w\varepsilon = 10$. The resistances of this equivalent T are found by plastic theory. *Clause D.4.5(1)* *Clause D.4.5(2)*

D.4.6 Classification for Vierendeel bending

Figure 14.6 shows the Vierendeel action effects above an opening, on members equivalent to those shown in Figure 14.4, with the addition of member CD to represent the shear connection. The position of its point of contraflexure is not shown, because it is only necessary to find it when the slab is stiff and within the scope of *Annex E*.

Figure 14.6. Action effects above an opening

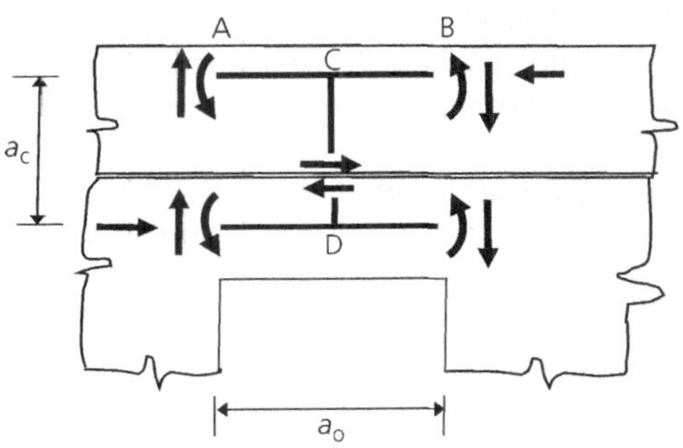

The local bending moments at the ends of the top T are of opposite sign, so the stress along its free edge is not uniform. For the rules that enable a Class 3 web to be treated as Class 2, reference is made to EN 1993-1-13, clause 7.5, which is more liberal than its clause 7.4. Its method takes no account of coexisting global compression. For $d_t/t_w\varepsilon \leq 14$, there is no reduction of cross-section until $a_{eff}/(t_w\varepsilon)$ exceeds about 46, where a_{eff} is the effective length of the opening ($= a_o$ for a rectangle). For wider openings, the qualifying value of $d_t/t_w\varepsilon$ reduces below 14 – for example, to 11.8 when $a_{eff}/t_w\varepsilon = 60$.

D.5 Ultimate limit states

D.5.1 Design rules

D.5.1.1 General

Comment on the clauses listed here is given in the relevant sections of this guide. ***Clause D.5.1.1(1)*** refers only to large web openings. The less extensive checks sufficient for beams with 'small' web openings are given in *clause D.2.2(3)*. *Clause D.5.1.1(1)*

Clause D.5.1.1(2) is an important reminder to follow the rules of EN 1993-1-13 wherever they are applicable. *Annex D* usually gives the cross-reference; for example, in its definition of $M_{NV,bT,Rd}$ in *clause D.5.1.4(2)*. *Clause D.5.1.1(2)*

In composite slabs, the ribs of the sheeting can prevent the use of longitudinal spacings of connectors less than about 0.3 m. ***Clause D.5.1.1(3)*** then prevents the use of composite Vierendeel action for rectangular openings shorter than about 0.6 m. *Clause D.5.1.1(3)*

Clause D.5.1.1(5) ***Clause D.5.1.1(5)*** appears to apply to all non-circular openings whether widely spaced or not: '*...or for other shapes of openings...*'. It refers to EN 1994-1-1, 8.6.3.3, where *Formula 8.12* enables the required minimum degree of shear connection to be reduced by a factor ρ_m^2 where the maximum design bending moment M_{Ed} is less than the resistance M_{Rd}, subject to some conditions.

Here, three further limits to ρ_m are added, involving the resistances N_{Rd} of the T-sections and the slab. These can depend on coexisting vertical shear and on the layout of the shear connectors, so this check cannot be made until late in the design process. If it then requires an increase in the shear connection, that is simple in a solid slab, but may be impracticable where profiled sheeting is present.

D.5.1.2 Global bending resistance

The resistances $N_{bT,Rd}$, $N_{tT,Rd}$ and $N_{oc,Rd}$ needed for applying *clause D.5.1.1(5)* are defined here.
Clause D.5.1.2(5) In ***clause D.5.1.2(5)***, 'moment-shear interaction' refers to the reduction in resistances M_{Rd} in T-sections (and in the web post) where the transverse shear V_{Ed} exceeds $V_{Rd}/2$. Further comment is given in *clause D.5.1.3*.

The share of the global axial compression above the centre of an opening that is resisted by the slab, $N_{oc,Ed}$, is transferred to it by the shear connectors between this cross-section and the nearer support of
Clause D.5.1.2(6) the beam. Failure of these at force $N_{oc,Rd}$ sets an upper limit to $N_{oc,Ed}$ (***clause D.5.1.2(6)***). Resistance $N_{oc,Rd}$ is also limited by the compressive strength of the slab. This check is made at the high-moment
Clause D.5.1.2(7) end of the opening, ***clause D.5.1.2(7)***. Its *Formula D.24* takes account of the small increase in $N_{oc,Ed}$ at that cross-section.

D.5.1.3 Shear resistance of composite beam at web opening

Clause D.5.1.3(2) The shear areas $A_{v,T}$ of the T's in ***clause D.5.1.3(2)*** are defined in *clause D.4.3(4)* by reference to EN 1993-1-1, 8.2.6(3). These areas consist of the web, the corner fillets and a small part of the flange.

Clause D.5.1.3(3) The shear resistance of the slab, $V_{oc,Rd}$, is the lesser of two values. The first is given by ***clause D.5.1.3(3)***, for flexible slabs only. It assumes shear failure of the studs directly above the opening. The height of the centre of the compression in the slab above the interface with the top T-section is d_c at the centre of the opening. In *Formula D.28* it is reduced because the check is at the higher-moment end of the opening, where the compressive force is greater.

In this clause n_r is the number of connectors per transverse row above the opening, as defined in *clause D.4.4*. This is inconsistent with the definition of n_r in *clause 3.2.2* ('number of shear stud connectors in one rib'), which does not apply here.

Clause D.5.1.3(4) The second limit to $V_{oc,Rd}$ is $V_{c,Rd}$, from ***clause D.5.1.3(4)***, which says that the resistance 'may be obtained' from EN 1992-1-1 in either of two ways. The second one is the quite complex strut-and-tie model given in its clause 8.5. That method allows for the global compression $N_{c,Ed}$, so its result depends on the position of the opening in the span.

In the T's, the reduction in resistance to bending caused by high transverse shear should be allowed
Clause D.5.1.3(5) for, ***clause D.5.1.3(5)***. It refers to EN 1993-1-1, 8.2.10(3), which uses symbol V_{Ed} for the total shear to be resisted. For a composite beam, V_{Ed} should be replaced by $V_{iT,Ed}$ for the T concerned. Where this exceeds half of the plastic resistance $V_{iT,Rd}$ of the T, the yield strength of the shear area is reduced by the factor $1 - \rho$, where

$$\rho = [2V_{iT,Ed}/V_{iT,Rd} - 1]^2$$

Alternatively, the effective thickness of the web may be reduced by the same factor, which can be the simpler method.

D.5.1.4 Shear resistance for Vierendeel bending

At this stage, a designer will have chosen values for the action effects $N_{bT,Ed}$ and $V_{bT,Ed}$, constant along the bottom T. Hence, the bending moments at its ends are $|M_{bT,Ed}| = V_{bT,Ed}\, a_o/2$. One would expect *Annex D* to require the check that $M_{bT,Ed}$ does not exceed the corresponding resistance. This is $M_{NV,bT,Rd}$, which can be found using EN 1993-1-13. The $_{NV}$ in this symbol means that account has been taken of the reduction in plastic resistance to bending caused by coexisting axial force and transverse shear.

For $M_{pl,V,Rd}$, the plastic resistance to bending after reduction for high shear, see *clause D.5.1.3(5)*. The reduction for axial force, given in EN 1993-1-13, clause 8.4 and applicable for all values of axial force, is

$$M_{NV,Rd} = M_{pl,V,Rd}(1-(N_{Ed}/N_{pl,Rd})^2) \quad (X14.2)$$

In place of a check on $M_{bT,Ed}$, ***clause D.5.1.4(2)*** gives *Formula D.30* for a resistance $V_{Vier,Rd}$, which must equal or exceed $V_{bT,Ed}$: Clause D.5.1.4(2)

$$V_{Vier,Rd} = (2\,(M_{NV,bT,Rd} + M_{NV,tT,Rd}) + k_a M_{vc,Rd})/a_{eq} \quad (D.30)$$

This assumes that resistances $M_{NV,Rd}$ have been reached at both ends of both T's (the plastic hinges shown in Figure 14.4) and that either the connectors above the opening have failed, or that the shear resistance of the slab, $V_{c,Rd}$ has been reached. This last contribution is subject to the reduction factor k_a, explained in this clause.

This limit to $V_{o,Ed}$, *Formula D.29*, assumes that five resistances have been reached at once. In practice, this assumed mode of failure is unlikely, as it is rare for the top T to be fully stressed. Then, $V_{Vier,Rd}$ will exceed $V_{Vier,Ed}$ because it assumes greater redistribution of action effects between the T's than that implied by the designer's choices.

Figure 14.7. Transfer of force ΔN_{oc} above an opening from concrete slab to top T

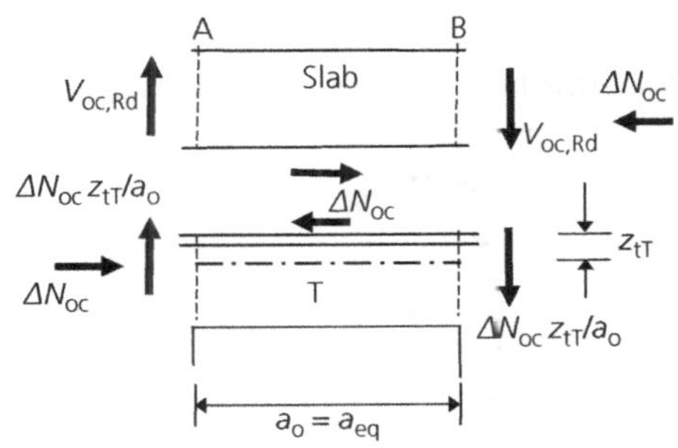

Clause D.5.1.4(3)) gives *Formula D.31* for $M_{vc,Rd}$, where Clause D.5.1.4(3)

$$M_{vc,Rd} = (V_{oc,Rd} + n_r P_{Rd} z_{tT}/s_x)\, a_{eq}$$

The use of symbol $M_{vc,Rd}$ is confusing because it is not a real bending moment in the slab. The formula shows that $M_{vc,Rd}$ is the sum of two couples acting on length a_{eq} above an opening. The concrete slab and the T are shown separately in Figure 14.7. The forces shown are assumed to be acting at a cross-section midway between A and B, where the Vierendeel model assumes that the bending moments in both elements are zero. The change in compression in the slab over length AB and the change in tension in the T each equal the force transferred by the shear connectors: $\Delta N_{oc} = n_r\, P_{Rd}\, a_{eq}/s_x$

The couple acting on the slab is $V_{oc,Rd}\, a_{eq}$, and that on the T is $(n_r P_{Rd}\, z_{tT}/s_x)\, a_{eq}$. Formula *D.31* gives their total.

D.5.1.5 Web buckling next to widely spaced openings

A bottom T resists more vertical shear at an opening than the adjacent parts of the I-section. The extra shear is transferred to it by compression at the top of the web adjacent to the vertical edge of the opening nearer to mid-span. For buckling at widely spaced openings, reference is made to EN 1993-1-13 clause 8.5.2 for regions too slender to be exempted by *paragraph (2)*. The compressive force $N_{w,Ed}$ is taken as the larger of the vertical shears resisted by the regions above and below the opening. The buckling resistance comes from the usual method of Eurocode 3, with relative slenderness based on h_o/t_w. The resistance does not depend on the width of the web, unlike the rules for closely spaced openings, for which *clause D.5.2.4* refers to EN 1993-1-13 clause 8.6.3.

D.5.2 Rules for closely spaced openings

Clause D.5.2.2(1) ***Clause D.5.2.2(1)*** on minimum degree of shear connection is applicable only to circular openings. It is based on the extensive research background to cellular beams and takes account of the different development of plasticity in the web post compared with other shapes.

The reduction factor ρ_m is explained in comment on *clause D.5.1.1(5)*. Here, a further reduction is given for minimum degree of shear connection below η from *clause 8.6.3.3(2)*. It can be significant – for example, for openings centred in the web and of depth 70% of that of the steel beam, the factor in parentheses in *Formula D.34* is 0.51. For long spans it is likely to reduce η to below the value η_{min} given in the same clause.

Clause D.5.2.3(1) ***Clause D.5.2.3(1)*** refers to EN 1993-1-13, 8.6.2(3) for bending of a web post between rectangular openings that is narrow enough for its width w_p to be relevant. The resistance to shear is given in EN 1993-1-13, 8.6.4 as the usual value, $w_p t_w f_y / \left(\sqrt{3}\gamma_{Mo}\right)$, which must not be less than $V_{wp,Ed}$.

Clause D.5.2.4(1) For web-post buckling, ***clause D.5.2.4(1)*** refers to EN 1993-1-13, 8.6.3. Figure 14.8 shows the action effects on bottom T's and the lower half of a web post, and the assumed hinges in the T's. The unchanged part of tension $N_{bT,Ed}$ is not shown. The moment $M_{wp,Ed}$ is shown in its positive direction and can be of either sign. Both it and the shear $V_{wp,Ed}$ cause compressive stress in the web post. EN 1993-1-13 requires its buckling to be checked for a compressive force given by

$$N_{wp,Ed} = V_{wp,Ed} + 2\,|\,M_{wp,Ed}\,|\,/h_o \qquad \text{(X14.3)}$$

Figure 14.8. Action effects for buckling of a web post between closely spaced openings

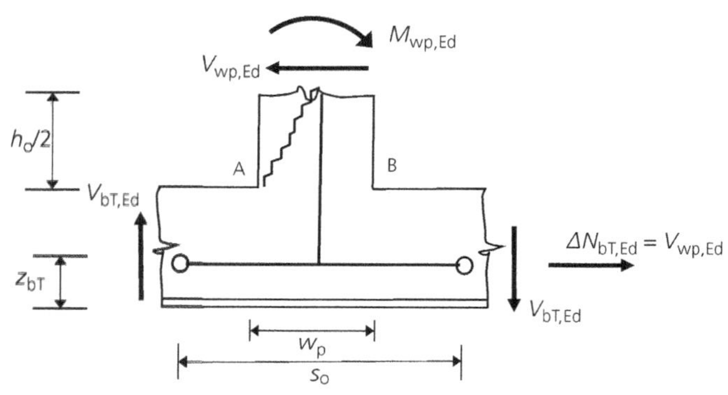

D.6 Serviceability limit states

Openings in a web increase the deflection of the beam for a given loading. Deflection can be a limiting factor in the design of cellular and castellated beams with high span/depth ratios. It is less significant for a standard rolled I-beam with a few openings for services. These subjects are treated separately in this clause, by giving predicted ratios of additional deflection to deflection without openings. They were obtained from extensive and quite complex algebraic elastic analyses and some finite element parametric studies, mainly for beams with composite slabs. They were confirmed by a few tests on beams.

The ratios are approximate, and this uncertainty is additional to that inherent in the values for beams without openings, which arises from external restraints, shrinkage and creep. As elsewhere in *Annex D*, their scope is limited to simply-supported beams of uniform depth and uniformly distributed loading. Second moments of area are usually calculated for the transformed ('equivalent') uncracked unreinforced composite section, using an effective modulus that allows for creep and neglecting effects of slip.

The extra deflection arises from

- loss of cross-section to resist global bending and shear at the openings
- curvatures from Vierendeel bending across openings

- shear deformation in the T's
- shear deformation in web posts between closely spaced openings.

Clause D.6(2) gives the additional deflection from a single opening, which increases with the distance of the opening from mid-span. The first of the two terms of *Formula D.35* gives the contribution from additional curvature at the opening. The second is a shearing deflection due to Vierendeel bending around the opening. No guidance is given for a beam with more than one opening. The extra deflections for a small number of openings can be accumulated for openings separated by web posts wide enough for their deformations to be negligible. Clause D.6(2)

Clause D.6(3) refers to beams that are either cellular or castellated. These have equally spaced identical openings throughout the span, giving additional deflections that are relatively greater than those from *paragraph (2)*. The formulae are based on finite element analyses of composite cellular beams loaded with 50% of the failure load. Clause D.6(3)

Clause D.6(4) gives the relative deflection from Vierendeel bending, $w_{v,add}$, for openings with $a_{eff} > 2h_o$. It is presumably intended for single openings. The deflection penalises wide openings, being proportional to a_{eff}^3. Clause D.6(4)

The cracks to which ***clause D.6(5)*** refers presumably arise from the local bending in regions near supports where global longitudinal compression is low. The cracks could, in theory, be on either surface of the slab. Where the second limit, $w_{add} \leq a_{eff}/200$, is exceeded, it is necessary to calculate the local bending moments and use the methods of Eurocode 2 to check whether crack control (meaning closer spacing of longitudinal reinforcement and/or smaller-diameter bars) is needed. Clause D.6(5)

Example 14.1: composite beam with web openings

A simply supported composite T-beam of span 8.6 m with a full-thickness concrete flange has been designed for distributed loading. It is proposed to make two rectangular openings in the web, close to one support, with the dimensions shown in Figure 14.9. The resistances of the beam at the openings are checked, taking account of the global action effects.

The following method has these main stages.

- At the opening nearer to the support, find the global compression in the slab, as limited by its shear connection.
- If the resistance of the slab to compression is sufficient, assume that the external moment M_{Ed} is resisted with zero compression in the top T.
- Find from Eurocode 2 the resistance of the slab to vertical shear, which may be reduced by the resistance to longitudinal shear of the studs above the opening.
- Assume this resistance is fully used and apply the rest of the global vertical shear to the two T's, either equally or in proportion to their shear areas.
- Check that the bending resistance of the bottom T, allowing for coexisting tension and shear, exceeds the Vierendeel bending moment caused by its vertical shear $V_{bT,Ed}$.
- Repeat the previous stages for the next opening; then check the web post between them for bending resistance and local buckling.

Data

The openings are 'closely spaced' and the slab is 'flexible'. Details of materials, loading and some resistances are as follows

- structural steel: $f_y = 355$ N/mm²; $f_{yd} = 355$ N/mm² unless noted otherwise; $E_a = 210$ kN/mm²
- reinforcement: $f_{sk} = 500$ N/mm²; $f_{sd} = 435$ N/mm²; E_s taken as 210 kN/mm²
- normal-density concrete: from EN 1992-1-1: $f_{ck} = 25$ N/mm²; $E_{cm} = 30.5$ kN/mm² (quartzite aggregate); free shrinkage (*Annex C*), 325×10^{-6}
- stud shear connectors: $f_u = 500$ N/mm²; $d_s = 19$ mm; $h_{sc} = 100$ mm; $\gamma_V = 1.25$; $\boldsymbol{P_{Rd} = 73.1}$ **kN** per stud.

- steel beam: 406 × 178 UB60; $A_a = 7650$ mm²; $I_a = 216 \times 10^6$ mm⁴
- composite beam: $n = 2n_o = 13.8$, $b_{eff} = 1.96$ m, $I = 692 \times 10^6$ mm⁴.

From *clause 5.1* for the design compressive strength of concrete, $\eta_{cc}\, k_{tc} = 0.85$, and so $\boldsymbol{f_{cd}} = 0.85 \times 25/1.5 =$ **14.2 N/mm²**.

Figure 14.9. Elevation and cross-section of a beam with openings near a support

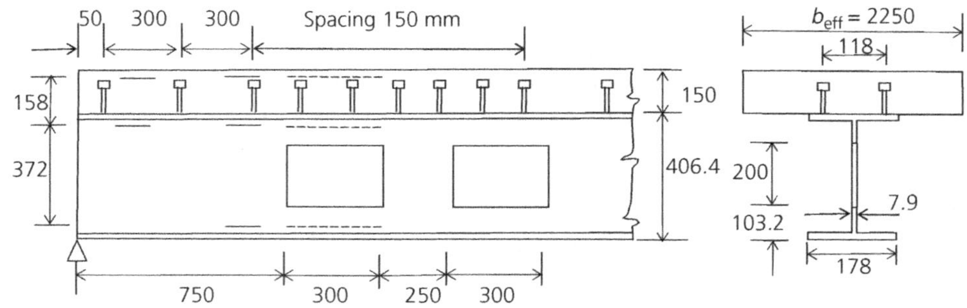

(a) Elevation, showing members as dashed lines (b) Cross-section at opening

Loading, shear connection and reinforcement

The beam is designed for propped construction. The design ultimate loading is $\boldsymbol{w_d}$ **= 64 kN/m**. The failure load with full shear connection is 85 kN/m. Partial shear connection ($\eta = 0.77$) was used, with pairs of studs at 300 mm spacing. The transverse reinforcement in the slab was 12-mm bars (top) and 8-mm bars (bottom), both at 200 mm spacing and with centres assumed to be 30 mm from the adjacent surface of the slab, so the effective depths for transverse bending, $d_{s,c,sag}$ and $d_{s,c,hog}$, are 118 and 120 mm. The proposed longitudinal reinforcement at the openings is 8-mm bars, top and bottom, at spacing 250 mm, giving A_s =402 mm²/m. At 0.27% this is just below the maximum at which a slab of this thickness is defined as 'flexible' by *clause E.2(3)*. From Figure 14.10, the effective depth $d = 110$ mm for both sagging and hogging bending. These details may need to be altered.

Figure 14.10. Proposed reinforcement of the deck slab at an opening

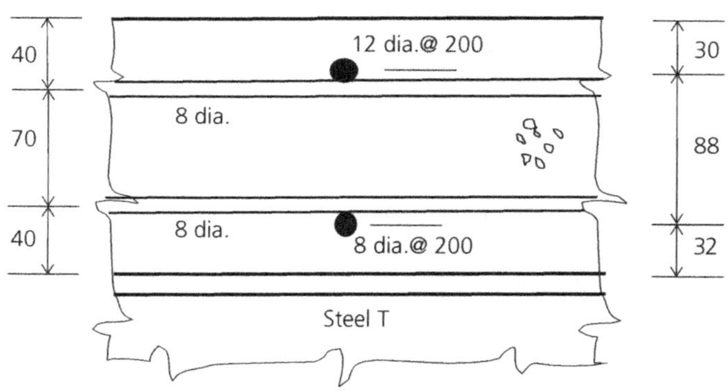

Properties of steel T-sections

It simplifies calculations to replace the corner fillets of the T's by an additional thickness of flange, so that the cross-section becomes a pair of rectangles. This flange thickness is $t_{f,eff} = 13.06$ mm, increased from 12.8 mm. For classifying the T-sections, the correct web outstand, c in EN 1993-1-1, is used. For Vierendeel action the top T is Class 2 to EN 1993-1-13. For global bending, EN 1993-1-13 requires the web of the T to be treated as an outstand flange in pure compression. The top T is in Class 3, so an effective web is used, with its depth of 80.2 mm reduced to the Class 2 limit, 64.3 mm. The global tension in the bottom T is sufficient to keep it in Class 2.

Some properties of these T-sections are as follows

$A_{a,bT} = 3035$ mm^2; $N_{pl,bT,Rd} = 1077$ kN; $A_{v,bT} = 940$ mm^2; $M_{pl,bT,Rd} = 15.5$ kN m; $z_{bT} = 18.6$ mm
$A_{a,tT} = 2909$ mm^2; $N_{pl,tT,Rd} = 1033$ kN; $A_{v,tT} = 814$ mm^2; $M_{pl,tT,Rd} = 11.6$ kN m; $z_{tT} = 15.3$ mm.

The design global action effects at the centres of the openings and web post, at cross-sections distance x from the left-hand support, are given in Table 14.1.

Table 14.1. Design global action effects

	Opening	Web post	Opening
x, m	0.9	1.175	1.45
V_{Ed}, kN	218	200	182
M_{Ed}, kN m	222	279	332

Figure 14.11. Analysis for longitudinal forces at the centre of an opening

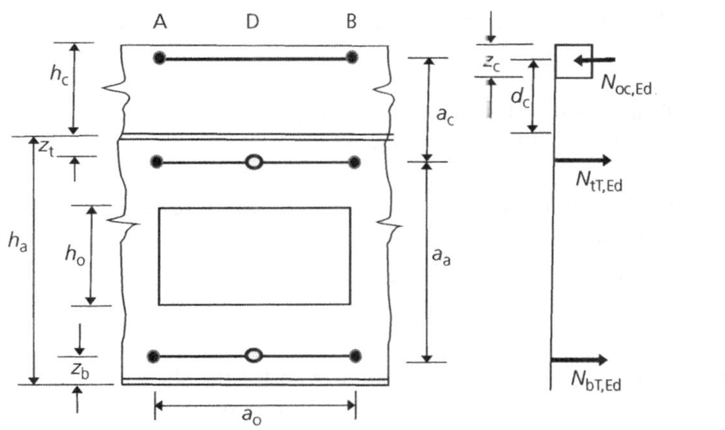

(a) Dimensions at an opening (b) Longitudinal action effects

Effective widths of the concrete slab at openings

With the bottom transverse reinforcement at depth 118 mm below the top of the slab, and the rows of studs at lateral spacing 118 mm

- for global and Vierendeel bending, $b_{eff} = 1.96$ m at $x = 0.9$ m from an end support
- for vertical shear, $b_{eff,V} = b_{ft} + 2h_{cs} = 178 + 2 \times 150 = 478$ mm.

Analysis of the cross-section at the centre of the first opening ($x = 0.9$ m)

This analysis is discussed under *clause D.4.2*. Direct calculation of the longitudinal forces involves algebra and a quadratic equation. The following method is a simpler way to find their values at the centre of an opening, shown in Figure 14.11. Dimension h_p depends on the direction of span of any profiled sheeting, and is zero here. The global sagging bending moment, $M_{o,Ed}$, is 222 kN m.

The initial spacing of pairs of studs, 300 mm, does not satisfy *clause D.5.1.1(3)*, which sets a limit of $0.5a_o$ or 150 mm here. To achieve this, the first pair of studs is placed at $x = 50$ mm, and the spacing of rows is reduced to 150 mm from $x = 650$ mm to $x = 1550$ mm, as shown in Figure 14.9. This provides eight studs before the centre of the first opening, and two pairs above each opening and above the web post.

The resistance of the slab to compression, $N_{oc,Rd}$, is governed either by shear failure of these eight studs: $N_{oc,Rd} = 585$ kN or by crushing of the slab. Crushing is unlikely; its resistance here is 4170 kN, based on the global effective width, $b_{eff} = 1.96$ m.

The initial assumption is that $z_c = N_{oc,Rd}/(b_{eff} f_{cd}) = 585/(1.96 \times 14.2) = 21$ mm.

Then, the lever arm to the bottom T is $a_a + a_c = 150 + 406.4 - 18.6 - 21/2 = 527$ mm, and with $N_{tT,Ed} = 0$ the compression in the slab is $N_{oc,Ed} = 222/0.527 = 421$ kN.

The correct result is $\boldsymbol{N_{oc,Ed} = N_{bT,Ed}} = M_{o,Ed}/(a_a + a_c) = 222/0.529 = \mathbf{419\ kN}$

The value 421 kN is too high because depth z_c is too high, being based on 585 kN, and is 15 mm when $N_{oc,Ed}$ is 419 kN.

If $N_{oc,Ed}$ had exceeded $N_{oc,Rd}$, then $N_{oc,Ed}$ would be chosen to equal $N_{oc,Rd}$. The remaining axial force would be assumed to be resisted by a force $N_{tT,Ed}$ in the top T, and the tension in the bottom T would be $N_{bT,Ed} = N_{oc,Rd} + N_{tT,Ed}$.

Finally, the bottom T is checked. It resists bending and shear as well as tension, but tension has priority and reduces the other resistances where necessary. From the list above, $N_{pl,bT,Rd} = 1077$ kN, which is sufficient. If it were too low, recalculation to *clause D.3.2(5)* would be needed.

Clause D.5.1.2(7) requires a further check on the resistance of the slab at the high-moment end of the opening. The additional compression is small, and obviously acceptable here.

Global shear forces

The shear resistance of the slab may be governed by failure of the four studs above the opening. The depth d_c is $150 - 419/(2 \times 1.96 \times 14.2) = 142.5$ mm

Formula D.28 reduces this depth by $N_{oc,Ed}/(2b_{eff} f_{cd}) = 7.5$ mm. Here, n_r is two studs per row and the spacing s_x is 150 mm, whence

$$V_{oc,Rd} = (2 \times 73.1/150)(142.5 - 7.5) = 132 \text{ kN} \le V_{c,Rd}$$

Resistance of the slab to vertical shear

For 'flexible' slabs, *clause D.5.1.3(3)* gives *Formula D.28* for the shear resistance of the slab at the opening. The result at the first opening is $V_{oc,Rd} = 91$ kN. It depends on the axial compression in the slab, so the resistance decreases with distance from the nearer support. The condition $V_{oc,Rd} \le V_{c,Rd}$ is now checked.

From *clause D.5.1.3(4)*, resistance $V_{c,Rd}$ 'may' be found from EN 1992-1-1:8.2. Its Formula 8.27 for the shear strength $\tau_{Rd,c}$ includes the following

- a term d_{dg} which is a size parameter for the roughness of the failure zone that depends on a parameter D_{lower}. As the choice of aggregate size within the concrete specification depends on the spacing and density of the reinforcement, the smallest value of D_{lower} permitted by EN 1992-1-1, 8 mm, is conservatively assumed here. This gives $d_{dg} = 24$ mm
- the effective area of longitudinal tensile reinforcement, 402 mm^2/m
- the depth to the tensile reinforcement. This is 110 mm. With $\gamma_V = 1.40$, $\tau_{Rd,c} = 0.54$ N/mm^2. This is a low value, and is found to be below the minimum given by Formula 8.20 in Eurocode 2, which is used here.

$$\tau_{\mathbf{Rdc,min}} = (11/\gamma_V)[(f_{ck}/f_{yd})\ (d_{dg}/d)]^{1/2} = (11/1.4)\ [(25/355)\ (24/110)]^{1/2} = \mathbf{0.97\ MPa}$$

Factor γ_V is an NDP, from Table 4.3 in EN 1992-1-1. The recommended value, 1.4, is used here.

Formula 8.18 in EN 1992-1-1 is $\tau_{Ed} = V_{Ed}/(b_w z)$. The breadth b_w is $b_{eff,V} = 0.478$ m. From clause 8.2.1(3) and Figure 14.10, $z = 0.9d = 99$ mm. Then, from Formula 8.18,

$$\boldsymbol{V_{c,Rd}} = 0.97 \times 0.478 \times 99 = \mathbf{46\ kN} = V_{oc,Rd} = V_{oc,Ed}$$

because the condition in *Annex D* governs. If it is assumed that the remainder of the shear $V_{oa,Ed}$ can be shared between the T's in proportion to their plastic shear resistances (*clause D.4.3(4)*), which are proportional to their shear areas, given above, the results (not the final values) are: $V_{tT,Ed} = 80$ kN; $V_{bT,Ed} = 92$ kN.

Resistances of T-sections

The bending resistance $M_{pl,Rd}$ is reduced for coexisting shear if $V_{Ed} > 0.5V_{Rd}$. For the T with the lower shear area A_v (814 mm^2, above), $V_{tT,Rd}$ = 167 kN, and V_{Ed}/V_{Rd} = 80/167 = 0.48, so no reduction is needed. Repeating the values given above with new symbols gives

- for the top T: $N_{V,Rd}$ = 1033 kN; $\boldsymbol{M_{V,Rd}}$ **= 11.6 kN m**
- for the bottom T: $N_{V,Rd}$ = 1077 kN; $M_{V,Rd}$ = 15.5 kN m

The bending resistance is reduced for coexisting axial force, which is present only in the bottom T. From Equation X14.2,

$$\boldsymbol{M_{NV,bT,Rd}} = M_{V,Rd}(1-(N_{Ed}/N_{V,Rd})^2) = 15.5(1-(419/1077)^2) = \mathbf{13.2\ kN\ m}$$

The vertical shear resisted by a T is limited by its resistance to bending moments at the ends of the length a_o, so the bottom T can resist 13.2/0.15 = 88 kN. The top T can resist 11.6/0.15 = 77 kN, both of which are below the allocations made above. The T's together are now assumed to resist 165 kN. Adding the resistance from the slab, 46 kN, gives a total of 211 kN.

There is further resistance from the transfer of the longitudinal shear force from the steel-concrete interface to the centre of the top T, a distance z_{tT} (15.3 mm). This increases V_{Rd} at the centre of the opening as follows. From *Formula D.31*,

$$M_{vc,Rd} = (46 + 2 \times 73.1 \times 15.3/150) \times 0.3 = 18.3 \text{ kN m}$$

From *clause D.5.1.4(4)*, with $k_a = 1$ as $a_o/h_{tT} < 5$ here, *Formula D.30* gives the sum of the contributions to the resistance to vertical shear at the centre of the opening

$$V_{Vier,Rd} = [2 \times (13.2 + 11.6) + 18.3]/0.3 = 226 \text{ kN}$$

which just exceeds the 218 kN required. This completes the check for the first opening.

Check of the cross-section at the centre of the second opening

At x = 1.45 m, M_{Ed} is 50% higher than before, at 332 kN m and V_{Ed} is 16.5% lower, at 182 kN. The properties of the T's and the method of checking are as for the first opening. The changes from the results for the first opening show the trends expected if the openings were continued towards mid-span.

The resistance of the slab to compression, $N_{c,Rd}$, increases faster than the action effect $N_{c,Ed}$ from the increasing bending moment because of the extra studs placed above the openings, so again there is no need to use the top T to resist compression. The increase in $N_{oc,Ed}$ increases the stress-block depth z_c faster than the increase in effective width b_{eff} reduces it. The lever arm a_c is 2 mm less, at 155 mm, and $\boldsymbol{N_{bT,Ed}}$ **= 630 kN**.

The shear resistance of the slab, $V_{oc,Rd}$, is based on $\tau_{Rdc,min}$ and is unchanged at 46 kN. The bending resistance of the bottom T is reduced to 10.2 kN m by the greater normal force, whence $V_{bT,Rd}$ = 68 kN. The shear resistance of the top T, 77 kN, is assumed to be fully used, leaving $V_{bT,Ed}$ = 59 kN for the bottom T.

Check of the web post

Figure 14.12, like Figure 14.8, shows the length s_o (= 550 mm) of the lower half of the steel beam. The design model assumes zero bending moment in the bottom T at cross-sections D and E. From *clause D.4.4(2)*, the design vertical shear for the bottom T is the average of those assumed at the centres of the openings

$$V_{av,bT,Ed} = (92 + 59)/2 = 75.5 \text{ kN}$$

The lever arms are: a_c = 0.155 m, a_a = 0.372 m. *Formula D.14* gives the web-post shear

$$V_{wp,Ed} = \Delta N_{oc,Ed} = 200 \times 0.55/(0.155 + 0.372) = 209 \text{ kN}$$

This differs slightly from the difference between the axial forces in the bottom T's at the two openings, 211 kN, because of changes in the depth of the compressive stress block for the concrete.

Figure 14.12. Action effects for check on web post (units: kN m)

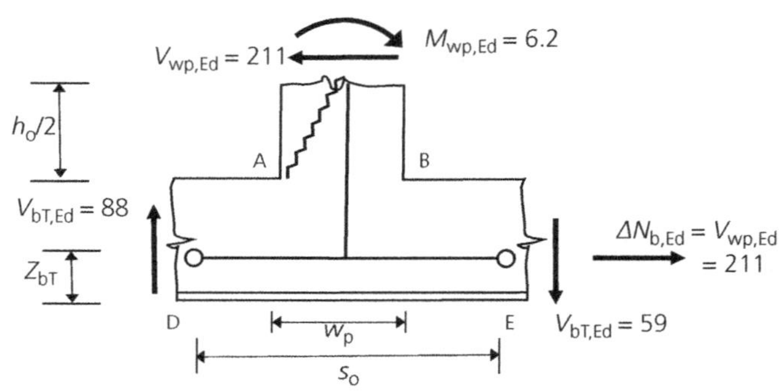

Equation X14.1 gives the web-post moment, with $V_{av,Ed} = (178 + 222)/2 = 200$ kN,

$$M_{wp,Ed} = (200 - 2 \times 59)(0.55/2) - 211 \times 0.155/2 = 6.2 \text{ kN m}$$

This is well below the elastic resistance to bending, 29 kN m.

The final check is on buckling, to clause 8.6.3(3) of EN 1993-1-13. Its Formula 8.30 gives the design compression as $\mathbf{N_{wp,Ed}} = 211 + 2 \times 6.2/0.2 = \mathbf{272\ kN.}$

The effective length to EN 1993-1-1 is 0.32 m. The slenderness $\bar{\lambda}_{wp} = 1.16$, and from buckling curve a (EN 1993-1-13, 8.6.3(5)), $\chi = 0.56$. Hence, with $w_p = 0.25$ m,

$$\mathbf{N_{wp,Rd}} = \chi_{wp} w_p t_{wp} f_{yT} / \gamma_{M1} = \mathbf{393\ kN}$$

which is sufficient. If it were too low, it might be possible to reduce $M_{wp,Ed}$ to zero by changing the allocation of vertical shear to the bottom T, subject to its ability to resist the resulting bending moment. Symbol s_o is used in EN 1994 for spacing of openings, with w_p for web-post width. The corresponding symbols in EN 1993-1-13 are s and s_o respectively.

Comment

The resistances of the two T's are not fully used in this example. It is always simpler to check openings for a given loading on the beam, rather than to predict their maximum resistance, because of the many interactions in the calculations.

Deflections

As allowed in *clause 7.4.2.2(11)*, the modular ratio is taken as $2n_o$ for all loading. With propped construction, the serviceability loading, assumed to be 40.7 kN/m, causes mid-span deflection 20 mm in the propped beam without openings. The two adjacent openings here are outside the scope of *Annex A*. Calculation for the right-hand one to *clause D.5(2)* gives the additional deflection as 0.2 mm. In practice, deflection from a small number of openings of moderate size will usually be negligible.

Where there are openings throughout a span, the extra deflection from *clause D.6(3)* should always be checked – for example, a cellular beam with $L/h_a = 20$ has 18 circular openings with $h_o = 0.7\ h_a$ and edge-to-edge spacing $w_p = 0.5\ h_o$. From *Formula D.36*, $w_{add}/w_b = 0.21$.

Example 14.2: web opening with profiled sheeting

The data for this example are identical with those for Example 14.1, except as follows.

The slab has transverse Comflor 70 profiled sheeting of net depth 55 mm, gross depth 70 mm and mean trough width 133 mm. This reduces the effective thickness of the slab from 150 mm to 80 mm. The shear resistance of the slab is much reduced, so the resistances $M_{NV,Rd}$ of the two T's must be checked for the additional vertical shear.

The longitudinal reinforcement is assumed to be 7-mm bars at 200-mm spacing, in two layers located at depths 40 mm and 76 m from the top of the slab, giving $\rho = 0.40\%$. The transverse reinforcement remains to be chosen. The slab is still 'flexible' under the other criterion of *clause E.2(3)* because the effective thickness is less than h_{tT}.

Longer studs are needed (125 mm), and the spacing of the pairs is increased from 150 mm to 300 mm, as this is the trough spacing. Two rows above an opening 0.3 m long may not be possible, as it depends on the position of the sheeting relative to the opening. Their location within a trough reduces the shear resistance per stud from 73.1 kN to 42.1 kN. In practice, only one row can be relied on. This governs the shear resistance of the slab and leads to so much shear in both the T's that their resistances $M_{NV,Rd}$ are not sufficient and the design fails.

Comment

The resistances of beams at large openings are strongly influenced by the layout of the local shear connectors and the limitations from the troughs of profiled sheeting.

Roger P Johnson and Stephen J Hicks
ISBN 978-1-83662-921-4
https://doi.org/10.1108/978-1-83662-920-720253018

CHAPTER 15
Annex E (Informative). Composite beams with web openings and a stiff slab

This chapter corresponds to *Annex E* of EN 1994-1-1, which has the following clauses.

- Use of this annex, with Note *Clause E.1*
- Scope and field of application *Clause E.2*
- Effective width for local bending and shear *Clause E.3*
- Analysis *Clause E.4*
- Additional checks at ultimate limit states for widely spaced openings with locally stiff slabs *Clause E.5*

Many of the clauses use the word 'may', so several design methods are possible. The symbols of *Annex E* are listed at the beginning of Chapter 14.

E.2 Scope and field of application

Clause D.5.1.1(4) refers to this *Annex E* for supplementary rules for 'widely spaced' openings (defined in *clause D.2.2*) under 'locally stiff' slabs (defined below). The limits to scope are as in *Annex D* (sagging moment regions of beams in buildings etc.). Shear connectors other than studs can be used if they are in category D2 or D3. There is comment under *clause D.2.1(5)*.

The share of the vertical shear at an opening that is resisted by the slab increases with its stiffness relative to that of the top T. For flexible slabs, it is found using *Annex D*. Above an opening, shear transfer across the concrete-steel interface causes tension in the connectors near the end of the opening closer to mid-span, and is assumed to become a potential failure mode where the slab is 'locally stiff'. This could be a brittle mode, *clause E.4(2)*. It leads to the replacement of the plastic resistance of the T-sections in *Annex D* by the elastic resistance in *clause E.4(5)*, and may require additional shear connection.

For simplicity, this progressive change of behaviour with increasing slab stiffness is replaced in EN 1994-1-1 by a step change of method that is significant, because the design rules for stiff slabs are complex and worth avoiding where possible. It is not known at present whether the UK's National Annex will enable the use of *Annex E*.

'Locally stiff' slabs as defined in ***clause E.2(3)*** are those that are not 'locally flexible'. The first condition, that the depth of the concrete that resists longitudinal compression exceeds the depth of the top T, is unlikely to be satisfied where the slab is composite. If it does apply, it can be difficult to fit in the extra connectors. This chapter is limited to full-depth ('solid') slabs. The other condition is that the longitudinal reinforcement ratio should exceed 0.3%. This ratio can be determined by detailing requirements. Where a slab has $h_c > h_{tT}$, the designer may be able to choose between making the slab flexible or stiff by altering the reinforcement. *Clause E.2(3)*

Most cellular beams, but not castellated beams, are deemed to be locally flexible in ***clause E.2(4)***, based on current practice. Locally stiff slabs above closely spaced non-circular openings are outside the scope of EN 1994-1-1 (***clause E.2(5)***), as shown in Figure 14.2. The reason is uncertainty in the distribution of bending moment in the slab, which is affected by interaction between the openings. *Clause E.2(4)* *Clause E.2(5)*

E.3 Effective width for local bending and shear

Clause E.3(1)

The width $b_{eff,b}$, defined in ***clause E.3(1)***, is similar to the length a_o of local openings, and much smaller than the width for global bending (defined in *clause D.3.3*). The width for vertical shear, $b_{eff,V}$, is also different from the value in *Annex D*, and depends mainly on $d_{s.c.ten}$, the effective depth under transverse bending. *Figure E.1* shows that this is sagging bending. The value $b_{eff,V}$ depends strongly on the detailing of the transverse reinforcement and would be small if parallel profiled sheeting is used. In UK practice, the transverse reinforcement is normally used for fire resistance purposes. This should not be a reason for ignoring it here, because 'fire' is a different limit state.

E.4 Analysis

Clause E.4(1)

Figure 15.1 and *Figure E.1* show the set of self-equilibrated local action effects in a length $3a_o$ of slab with the nearer support on the left, that is assumed in ***clause E.4(1)***, and part of a rectangular opening. (This account is limited to rectangular openings with $a_{eq} = a_o$, and symbol a_o will be used here.) The design model for these effects is based on the bending moments in the slab at cross-sections A and B, with action effects $M_{A,Ed}$ and $M_{B,Ed}$ not exceeding their resistances $M_{A,Rd}$ and $M_{B,Rd}$. Their values are both taken as positive, although the shear deformation causes bending moment M_A to be hogging and M_B to be sagging.

The bending moments are assumed to be caused by distributed vertical forces from the top T over lengths a_o each side of the opening. Thus, the compressive force to cause the sagging moment M_B is $F_B = 2M_B/a_o$. The force is opposed by the tensile force $P_{ten,Ed}$ (defined in *Formula E.8*) from shear connectors near section B. Force $P_{ten,Ed}$ exceeds F_B by the amount of global vertical shear assumed to be transferred across the opening in this way. Similarly, there are forces F_A and $P_{comp,Ed}$ (defined in *Formula E.6*) near section A.

Figure 15.1. Action effects in a stiff slab above an opening

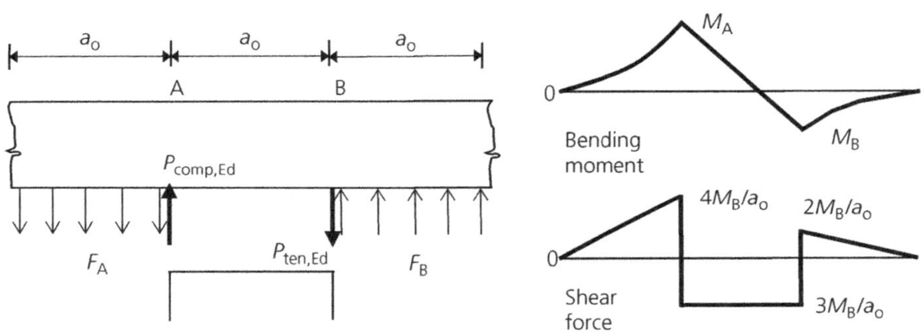

(a) Action effects on a stiff slab from the top T

(b) Bending moment and shear force in the slab (example)

These forces are additional to the forces on this length of slab considered earlier that led to the term $M_{vc,Rd}$ in *Formula D.31*. These are not shown.

Clause E.4(2)

The forces $P_{comp,Ed}$ and $P_{ten,Ed}$ are proportional to the bending moments $M_{A,Ed}$ and $M_{B,Ed}$, respectively, in the slab, which are chosen by the designer. ***Clause E.4(2)*** explains why they should not be too high to reduce the risk of a brittle shear failure of the slab. The check on this comes later, *clause E.5.3(1)* and *(2)*. The analysis that follows is approximate, and one reason why this Annex is Informative, with 'advanced analysis' as an alternative.

Clause E.4(3)

The global axial forces at the centre of the opening will have been determined earlier. In finding them, shear resistance of some of the connectors is neglected, ***clause E.4(3)***. This is because they are assumed to be resisting vertical tension.

Clause E.4(4)

Resistances $M_{A,Rd}$ and $M_{B,Rd}$ are found using EN 1992-1-1, ***clause E.4(4)***, taking account of the co-existing global compressive force in the slab. The line of action of this force is slightly different at cross-sections A and B, and is assumed to be at mid-height of the slab.

Areas for the top and bottom longitudinal reinforcement have to be assumed, and may be different. Then, $M_{A,Rd} \neq M_{B,Rd}$. It may be difficult to provide sufficient connectors to resist the tension $P_{ten,Ed}$, which is proportional to $M_{B,Ed}$. Then, it may help to assume $M_{A,Rd} > M_{B,Rd}$.

Clause E.4(5) gives *Formula E.3* for 'The maximum possible shear in the slab…'. It is an equilibrium equation, giving the value $V_{slab,Ed}$ when the steel T's and the shear connectors are fully utilised in resisting the external shear V_{Ed}. There remain the contributions from composite action and the moments M_A and M_B, which must reach this value V_{slab}. *Clause E.4(5)*

It is assumed that the steel T's are still elastic. Its purpose is to reduce the shear strain in the slab, to avoid brittle shear failure of the concrete slab. The term $n_r\ P_{Rd}\ d_c/s_x$ is an approximation, with a limit on d_c/s_x. The more accurate *Formula E.14* takes account of the length of the shear connection and its lever arm to the top T.

If these two contributions to resistance to external vertical shear in the slab at the opening, $V_{oc,Ed}$, together exceed the shear resistance $V_{c,Rd}$ of the slab to Eurocode 2, the lower value governs (*clause E.6.3(2)*), and the resistance of the connectors is not fully used. In Example 15.1, below, $V_{slab,Ed}$ is found to be much less than the limit from *Formula E.3* because it is determined by the limited space available for the studs that resist vertical tension. That can be a more severe condition.

Example of choice of action effects $M_{A,Ed}$ and $M_{B,Ed}$

Let us assume that $M_{A,Ed}$ and $M_{B,Ed}$ equal the resistances $M_{A,Rd}$ and $M_{B,Rd}$, with $M_{A,Rd} = 2M_{B,Rd}$. Then, from *Formulae E.2* and *E.3*,

$$V_{slab}a_{eq} = M_{A,Rd} + M_{B,Rd} \qquad \text{(X15.1)}$$

From *Formula E.7*, $a_{eq,A} = 2\ a_o/3$, as shown in Figure 15.1(b).

From *Formulae E.6* to *E.9*, $P_{ten,Ed} = 5\ M_{B,Rd}/a_o$ and $P_{comp,Ed} = 7\ M_{B,Rd}/a_o$

The bending-moment and shear-force diagrams for this example are shown to scale in Figure 15.1(b). The vertical shear $V_{slab,Ed}$ transferred across the opening is $3M_{B,Rd}/a_o$, from *Formula E.4*.

Clause E.4(5) sets a lower limit to the value for spacing of rows (s_x) that can be used in *Formula E.3* ('d_c/s_x should not be taken as more than 0.5…' with $d_c = h_c/2$ for a solid slab). This does not limit the actual spacing, which could be closer than h_c in a solid slab. It also applies, with the same wording, to *Formulae E.10* and *E.14*. The intention is to limit the steepness of the direction of thrust from the studs into the slab.

Clause E.4(5) suggests using *Formula E.3* as a starting point for an analysis and leads to the following design process. Firstly, find V_{slab} from *Formula E.3*, then choose $M_{A,Ed}$ and $M_{B,Ed}$ to satisfy *Formulae E.4* and *E.5*. Then check that the longitudinal reinforcement is such that in each case $M_{Rd} \geq M_{Ed}$. Finally, find $P_{ten,Ed}$ from *Formula E.8*. The comments above on *Formula E.3* suggest that may not be the best method. It is not used in Example 15.1.

Clause E.4(6) refers to a known 'sequence of construction', which is misleading. The clause applies only to unpropped construction. *Clause E.4(6)*

A design may be governed by fitting in the number of studs needed to resist force $P_{ten,Ed}$. For shear connection, this force has to be distributed along the interface, and ***clause E.4(10)*** suggests a length for this purpose. These connectors will also be resisting longitudinal shear. Their design is discussed in Chapter 18 on *Annex H*. *Clause E.4(10)*

The words in this clause '*should include a maximum group of two lines of shear connectors*' are unclear. They probably mean that a row across the width of a steel flange should not have more than two studs. The tension pull-out cones for two studs near the edges of a typical flange would overlap, so a third stud placed between them would add little to the group's resistance to tension.

E.5 Additional checks at ultimate limit states for widely spaced openings with locally stiff slabs

Clause E.5.2
Clause E.5.3(1)
Clause E.5.3(2)

In general, *Annex D* is followed, except as given here. For global bending resistance ***clause E.5.2*** refers to the longitudinal forces found using *clause E.4*. *Formula E.10* in ***clause E.5.3(1)*** gives the total vertical shear to be resisted by the concrete slab, and repeats the rule concerning d_c/x_s referred to above. ***Clauses E.5.3(2)*** and ***(3)*** are the same as in *clause D.5.1.3* of *Annex D*.

Clause E.5.3(5)

Clause E.5.3(5) refers to Part 4 of EN 1992, 'Design of fastenings for use in concrete', Annex A.2: 'Verification of the shear resistance of the concrete member'. This concerns the anchorage of the connectors that resist force $P_{ten,Ed}$, and is partly superseded by *Annex H*, 'Design tension resistance of headed studs'. Calculations are given in Example 15.1.

E.5.4 Resistance to Vierendeel bending

Clause E.5.4(3)

In ***clauses E.5.4(3)*** and ***(4)***, *Formulae E.12* to *E.14* give the sum of the four contributions to $V_{Vier,Rd}$, which limits the vertical shear that can be transferred across an opening. It is now compared with the same resistance where the slab is 'flexible', given by *Formulae D.28, D.30* and *D.31*. Combining these three,

$$a_{eq}V_{Vier,Rd,\,flexible} = 2(M_{NV,bT,Rd} + M_{NV,tT,Rd}) + k_a a_{eq}(n_r P_{Rd} / s_x)[d_c + z_{tT} - N_{oc,Ed} / (2b_{eff} f_{cd})]$$

For stiff widely spaced slabs, *Formula E.12* gives the same contribution from the T-sections, but there are four differences, as follows.

- The reduction factor k_a is omitted (i.e. is taken as 1.0), presumably because the stiffer slab reduces the effect of local deflection at the opening and tension in the studs is being allowed for.
- In *Formula E.14* for $M_{vc,Rd}$, length a_{eq} is reduced by $h_c/2$ to allow for the length of opening where the connectors are assumed to be resisting vertical tension.
- The reduction in d_c in *Formula D.28* to allow for the difference between the levels of the global compression at the ends of the opening, $N_{oc,Ed}/(2b_{eff}f_{cd})$, is omitted because of the different assumption made in *clause E.4(4)* about the lines of action of these compressive forces.
- The terms $M_{A,Ed}$ and $M_{B,Ed}$ are added because these are not in the theory for 'flexible' slabs.

Where, from *Formula E.10*, $V_{oc,Ed}$ is found to exceed $V_{c,Rd}$ and (from *clause E.5.3(2)*) has to be limited to that value (which occurs in Example E.1), the moments $M_{A,Ed}$ and $M_{B,Ed}$ have accordingly been reduced here, to provide consistent calculations. *Annex E* does not say that this should be done. These lower values are used in *Formula E.13* for $M_{c,Rd}$ and, therefore, affect resistance $V_{Vier,Rd}$ (*Formula E.12*).

E.5.5 Resistance of the shear connection at the opening edges

This clause provides a direct method for calculating n_t, the number of connectors to be provided within the length $2(h_c + h_p)$ to resist the vertical tension $P_{ten,Ed}$ given in *clause E.4(9)*. The number depends on the resistance to tension of a connector. The only type for which rules are given, in *Annex H*, is the headed stud, so the scope of *Annex E* is effectively limited to their use. *Annex H* is the subject of Chapter 18.

E.5.6 Resistance to web buckling

The compressive force $P_{comp,Ed}$ shown in Figure 15.1 adds to the compression in the web adjacent to the side of the opening below. This term is, therefore, added to the compression that must not exceed the buckling resistance of the web. Otherwise, this clause is the same as *clause D.5.1.5(1)*. The concession (omission of buckling check) in *clause D.5.1.5(2)* is not repeated, but presumably applies.

E.5.7 Resistance of transverse reinforcement to local loads

At an edge of an opening, the vertical shear $V_{oc,Ed}$ is treated here as a 'local load' on the slab, requiring a bottom transverse reinforcement ratio ρ_t given by *Formula E.19*. The ratio is calculated using the effective depth of this reinforcement, even where transverse sheeting is present. If the sheeting is 'adequately anchored', its area may presumably be included, with the designer left to decide whether its anchorage is 'adequate'. As this is bottom transverse reinforcement for 'local loads', it is presumably needed above the end of the opening nearer to mid-span, where transverse bending is sagging.

Example 15.1: web opening with a stiff slab

Annex E includes nineteen formulae, some of which are for options, and design is complex. The sequence followed here, one of the several possible methods, is to find first those resistances that are independent of other parameters. To simplify presentation, it is assumed here that the longitudinal reinforcement at an opening is such that $M_{A,Rd} = M_{B,Rd}$ (Figure 15.1). Then, $a_{eqA} = a_{eqB} = a_o/2$. Also, each stud connector is assumed to be resisting either tension or longitudinal shear, so that interaction *Formula E.16* is not needed.

Data

The initial data for this example are identical with those for Example 14.1 (Figure 14.9), except as follows.

- The width of the web post is increased from 250 mm to 600 mm by moving the second opening towards mid-span, as shown in Figure 15.2.
- The initial longitudinal reinforcement is 8-mm bars at 200-mm spacing, in two layers located at depths 40 mm and 110 mm from the top of the slab. For both layers and with $h_c = 150$ mm, $\rho = 0.33\%$. The transverse reinforcement is determined later. Assuming 10-mm bars, the effective depth for sagging bending is $d_{s,c,ten} = 119$ mm.
- The spacing of the pairs of 19 × 100-mm stud connectors (initially 300 mm) is revised, as explained below.

The conditions for the openings to be widely spaced and the slab to be stiff are satisfied. They are: $s_o \geq \max\{a_o, 2h_o\}$, $h_c \geq h_{tT}$, and longitudinal reinforcement ratio $\rho > 0.3\%$.

Effective widths from *clause E.4*, $b_{eff,b} = 118 + 0.83 \times 300 = 367$ mm

$$b_{eff,V} = 0.5 \times 118 + 7 \times 119 = 892 \text{ mm}$$

Resistances and analysis

Resistance to vertical tension, $P_{ten,Rd}$

It will be difficult to provide sufficient stud connectors to transfer the force $P_{ten,Ed}$ shown in Figure 15.1. The length of flange available, $2h_c$ from *clause E.4(10)*, is 300 mm, of which 75 mm is above the opening. The resistance of a stud to tension is governed by cone failure of the concrete, so there is no advantage from using closely spaced studs.

Figure 15.2. Arrangement of shear connectors with stiff slab above an opening

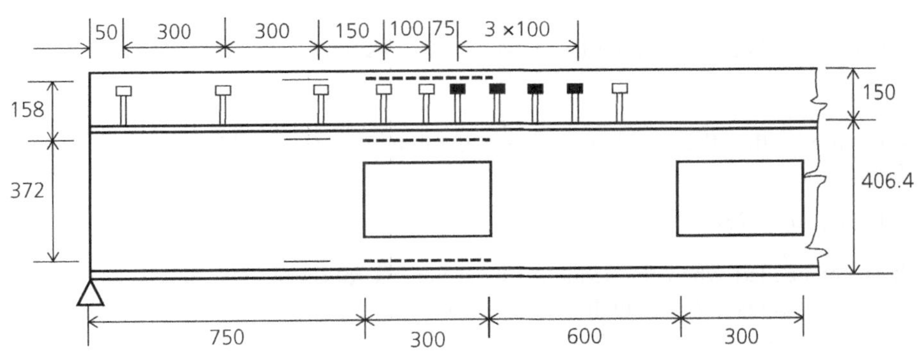

It is proposed to use four rows at 100 mm, those shown with black heads in Figure 15.2. Each row has two studs at 118 mm transverse spacing, so eight studs resist vertical tension. For the group, from Example 18.1, $\boldsymbol{P_{ten,Rd}}$ **= 51.5 kN**. It cannot be increased by adding more studs.

For longitudinal shear, from Example 14.1, pairs at 300 mm spacing are needed. Six rows are shown in the figure, with average spacing 1325/5 = 265 mm. In reality, studs near the opening will resist a combination of tension and shear. The spacing near the opening satisfies the limit of *clause D.5.1.1(3)*, 150 mm.

Resistance of the slab to vertical shear

With 0.23% from one layer of longitudinal reinforcement, this resistance is governed by the value $\tau_{Rd,min}$ from EN 1992-1-1, clause 8.2.2(4), Formula 8.20. It is found to be 0.974 MPa. From Formula 8.18,

$$V_{c,Rd} = V_{oc,Rd} = \tau_{Rd,min} b_{eff,V}(0.9\ d) = 0.974 \times 892 \times 0.9 \times 0.11 = \mathbf{86.0\ kN}$$

Bending action effects for the slab above the first opening

It is assumed that for the studs in tension, $P_{ten,Ed} = P_{ten,Rd}$. Then, from *Formula E.8*, with $a_{eqB} = a_o/2$,

$$M_{B,Ed} = M_{A,Ed} = P_{ten,Rd} a_o / 4 = 51.5 \times 0.3 / 4 = \mathbf{3.86\ kN\ m}$$

Maximum applied shear in the concrete slab, $V_{oc,Ed}$

This is given by *Formula E.10*, for which $n_r = 2$, $P_{Rd} = 73.1$ kN and d_c/s_x is taken as 0.5, based on the average of the non-uniform spacings of the studs,

$$V_{oc,Ed} = 2 \times 3.86 / 0.3 + 73.1 = 99\ \text{kN}$$

From *clause E.5.3(2)*, $V_{oc,Ed}$ cannot exceed $V_{oc,Rd}$ (86 kN), so the action effects assumed in *Formula E.10* must be reduced. Resistance P_{Rd} was taken above as the resistance to shear, not to tension, because this clause is for global shear, not local effects at the opening. It is now reduced from 73.1 kN to 60.1 kN, giving $V_{oc,Ed} = 86$ **kN**, with the slab at shear failure and the studs not fully loaded.

Shear resistance $V_{Vier,Rd}$

This is given by *Formula E.12* as the sum of four terms. It is now calculated. From *Formula E.14* for $M_{vc,Rd}$, with

$$n_r = 2,\ h_c = 150\ \text{mm},\ d_c = 150/2\ (\textit{clause E.2(3)}),\ s_x = 150\ \text{mm},\ z_{tT} = 15.3\ \text{mm},$$
$$M_{vc,Rd} = 2 \times 60.1(0.3 - 0.075)(0.5 + 15.3/150) = \mathbf{16.3\ kN\ m}$$

From *Formula E.13*, $M_{c,Rd} = 2 \times 3.86 = \mathbf{7.72\ kN\ m}$

These plastic resistances of T's are from Example 14.1

- top T: $M_{NV,tT,Rd} = \mathbf{11.6\ kN\ m}$
- bottom T: $M_{V.bT,Rd} = 15.5$ kN m, reduced by axial force to $M_{NV,bT,Rd} = \mathbf{13.2\ kN\ m}$.

From *Formula E.12*,

$$V_{Vier,Rd} = [2(13.2 + 11.6) + 16.3 + 7.72] / 0.3 = \mathbf{245\ kN}$$

This exceeds $V_{o,Ed}$ (218 kN from Table 14.1), which satisfies *Formula E.11*.

Applied moments in the slab

The decision, above, that the equal resistances of moments M_A and M_B are fully used simplifies *Formula E.4* to

$$V_{slab} = 2M_{A,Rd} / a_o = \mathbf{25.7\ kN}$$

This result is not needed when the sequence used here is followed. The only other appearance of symbol V_{slab} is in *Formula E.3*, which gives its 'maximum possible value'. Elastic analysis of the T-sections finds $M_{elN,tT} = 5.89$ kN m and $M_{elN,bT} = 4.57$ kN m.

Then from *Formula E.3* with $P_{Rd} = 60.1$ kN,

$$V_{slab,max} = 218 - 2(5.89 + 4.57)/0.3 - 2 \times 60.1 \times 0.5 = \mathbf{88.2\ kN} > 25.7\ \text{kN}$$

Resistance to web buckling

This does not govern here. It is included in the check on the web post in Example 14.1.

Reinforcement

Flexural reinforcement for the slab

The longitudinal reinforcement has not yet been checked. Design to EN 1992-1-1 finds that 8-mm bars at 200 mm spacing, top and bottom, provide resistance $M_{A,Rd} = M_{B,Rd} = 4.44$ kN m, which is more than sufficient. The depth of the compressive-stress block is 8 mm.

The slab is also resisting global axial compression of 419 kN (Example 14.1) which, from a simplification given in *clause E.4(3)*, has been assumed to act at the centre of the slab, with a stress block of depth 15 mm. If it is thought necessary to check compressive strain in concrete, the two stress blocks should be adjacent, giving a neutral-axis depth of 23 mm. This is less than 25% of the effective depth, so the compressive strain is quite low.

Bottom transverse reinforcement above the opening

Clause E.5.7 gives the minimum transverse reinforcement ratio, ρ_t. The relevant effective depth, $d_{s,c,b}$, is 119 mm and ρ_{min} is given as 0.2%. From *Formula E.19* with $F_{tr,Rd} = V_{oc,Ed} = 86$ kN,

$$\rho_t = F_{tr,Rd}\rho_{min} / (0.006\ h_c d_{s,c,b} f_{sd}) = 86 \times 1000 \times 0.002 / (0.006 \times 150 \times 119 \times 435) = 0.0037$$

Hence, $A_s \geq 0.0037 \times 119 \times 1000 = 440$ mm^2/m

This can be provided by 10-mm bars at 175 mm or 12-mm bars at 250 mm.

Summary

The calculations take account of the coexisting global bending. The method uses the full resistance of the following components

- eight studs that resist the vertical tension $P_{ten,Ed}$, which determines $M_{A,Rd}$ and so on
- the concrete slab in vertical shear, $V_{c,Rd}$
- the two steel T-sections at their plastic resistances $M_{NV,T,Ed}$.

The components not fully used are the bending resistance of the slab (3.86 kN m compared with 4.44 kN m) and the longitudinal shear resistance of the studs (60.1 kN compared with 73.1 kN).

Consequently, the vertical shear resistance $V_{o,Rd}$ exceeds $V_{o,Ed}$ (245 kN compared with 218 kN).

Roger P Johnson and Stephen J Hicks
ISBN 978-1-83662-921-4
https://doi.org/10.1108/978-1-83662-920-720253017

CHAPTER 16
Annex F (Normative). Headed studs that cause splitting forces in the direction of the slab thickness

This chapter corresponds to *Annex F* of EN 1994-1-1, which has the following clauses.

- Use of this annex — *Clause F.1*
- Scope and field of application — *Clause F.2*
- Design resistance and detailing — *Clause F.3*
- Fatigue strength — *Clause F.4*

Annex F gives design rules for the detailing and resistance of shear studs that are close to an edge of a concrete slab, as shown in *Figures 8.16, F.1* and *F.2*, and here in Figures 16.2 and 16.3. Details of this type can occur in composite bridges, and EN 1994-2:2005 included a set of these rules as Annex C. In buildings, the side of a deep haunch is also a potential region of risk. The rules for haunches, *clause 8.6.10.4*, pre-date the research that led to *Annex F*. It is a matter of judgement whether a design in accordance with those rules creates 'splitting forces in the direction of the haunch width' such that *Annex F* is applicable. This is discussed later, in the section 'Applicability of *Annex F*'.

The rules were developed from research at the University of Stuttgart (Kuhlmann and Kürschner, 2006). There were extensive push tests and finite element analyses, which showed that to avoid premature failure by splitting of the slab and to ensure ductile behaviour, additional rules for the details are needed. They are partly based on elaborate strut-and-tie modelling. It was not possible to find rules that are dimensionally consistent, so the units to be used are specified. The simplified and generally more conservative rules given in *clause 8.6.8.1(3)* do not cover interaction between longitudinal and transverse (e.g. vertical) shear or resistance to fatigue.

It will be found that these so-called 'lying studs' have to be longer than usual, and that the minimum slab thickness to avoid a reduction in the shear resistance per stud can exceed 250 mm. The comments that follow are illustrated in Example 16.1, and in Figure 16.3 where the longitudinal shear acts normal to the plane of the figure and vertical shear acts downwards from the slab to the steel plate.

F.3 Design resistance and detailing

Clause F.3(1)

Clause F.3(1) gives, for this mode of failure, the static resistance of a stud to longitudinal shear. The minimum length h_{sc} of the stud and the reinforcement details are intended to be such that splitting of the slab is followed by fracture or pulling out of the stud, giving a ductile mode of failure.

Comparison of resistances from *Formula F.1* with those from *Formulae 8.23* and *8.24* shows that the lowest of the three resistances is likely to depend on the thickness of the slab, which determines a_{rp}, the effective edge distance, and on the ratio of the stirrup spacing s to the stud spacing s_x.

As an example, resistances P_{Rk} from *Formula F.1* for 19-mm studs in C30/37 concrete are shown in Figure 16.1 for two dimensions a_{rp}. They are compared with the result from *Formula 8.24* for concrete failure: $P_{Rk} = 102$ kN.

A condition for the use of *Formula 8.24* in *clause 8.6.8.1(3)* is $a_{rp} \geq 6d$. For a slab with 10-mm stirrups and 30 mm cover, its minimum thickness is

$$2\,a_{rp} + 60\text{ mm} = 2(6 \times 19 + 30) = 288\text{ mm}$$

The other rule in this clause, $v \geq 14d$, requires very long studs (*Figure 8.16*). Hence, *Annex F* will normally be used, as in Example 16.1.

Clause F.3(1) requires $a_{rp} \geq 50$ mm, so the minimum possible slab thickness for the use of *Annex F* in this example is 160 mm. Then, P_{Rk} is only 85 kN for $s/s_x = 1$ (Figure 16.1). To obtain a resistance of 102 kN, the stirrup spacing has to be reduced to half that of the studs. The other curve in Figure 16.1 is for $a_{rp} = 80$ mm, which corresponds to a slab thickness of 220 mm. For this and thicker slabs, *Formula F.1* does not govern the shear resistance. These results are for an edge position (*Figure F.1(c)*). The influence of grade of concrete on this comparison is small until it becomes high enough for *Formula 8.23* (steel failure) to govern.

Figure 16.1. Comparison of resistances of studs from *clauses 8.6.8.1* and *F.1*

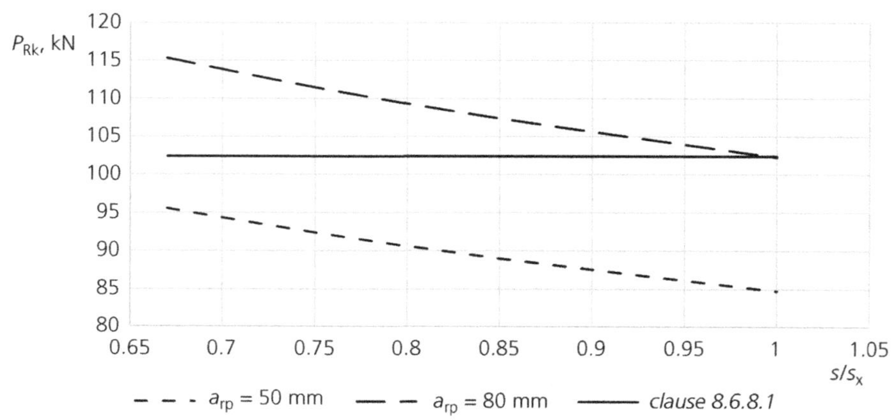

Formula F.1, repeated in Example 16.1, uses factor k_v to distinguish between two situations. The more favourable, where $k_v = 1.14$, applies where the slab is connected to both sides of a steel web and so must resist hogging bending – an 'internal position'. This requires reinforcement to pass continuously above the web, as shown in *Figure F.1(a)*. Some shear is then transferred by friction at the face of the web. Where the web is on one side of the slab, an 'edge position', k_v has the lower value 1.0. In *Figure F.1(a)* and *(c)*, the vertical dimensions are to the centre of the reinforcing bar, but the horizontal dimension c_h to stirrups is to the external face.

Vertical shear

Clause F.3(3)

Resistance to vertical shear is given by *Formula F.4* in ***clause F.3(3)***. Minimum diameters are given for both the stirrups and the longitudinal bars passing through them. The effective edge distance $a_{rp,o}$ is defined in the same way as a_{rp}, the subscript o meaning above (from the German, *oben*). The upper limit to the spacing of stud connectors is 250 mm, lower than the 440 mm for longitudinal shear.

The vertical shear resistance given by *Formula F.4* is typically less than 40% of the longitudinal shear resistance, being governed mainly by the upper edge distance $a_{rp,o}$. Application of vertical shear to lying studs is best avoided. The use of *Formula F.3* for the interaction between longitudinal and vertical shear is straightforward.

F.4 Fatigue strength

This clause is more comprehensive than it was in the previous EN 1994-2. It finds application mainly in bridges and is not discussed here.

Applicability of *Annex F*

Annex F does not state how close to a free concrete surface a stud must be for its rules for resistance and detailing to apply.

Figure 16.2. Cross-sections at an edge of a slab: (a) *clause 8.6.10.3*; (b) *Annex F*

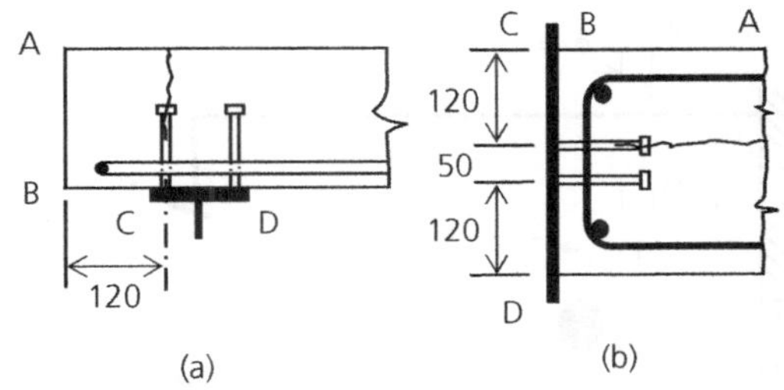

Adjacent edges

Clause 8.6.10.3 refers to a '*longitudinal edge*', not a top surface, but like *Annex F* is concerned with splitting parallel to a free surface. Where the edge distance (i.e. a_r in *clause F.3(1)*) is less than 300 mm, it specifies an edge distance $a_r \geq 6d$, and 'U-bars' (i.e. stirrups) of diameter $\phi_s \geq 0.5d$ that pass around the studs (Figure 16.2(a)), whereas in *Annex F* they pass between them (Figure 16.2(b)). This is because the surface parallel to the plane of splitting, AB in Figure 16.2, is parallel to the plane of the slab in one case, not in the other. This can determine which clause is applicable.

The '6*d*' rule of *clause 8.6.10.3* normally gives a_r as about 120 mm, as shown, corresponding to $a_{rp} \approx 90$mm. Figure 16.1 shows that *Annex F* is then unlikely to govern the resistance of the studs.

Example 16.1: design of lying studs

A floor slab 250 mm thick is to be connected along its sides to steel webs. The final detail is shown in Figure 16.3. The design longitudinal shear at each edge is $V_{L,Ed} = 600$ kN/m. The vertical shear is at first assumed to be negligible. The effect of adding $V_{V,Ed} = 60$ kN/m is then found. The strengths of the materials are

$$f_{yd} = 500/1.15 = 435 \text{ N/mm}^2, f_u = 500 \text{ N/mm}^2 \text{ and } f_{ck} = 40 \text{ N/mm}^2$$

Resistances in *Annex F* are given in terms of f_{ck}, not f_{cd}, because the partial factor γ_V allows for both the stud material and the concrete. The minimum cover is 30 mm.

There is space only for a single row of 19-mm studs at mid-depth. The studs have a height greater than four times their shank diameter, and their shear resistance to *Formula 8.24* is

$$P_{Rd} = 0.29 \times 19^2 (40 \times 34\ 500)^{0.5} / (1.25 \times 1000) = 98.4 \text{ kN/stud} \tag{X16.1}$$

and to *Formula 8.23* is $\mathbf{P_{Rd}} = 0.8 \times 500 \times \pi \times 19^2 / (4 \times 1.25) = \mathbf{90.7\ kN/stud}$ (X16.2)

This gives a maximum spacing of 90.7/600 = 0.151 m, so a spacing s_x = 150 mm is assumed, with stirrups also at 150 mm. Their required diameter ϕ_s can be estimated from *clause F.3(5)*, *Formula F.6*.

$$T_{Ed} = 0.3\ F_{L,Ed} \text{ and } T_{Rd} = (\pi\phi_s^2/4) \times 500/1.15 \geq 0.3 \times 600 \times 1000 \times 0.15$$

This gives $\phi_s \geq 8.9$ mm, so 10-mm stirrups are assumed for finding a_{rp}.

With cover of 30 mm, a_{rp} = [250 – 2(30 + 5)]/2 = 90 mm.

From *Formula F.1* with $k_v = 1.0$ for an edge position and $s_z = s$,

$$P_{L,Rd} = 1.4\ k_v (f_{ck}\ d\ a_{rp})^{0.4} (s_x/s)^{0.3} / \gamma_V = 1.4(40 \times 19 \times 90)^{0.4} / 1.25 = 96.2 \text{ kN/stud} \tag{F.1}$$

This is greater than result X16.2, so the shear resistance is governed by *Formula 8.23*. Where this occurs, the term $P_{L,Rd}$ in *Formula F.3* should be replaced by the governing shear resistance.

Figure 16.3. Notation and dimensions for lying studs in Example 16.1

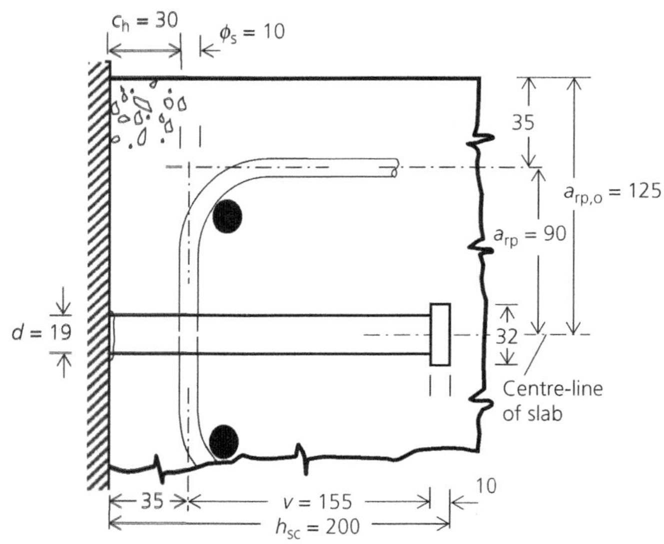

For studs at 0.15 m spacing, $\boldsymbol{V}_{\mathbf{Rd,L}} = 90.7/0.15 =$ **605 kN/m**, which is sufficient. The length of the studs is governed by the anchorage length v from *clause F.3(2)*

- for uncracked concrete, $v \geq \max\{110, 1.7 \times 90, 1.7 \times 75\} = 153$ mm
- for cracked concrete, $v \geq \max\{160, 2.4 \times 90, 2.4 \times 75\} = 216$ mm.

The minimum stud height after welding is $v + c_h + \phi_s/2$ plus the thickness of the head of the stud, 10 mm, giving heights h_{sc} of 193 mm for uncracked concrete and 256 mm for cracked concrete. Here, for uncracked concrete, $h_{sc} = 200$ mm is used.

The simpler and usually more conservative rules of *clause 8.6.8* would require a_{rp} to be increased from 90 mm to $6 \times 19 = 114$ mm. It would then be necessary for the slab thickness to be increased locally from 250 mm to at least 298 mm, and longer studs would be required.

Interaction with vertical shear

The design vertical shear is $F_{V,Ed} = 0.15 \times 60 = 9$ kN/stud.

It is found that the assumed diameters of stirrups and longitudinal reinforcement are too small. They are increased to the maxima allowed by *clause F.3(3)*: 12 mm for stirrups and 16 mm for longitudinal reinforcement. From *Formula F.4* with $k_v = 1.0$ and $s_x = s$,

$$\begin{aligned} \boldsymbol{P}_{\mathbf{V,Rd}} &= 0.012\ (f_{ck}\phi_{\ell})^{0.5}\ (d\ s_x / s)^{0.4}\ (\phi_s)^{0.3}\ (a_{rp,o})^{0.7}\ k_v / \gamma V \\ &= 0.012\ (40 \times 16)^{0.5} (19)^{0.4} (12)^{0.3} (125)^{0.7} \times 1.0 / 1.25 = \mathbf{48.8\ kN / stud} \end{aligned} \qquad (F.4)$$

From interaction expression *Formula F.3*,

$$(600 \times 0.15 / 90.7)^{1.2} + (60 \times 0.15 / 48.8)^{1.2} = 1.12 \text{ (but } \leq 1.0)$$

which is too large. Changing the spacing of the studs and stirrups from 150 mm to 125 mm would reduce result 1.12 to 0.90, which satisfies the interaction condition.

Roger P Johnson and Stephen J Hicks
ISBN 978-1-83662-921-4
https://doi.org/10.1108/978-1-83662-920-720253019

CHAPTER 17
Annex G (Informative). Design resistance of headed studs used with open trough profiled steel sheeting in buildings with ribs transverse to the supporting beams

This chapter corresponds to *Annex G* of EN 1994-1-1, which has the following clauses.

- Use of this annex — *Clause G.1*
- Scope and field of application — *Clause G.2*
- Shear resistance — *Clause G.3*

Annex G is new in Eurocode 4. The comments on *clause 8.6.9.2*, which covers the same subject, explain why its scope had to be reduced. The more recent method of *Annex G* has wider scope, is longer, more complex and, in some situations, more conservative. To avoid presenting alternative methods, the use of *Annex G* is limited to situations outside the reduced scope of *clause 8.6.9.2*. Even its wider scope does not cover the whole of current practice, as it is limited by the available results from research. The depth of steel profile is limited to 135 mm, but sheeting deeper than 200 mm is in use. Background information is available in Vigneri *et al.* (2022).

The rules of ***clause G.2(1)*** that prevent the use of *clause 8.6.9.2* are the limit on embedment depth of the stud, $h_A < 2.7d$, and the distance from the edge of the concrete rib to the nearest stud connector (*Figure 8.19*), which must satisfy 25 mm $< e_k \leq$ 60 mm. The limits of *clause 8.6.9.2*, $h_A \geq 2.7d$ and $e_k >$ 60 mm, are shown in Figure 8.20. *Clause G.2(1)*

The scope of *Annex G* is more restrictive than that of *clause 8.6.9.2* in other ways. The maximum concrete strength is C50/60 (compared with C70/85), and the minimum density is 2200 kg/m^3 (compared with 1750 kg/m^3).

Clause G.2(2) requires the ductility category of the studs to be verified by tests to *Annex B*, which is not required (*clause 5.4.2.1(3)*) for studs within the scope of *clause 8.6.9.2*. Presumably, this reliance on testing means that the rule 'a reinforcement layer is placed underneath the head of the stud' in *clause 8.6.9.2(3)* need not apply here. *Clause G.2(2)*

Shear resistance

Shear resistance to *clause 8.6.9.2* for studs in transverse ribs is the lower of one resistance based on stud failure and another based on concrete failure, both multiplied by a reduction factor k_t. Similarly, shear resistance to ***clause G.3*** is the lower of two resistances. The 'concrete' resistance, *Formula G.2*, depends on five parameters that appear in *clause 8.6.9* (d, b_0, h_{sc}, h_p and n_r), but not in the same ways, and on the following new ones *Clause G.3*

- the tensile strength of the concrete, which replaces $(f_{ck}E_{cm})^{1/2}$
- the transverse spacing of studs, where there are two in a trough
- the width of the top of the concrete rib
- the bending resistance of the stud, based on its strength f_u
- the position of the stud or studs within a trough, whether central, staggered, in series or parallel.

Comparison of these methods is only meaningful for layouts close to the interface between them. This has been done in Example 8.2. There, the characteristic resistance to *Annex G* is 71 kN/stud, compared with 87 kN/stud to *clause 8.6.9.2*. This is for a profile with $t_p = 1.0$ mm. The difference would be greater for thinner profiles. It is expected that the UK's National Annex will either enable the use of *Annex G* or prefer an alternative that has been proposed (Degtyarev and Hicks, 2023).

Roger P Johnson and Stephen J Hicks
ISBN 978-1-83662-921-4
https://doi.org/10.1108/978-1-83662-920-720253020

CHAPTER 18
Annex H (Normative). Design tension resistance of headed studs

This chapter corresponds to *Annex H* of EN 1994-1-1, which has the following clauses.

- Use of this annex *Clause H.1*
- Scope and field of application *Clause H.2*
- Design tension resistance for headed studs *Clause H.3*

From ***clause H.2(2),*** the scope does not include studs covered by *Annex G* because there is no background information on combined shear and tension resistance of headed studs welded within the ribs of modern decking. *Clause H.2(2)*

Much of this annex is extracted from Eurocode EN 1992, 'Design of concrete structures', Part 4, 'Design of fastenings for use in concrete', which has wide scope and does not refer directly to the situation here: a rectangular array of headed studs resisting direct tension, assumed to be shared equally between them. The presentation in *Annex H* is more helpful than a reference to the whole of EN 1992-4. However, ***clause H.2(3)*** does require the use of its clause A.2, 'Verification of the shear resistance of the concrete member', which covers local verification for the loads applied by each 'fastening'. *Clause H.2(3)*

Clause 8.6.8.2(4) says '*The tensile resistance of headed studs, $P_{ten,Rd}$, shall be calculated in accordance with Annex H for the studs within its scope.*' This use of 'shall' does not alter the meanings of the words 'may' and 'should' used in *Annex H* (e.g. '*tension resistance may be applied*' in *clause H.1(1)*).

H.3 Design tension resistance for headed studs

Calculations for the three modes of failure listed in ***clause H.3(1)*** are given in Example 18.1. Direct tension is the simplest mode. It is assumed that the studs resist no bending or shear. The resistances $N_{s,Rk}$ given in *Table H.1* are calculated for a tensile strength of 450 N/mm^2. This is consistent with the condition that the ultimate tensile strength of the material (R_m) should be at least 450 N/mm^2 for studs of types SD1 and SD3, given in Table 2 of EN ISO 13918:2018, 'Welding – Studs and ceramic arc ferrules for arc stud welding' (BS EN ISO 13918, 2018). A stud connector of type SD2 could have a strength as low as 400 N/mm^2. They would be outside the scope of *Annex H*. It is understood that they are not used in practice. *Clause H.3(1)*

Pull-out failure of a stud is resisted mainly by compression in concrete under its enlarged head. The characteristic values for $N_{p,Rk}$ and ψ_p given in *Tables H.2* and *H.3* of ***clause H.3(3)*** correspond to a pressure under the head of many times the cylinder strength. In *Table H.2* the head diameter is given for 16-mm studs and not for others. It is not a value needed for design. The reason is because EN ISO 13918, Table 10, permits a reduction to the head diameter for 16-mm studs in certain applications, and ensures compliance with the minimum required by *clause 8.6.10.7(2)* of 1.5*d*. *Clause H.3(3)*

With the partial factors proposed here, $N_{p,Rd}$ is lower than $N_{s,Rd}$ for grade C30/37 and weaker concretes, and $N_{s,Rd}$ is lower for stronger concretes.

Concrete cone failure is the mode most likely to govern. *Formula H.5* in ***clause H.3(4)*** gives the resistance $N_{c,Rk}$ in symbols that are now discussed in turn. ***Clause H.3(5)*** defines the maximum characteristic resistance, $N^0_{c,Rk}$, in *Formula H.6*. *Clause H.3(4)* *Clause H.3(5)*

The value of constant k_1 is based on the assumption that the surrounding concrete is cracked. Its definition refers to a 'large web opening', a term which is used in *Annexes D* and *E* in EN 1994 and not explained or defined. The title of EN 1993-1-13, 'Beams with large web openings', provides a clue, but there is no definition there. It clearly means all openings within the scope of its rules. 'Small web opening' is not defined either. Clause 8.2.2.11 of EN 1993-1-1 provides the guidance needed: 'Holes for fasteners may be neglected, but larger openings should be taken into account.'

Clause H.3(6)

The ratio $A_{c,N}/A^0_{c,N}$ in ***clause H.3(6)*** is concerned with the overlap of failure cones for individual studs. A cross-section through the failure surface for an isolated stud is shown in Figure 18.1(a), assuming (as is likely here) that it does not intersect a free edge of the slab. Resistance is assumed to be proportional to the base area, which is given as $9h_{ef}^2$, so the surface is in fact a pyramid, not a cone. This anomaly occurs also in EN 1992-4, and enables the method to be simplified. The minimum thicknesses of the head of a stud of height h_{sc} are given in EN ISO 13918 as approximately half of the stud diameter d_s, so from *Formula H.7*, $h_{ef} \approx h_{sc} - 0.5d_s$.

Figure 18.1. Base areas for concrete cone failure

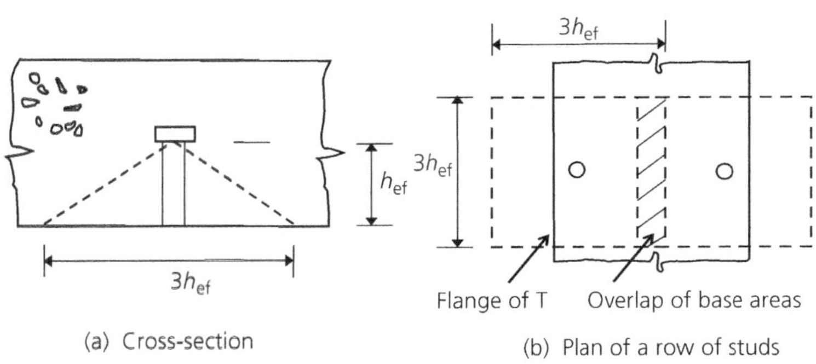

The effective base area per stud is reduced below $9h_{ef}^2$ where the centre-to-centre spacing of studs is less than $3h_{ef}$ in either direction, so that their areas overlap, Figure 18.1(b). *Formula H.8* allows for this. For example, in the notation of *Formula H.8*, the base area per stud is found for a group of many rows of studs with $n_s = 2$ and transverse and longitudinal spacings $2h_{ef}$. Then, $s_x = b_x = 2\ h_{ef}$, and $s_y = 2h_{ef}$. The formula gives $A_{c,N}/A^0_{c,N} = [2h_{ef}(3h_{ef} + 2h_{ef})]/9h_{ef}^2 n_s = 10/18$.

Multiplying by $9h_{ef}^2$ gives the correct result: each stud has a base area of $5h_{ef}^2$.

Worked examples find that, except where the top T has a wide flange, this overlap of failure pyramids leads to low values for the resistance of the group to tension. This and the short length of span available ($2h_c$ for a solid slab, *clause E.4(10)*) can make it impossible to fit in the studs required.

For example, suppose that the steel flange is quite wide, at 300 mm. The closest that 19 × 100 mm studs can be placed to the edge of the flange is 20 + 10 = 30 mm. Their effective height is about 85 mm, so their spacing, at 240 mm, is below $3h_{ef}$. The resistance of the group is not increased by adding a third stud at the centre of the row, although the stud spacing still exceeds the minimum of $2.5d_s$ given in EN 1994-1-1.

Clause H.3(7)

In ***clause H.3(7)***, a transverse-spanning composite slab with typical troughs will be penalised by factor $\psi_{s,N}$ – for example, the sheeting shown in Figure 10.5(a) has a relatively wide trough, with $b_{bot} = 136$ mm. For 19-mm centrally placed studs 95 mm high, $h_{ef} = 85$ mm. *Formula H.9* gives $\psi_{s,N} = 0.7 + 0.3 \times 68/(1.5 \times 85) = 0.86$.

Clause H.3(8)

The factor $\psi_{re,N}$ in ***clause H.3(8)*** would apply for studs with length after welding below about 110 mm, but not to longer ones, and where the local reinforcement is dense. The rules for reinforcement are copied from a clause in EN 1992-4 which adds 'the conditions a) or b) shall be fulfilled in both directions in case of reinforcement in two directions'. In *Annex H*, this can be inferred from the words '*reinforcement between which the studs are installed*'. In practice, it should be possible to detail reinforcement or stud connectors such that $\psi_{re,N} = 1$.

Clause H.3(9) gives factor $\psi_{ec,N}$ for the eccentricity e_N of the resultant tension force on a group of connectors. *Formula H.11* shows that there is a penalty for any value of the eccentricity, increasing for closely spaced studs. *Figure E.2* shows that in the longitudinal direction e_N can be taken as zero. In the transverse direction, e_N is also zero where the centre of the group of studs is above the steel web. *Clause H.3(9)*

Example 18.1: resistance of stud connectors to direct tension

The connectors that resist tension are as for Example 15.1: four rows of 19-mm studs in pairs 118 mm apart, with longitudinal spacing 100 mm (Figure 15.2), in a solid slab 150 mm thick with C25/30 normal-density concrete. Recommended values of NDPs are used here. From Example 14.1, $P_{Rd} = 73.1$ kN.

From *clause H.3(2)*, the design resistance to tension is 128/1.55 = 82.6 kN. From *clause H.3(3)*, the design resistance to concrete pull-out failure is 1.2 × 75/1.5 = 60.0 kN, both far above the resistance to cone failure, found below.

The length after welding (LAW) of the studs is 95 mm; the head is about 10 mm thick, so from *clause H.3(5)*, the effective embedment depth h_{ef} is 85 mm. From *Formula H.6*,

$$N^0_{c,Rk} = (8.5 \times (25)^{1/2} \times 85^{1.5})/1000 = \mathbf{33.3\ kN}$$

For the correction factor for base area, $b_x = 100$ mm, $n_s = 2$, $s_y = 118$ mm, and *Formula H.8* gives $A_{c,N}/A^0{}_{c,N} = \mathbf{0.29}$, a severe penalty.

Factor $\varphi_{s,N}$ is 1.0 from the Note below *Formula H.9*, assuming this is an 'internal composite beam with a solid slab'. Without this concession, $\varphi_{s,N}$ would be 0 71. 'Internal' is not defined. It probably means beams with concrete flanges on both sides of the web. Then, *Formula H.9* should be used for composite edge beams.

Factor $\varphi_{re,N}$ is 1.0 as the reinforcement spacings exceed 150 mm (Example 15.1). Factor $\varphi_{ec,N}$ is 1.0 as the eccentricity of $P_{ten,Ed}$ is zero. Hence, from *clause H.3(4),*

$$N_{c,Rd} = 33.3 \times 0.29 \times 1/1.5 = \mathbf{6.44\ kN\ per\ stud}$$

For eight studs, $\mathbf{P_{ten,Rd}} = 6.44 \times 8 = \mathbf{51.5\ kN}$. It does not seem to be possible to increase this rather low result – for example, use of longer studs (LAW = 120 mm) reduces $A_{c,N}/A^0{}_{c,N}$ from 0.29 to 0.21, which is counterintuitive. There is further comment in Example 15.1.

Roger P Johnson and Stephen J Hicks
ISBN 978-1-83662-921-4
https://doi.org/10.1108/978-1-83662-920-720253022

CHAPTER 19
Annex I (Normative). Additional rules for shallow floor beams

This chapter corresponds to *Annex I* of EN 1994-1-1, which has the following clauses.

- Use of this annex — *Clause I.1*
- Scope and field of application — *Clause I.2*
- Structural analysis — *Clause I.3*
- Ultimate limit states — *Clause I.4*
- Serviceability limit states — *Clause I.5*

Structures of this type, also known as 'slim-floor construction', are a new subject for Eurocode 4. The 2012 edition of this guide stated that their flexural behaviour was 'quite well understood', but there were no rules then for shear resistance or partial shear connection. It was not clear whether the properties of stud shear connectors were compatible with shear resistance from partial encasement in concrete. Their use is now permitted. The use of hollow-core slabs in shallow floors is illustrated in Example 19.1.

I.2 Scope and field of application

Beams within the scope of *Annex I* have a wide variety of cross-sections, illustrated in *Figures I.1* and *I.2* and *Table I.1*, and here in Figures 19.1 and 19.2. The *Figures* in EN 1994-1-1 include examples that have no in situ concrete above the top flange of the steel section. EN 1994-1-2 requires reinforced topping at least 30 mm thick for fire resistance periods of 60 min or above. In addition, there may be bottom longitudinal reinforcement, which will be essential if stirrups are required for vertical shear.

The essential difference from other types of composite floor structures is that in situ concrete encases the full depth of the steel web or webs. The deck slab may be concrete (Figure 19.1) or composite with profiled sheeting that may be over 200 mm deep (Figure 19.2). The only additional depth of the structure is typically the thickness of the steel bottom flange or of a steel plate attached to it.

Figure 19.1. Cross-sections of shallow floor beams in Types I to III

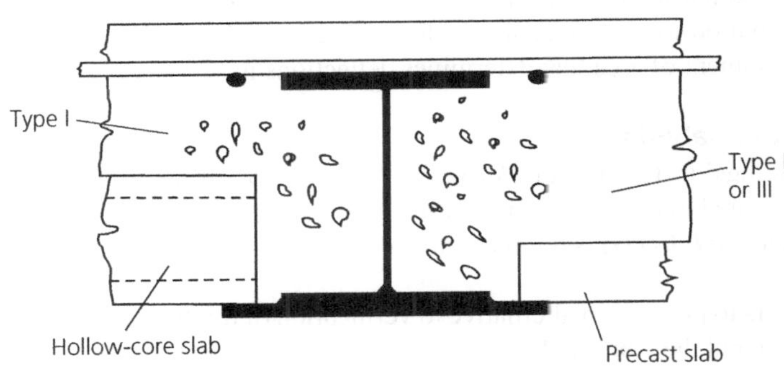

This plate supports permanent formwork, which may be precast hollow-core slabs of thickness that may be a high proportion of the total depth, or precast solid slabs with thickness and resistance just sufficient to support the wet concrete without excessive deflection, taking account of any propping that may be used. *Table I.1* refers to floors with the thick precast slabs as Type I, meaning that the flexural stiffness of the concrete above the slabs is not a 'significant' part of the total flexural stiffness. Type II slabs are those where it is 'significant', which it will be where solid precast slabs are used. Where the

Figure 19.2. Asymmetric steel beam and composite slab with deep decking (courtesy of the Steel Construction Institute)

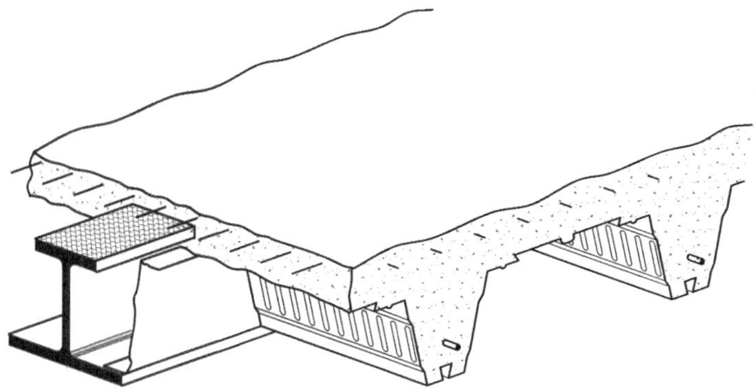

steel section is a shallow box of depth less than 60% of the overall depth, its flexural stiffness may be negligible compared with that of the slab. Such sections are named as Type III. Design for vertical shear (*clause I.3.2*) depends on the type of the section.

These types of concrete encasement are additional to the partial encasement treated in *clause 8.3*. Steel members of closed cross-section, with or without concrete infill, are included in *Figure I.1*. They do not appear elsewhere in *EN 1994-1-1* for beams. The much larger box girders are treated in EN 1994-2.

Clause I.2(3)

Steel sections in Class 4 are unlikely to be used in a shallow floor, and Class 3 can be treated as effective Class 2, so from ***clause I.2(3)***, their slenderness is not a limitation.

Clause I.2(5)

Clause I.2(5) places the design of the concrete slab outside the scope of *Annex I*, for reasons explained in its two Notes. When the slabs are erected, the supporting forces can be concentrated near their corners, making the assumption of uniform support unsafe in both normal and fire conditions. Existing guidance recommends limiting the design shear force at the ends of the hollow-core units (ECCS/IPHA, 1998).

Lightweight concrete is outside the scope of *Annex I*. So is fatigue loading, which would probably exclude floors designed for fork-lift trucks. Some other limitations are referred to below.

No guidance is given in *Annex I* on whether precast slabs can be assumed to contribute to the resistances of a shallow-floor beam. The reasons may be lack of evidence from research, and uncertainty over how much longitudinal stress is transferred across the joints between the long edges of the slabs. At present, no contribution to resistance to bending or vertical shear should be assumed. The slabs increase the second moment of area that determines deflections in service, *clause I.5(4)*.

I.3 Structural analysis

Clause I.3.1

Clause I.3.1(1)

The words 'for buildings' in the title of ***clause I.3.1*** do not appear elsewhere in *Annex I*. Its other clauses, though not limited in this way, were drafted with buildings in mind. The use of the rigid plastic global analysis (***clause I.3.1(1)***) is not recommended, because of problems with rotation capacity.

Clause I.3.1(2)

Clause I.3.1(2) appears to provide an alternative to verification of rotation capacity, where rigid plastic global analysis is used, but the wording for Type I sections is obscure. The 'condition in 7.4.4(6)' is for steel grades above S355, made more severe here by application to all steel grades but with a 'where it can be demonstrated' condition. The words appear to mean that the 'all grades' rule does not apply where that condition cannot be met. That cannot be the intention.

The first paragraph of *clause I.3.1(2)* does not explain how the effects it lists should be 'taken into account'. Its first bullet point requires demonstration that the cross-section 'reaches' $M_{pl,Rd}$ in both sagging and hogging bending. Redistribution for a given design ultimate loading may assume $M_{Ed} = M_{pl,Rd}$ at an internal support or supports, and the resulting sagging moment M_{Ed} can then be found. It is not clear whether 'demonstrate' means showing that this bending moment does not exceed sagging $M_{pl,Rd}$.

A conservative interpretation of these rules for Type I sections is to use elastic global analysis with no redistribution for Class 3 and 4 sections and steel grades above S355. For redistribution for Class 1 and 2 sections, use the rules given in *clause 7.4.4(6)* for class 2 sections, but only for steel grades not exceeding S355.

For Type II and Type III sections, rules in EN 1992-1-1 apply. These are too complex to repeat here. They take account of the type of reinforcement, the depth to the neutral axis at ULS, the effective depth, and strain ε_{cu}, a property of the concrete.

The rules on classification of the steel section in ***clause I.3.2*** depend on the extent to which out-of-plane deformation of each steel plate in compression is restrained by adjacent concrete. Where the rules are not complied with, the reference to EN 1993-1-1 means that restraint is assumed to be negligible. *Clause I.3.2*

The rule for compression flanges in ***clause I.3.2(1)*** is not relevant for regions of hogging bending, where *clause I.3.2(4)* applies. The reference in ***clause I.3.2(2)*** to concrete between flanges being 'fixed to the web' (see examples in *clause 7.5.3(2)*) is to prevent a web buckling away from the side where casing is present. *Clause I.3.2(1)* *Clause I.3.2(2)*

Closed sections without concrete infill are given no benefit from the concrete on the outside of the section (***clause I.3.2(5)***) because the webs could buckle inwards. ***Clause I.3.2(6)*** for outstand elements of closed sections refers to rules in *clause 7.5.3* that are more liberal than those in EN 1993-1-1. *Clause I.3.2(5)* *Clause I.3.2(6)*

I.4 Ultimate limit states

Resistance to bending

The reduction in bending resistance caused by transverse bending of the bottom plate, ***clause I.4.1.2(2)***, depends on the assumed position of the shear force from the floor slab. Midway along the bearing length of a precast slab is a reasonable assumption, for both precast and in situ slabs. Use of the flange tip is conservative. *Clause I.4.1.2(2)*

Precast concrete installers have difficulty in ensuring that bearing lengths are equal at the ends of a unit, so there is a risk of 'fall-through' during erection if the tolerances are adverse. This is discussed in Rackham *et al.* (2006).

Formula I.1 for the ratio $f_{y,red}/f_y$ can be conservatively replaced by the much simpler $f_{y,red}/f_y = 1 - \eta_m^2$. Its results are within 2% for $\eta_m \leq 0.4$, with a maximum reduction in $f_{y,red}/f_y$ (from 0.43 to 0.31) at $\eta_m = 0.82$.

The stiffness in torsion of a closed cross-section is likely to be sufficient to prevent the eccentricity of the force from the slab causing rotation of an edge beam (***clause I.4.1.3(2)***). For an open cross-section, *Figure I.3(b)* shows the tension F_t in an anchor bolt that reduces the clockwise rotation of the steel beam caused by the eccentricity of loading on its bottom flange. The rules for provision of robustness may require connection of the type shown, and could be used for its design. *Clause I.4.1.3(2)*

Resistance to vertical shear

The method of ***clause I.4.2(1)*** for Type I cross-sections is to neglect any contribution from the concrete slab. The reference to a 'solid web' means that beams with web openings are excluded. Holes for reinforcing bars are permitted. *Clause I.4.2(1)*

Clauses I.4.2(2) to ***(5)*** concern sharing the resistance to shear between the slab and the steel section. This will usually be necessary for Type III sections; less so for Type II. There is background information in Schäfer *et al.* (2018). *Clause I.4.2(2)*

Paragraph (3) refers to the provision of stirrups. This is practicable in sections of Types II and III, but adds to cost. For its shear resistance, the in situ concrete is treated as a T-beam with web of width given by *Formula I.2*, which refers to 'the width of the bottom plate of the steel section'. This should include any extension of the steel flange by a welded plate.

'The slab reinforcement in tension' here refers to the in situ construction, not the precast slab. Where bottom reinforcement is provided, the second term in the formula, $7d$, can lead to a significant shear resistance for the concrete component, from the method of Eurocode 2.

In beams of Types II and III, the majority of the vertical shear is likely to be transferred to a supporting steel column by way of the steel component. *Clause I.4.2(4)* requires its shear resistance to be sufficient for the whole vertical shear V_{Ed}. However, part of shear V_{Ed} will be resisted by the concrete, and its sudden transfer into the steel component risks failure in punching shear. This is why a minimum shear resistance for the concrete slab is required here.

Clause I.4.3.2(1) **_Clause I.4.3.2(1)_** on the use of partial shear connection refers to *clause 8.6.3*. This requires the cross-sections to be in Class 1 or 2 and the shear connectors to be in ductility category D2 or D3. This can rule out reliance on transverse bars through webs, in category D1 (*clause I.4.3.3(1)*). *Clause 8.6.3.3* includes a rule for steel sections with flange-area ratios (bottom/top) of three. For sections with a ratio
Clause I.4.3.2 (2) greater than three, **_clause I.4.3.2(2)_** repeats that formula, but reduces its scope for spans up to 18 m, rather than the previous 20 m.

Clause I.4.3.3(3) The shear resistance in **_clause I.4.3.3(3)_** assumes that the transverse bars are plastic in shear, so they should be fully anchored, to EN 1992-1-1, *clause I.4.3.3(7)*. The method takes no account of the dowel effect of the concrete at the hole in the steel web. Research on this effect has found it to be significant. This extra resistance is allowed for in the Eurocode for composite dowels (BSI, 2026, CEN/TS 1994-1-102). This can be assumed to be an 'advanced method' (**_clause I.4.3.3(4)_**).

I.5 Serviceability limit states

Clause I.5(1) Reference is made in **_clause I.5(1)_** to *clause 9* for the calculation of stresses in the steel section, where, for beams within the scope of this Annex, *clause 9.2.2(3)* gives limits to stress only for concrete and reinforcement. The rule for the steel section given here is relevant to the combined stress in its bottom flange plate from longitudinal and transverse bending. Elastic theory should be used for their combination, unlike the corresponding rule in *clause I.4.1.2(2)* for ultimate limit states.

The rest of this clause concerns the second moment of area of the cross-section, for use in predicting
Clause I.5(4) deflections. For shallow floor beams of Type I, **_clause I.5(4)_** refers to *clause 9.3.1* which refers to *clause 7*. It will be necessary to decide whether to assume the section is 'cracked' or not. One could find I for both the uncracked and cracked reinforced sections and use an intermediate value based on judgement. The precast slabs must make some contribution to stiffness.

Clause I.5(5) **_Clause I.5(5)_** gives a complex method for finding the longitudinal stiffness of sections of Types II and III (Kuhlmann and Hauf, 2015). Its scope is a steel section composite with what the clause calls 'the concrete slab' – that is, an in situ concrete T-section plus precast or composite slabs. The clause has the following problems.

- It is not clear whether symbols $A_{1,L}$, $I_{1,L}$ and $S_{i,L}$ are calculated for cracked or uncracked sections, and whether account need be taken of the method of construction, where either or both of the slabs and the beam could be propped.
- Calculation of M_{crack} would be difficult in practice and depend on the method of construction.
- It is not clear why there is an upper limit to areas of reinforcement, when more reinforcement increases stiffness.

Example 19.1: shallow floor with precast hollow-core slabs

A shallow floor structure for a building consists of composite beams of span 8.0 m that support precast prestressed hollow-core slabs of span 7.0 m with a topping of in situ normal-density concrete. Both spans are to be designed as simply supported. The dimensions assumed here (Figure 19.3) are based on much preliminary calculation. It can be difficult to predict which checks will govern design of a shallow floor. It could be torsion during construction, or deflection, or even the detailing of the region where the beam supports the slabs. Routine calculations are not given in full detail. This Example does not include resistance to fire or a check on robustness.

Characteristic loading

- Imposed load: 4.0 kN/m^2 plus an allowance of 1 kN/m^2 for partitions
- Construction loading during concreting: 0.75 kN/m^2, with $\gamma_F = 1.5$ for it and for wet concrete at density 25 kN/m^3

- Construction loading after concreting: 1.0 kN/m^2 with $\gamma_F = 1.35$
- Permanent loading: concrete at 24 kN/m^3, structural steel and reinforcement at 78 kN/m^3, and 0.40 kN/m^2 for ceiling and floor finishes.

Materials

- Normal-density concrete, class C30/37, with $f_{cd} = 0.85 \times 30/1.5 = 17$ N/mm^2
- Creep: from EN 1992-1-1, the short-term elastic modulus is $E_{cm,28} = 33.5$ kN/mm^2. From *clause 7.4.2.2(11)* the modular ratio can be taken as twice the short-term value for all loading, so $\boldsymbol{n} = 2 \times 210/33.5 = \mathbf{12.6}$
- Reinforcement with design strength $f_{sd} = 500/1.15 = 435$ N/mm^2
- Structural steel grade S355 with $f_{yd} = 355$ N/mm^2.

Hollow-core slabs

Data from the manufacturer: slabs 1.2 m wide, 250 mm thick, with nine cores, prestressed by six strands and four wires. Tabulated properties per slab converted to properties per metre width are as follows. There is no information on camber or creep deflection.

- Area of cross-section: 0.140 m^2/m; (weight at 24 kN/m^3: 3.36 kN/m^2)
- $10^6 I_y = 1021$ m^4/m in 'concrete' units (or 81.0 m^4/m in 'steel' units, with $n = 12.6$)
- Elastic neutral-axis depth: 0.128 m below top of slab
- Section modulus: $10^6 W_{bot} = 8370$ m^3/m
- Bending moment (serviceability): $M_S = 87.1$ kN m/m (from the manufacturer)
- Bending resistance (ULS) $M_{Rd} = 129.9$ kN m/m
- Shear resistance: $V_{Rd} = 156.4$ kN/m.

Structural steel

From *clause I.1(3)*, *Annex I* is applicable only to steel cross-sections in Class 1, Class 2 or effective Class 2. The top flange of the steel may be fully in compression during construction. The lightest UC column section in the '305' group with flanges in Class 2 to EN 1993-1-1 was chosen: 305 × 305 UC 118, weight 1.157 kN/m. For calculations, the web fillets are replaced by extra flange thickness of the same area, increasing thickness t_f from 18.7 mm to 19.0 mm.

The properties of the UC section are: $A_a = 0.0150$ m^2; $10^6 I_y = 276.7$ m^4; and as shown on Figure 19.3.

Figure 19.3. Cross-section of beam, showing I-section equivalent to 305 × 305 UC 118, modelled as three rectangular components (dimensions in mm)

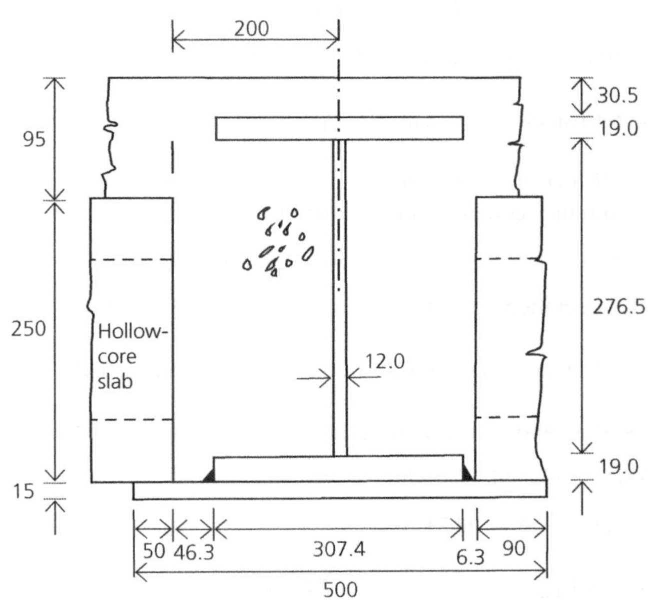

The added bottom-flange plate needs sufficient resistance to bending for the load from the slab. Its dimensions are $b_{pl} = 500$ mm, $t_{pl} = 15$ mm, after an initial trial with 10-mm plate. The distance between the centres of area of the two flanges, h_f, is 0.305 m. The new area of the steel cross-section is 0.0225 m^2. For calculation of stresses from torsion, the combined bottom flange section is replaced by a single rectangle (Figure 19.6), with the same total area and minor-axis moment of area, at a level such that $h_f = 0.305$ m.

Design loadings and action effects for construction and in service

It is assumed that the slabs are erected unpropped. The beam is propped, with the assumption that propping provides no resistance to twisting of the beam. The maximum torsion in the beam occurs where precast slabs are erected on one side only (factor for ULS, 1.35), with the topping as wet concrete, plus construction loading (factor 1.5). The design (factored) loads, with symbols q (imposed), g (permanent) and w (total), are

- Weight of 95-mm in situ topping: 2.375 kN/m^2 (wet) and 2.28 kN/m^2 (dry)
- Assumed weight of reinforcement: 0.04 kN/m^2
- Construction (SLS, wet), slab: $q = 0.75 + 2.375 = 3.125$ kN/m^2; $g = 3.36 + 0.04 = 3.40$ kN/m^2
- Construction (SLS, dry), slab: $q = 1.0 + 2.28 = 3.28$ kN/m^2
- Construction (ULS, wet), slab: $w = 3.125 \times 1.5 + 3.4 \times 1.35 = 9.28$ kN/m^2
- Construction (ULS, dry), slab: $w = 1.0 \times 1.5 + (3.4 + 2.28) \times 1.35 = 9.17$ kN/m^2
- In service (SLS), slab: $q = 1.0 + 4.0 = 5.0$ kN/m^2; $g = 3.36 + 0.04 + 2.28 + 0.40 = 6.08$ kN/m^2
- Final (ULS), slab: $w = 5.0 \times 1.5 + 6.08 \times 1.35 = 15.7$ kN/m^2

The nominal bearing length of the slabs on the steel flange plate is assumed to be 70 mm, with an allowance of ± 20 mm for tolerances. The two extreme positions are shown on the left and right sides of the web in Figure 19.3. For maximum torsion, the load on the beam is assumed to act at the centre of the 50-mm minimum bearing length (a span of 6.55 m). The length of each slab, for load calculation, is $6.55 + 2 \times 0.025 = 6.6$ m.

Hence, the load from the slab (ULS) is $w_1 = 9.28 \times 3.3 = 30.6$ kN/m. The eccentricity e_1 from the web is $(7 - 6.55)/2 = 0.225$ m. The torque is 6.89 kNm/m, if there is no load on the other side of the beam (Figure 19.6).

There is additional weight from the loads on width 0.2 m on one side of the web. For torsion during construction, these design loads per unit length of beam (ULS), from width 0.2 m and depth about 0.32 m are

$$w_2 = (0.2 \times 0.32 \times 25 + 0.2 \times 0.75) \times 1.5 = 2.62 \text{ kN/m}$$

The eccentricity e_2 is 0.10 m, so the torque is 0.26 kN m/m (loading on one side of the beam).

The total torque for the 8-m span of the beam is $\boldsymbol{T_{Ed}} = 8 \times (6.89 + 0.26) = \mathbf{57.2\ kN\ m}$

For this Class 2 section, ULS verification can be based on plastic resistances to the total loading, ignoring stresses remaining from the sequence of construction. The loading from one 0.2-m width is

$$0.2 \times ((0.32 \times 24 + 0.04 + 0.4) \times 1.35 + 5 \times 1.5) = 3.69 \text{ kN/m}$$

The design weight of the steel member is $0.0225 \times 78 \times 1.35 = 2.37$ kN/m.

The corresponding characteristic loads, for deflections, are 2.62 and 1.76 kN/m, respectively.

From the loading for the slab, above, the final design load for the beam in service is

$$2 \times (15.7 \times 3.3 + 3.69) + 2.37 = 113.4 \text{ kN/m}$$

At mid-span, $\boldsymbol{M_{Ed}} = 113.4 \times 8^2/8 = \mathbf{907\ kN\ m}$

The design vertical shear is $\boldsymbol{V_{Ed}} = 113.4 \times 8/2 = \mathbf{454\ kN}$

Elastic properties of cross-sections

Structural steel section: neutral axis, 0.243 m below top of the slab; $10^6 I_{at,y} = 413$ m^4

Cracked composite section in sagging bending, full interaction,

- effective flange width, $L/4 = 2.0$ m
- neutral axis, 0.158 m below top of the slab; $10^6 I_{comp} = 776$ m^4 in 'steel' units.

The topping is 0.095 m thick, so the concrete assumed to be uncracked extends 63 mm into the hollow-core slabs, which may include part of the cores. This is not allowed for. It is assumed that the joints between slabs are grouted, and are as stiff as concrete.

Deflection, hollow-core slabs

It is assumed that the concrete above the slabs hardens with its 'wet' density. Without construction load, the loading for deflection is $2.375 + 3.36 + 0.04 = 5.78$ kN/m^2. The span between assumed support points is 6.55 m. The mid-span deflection is

$$5 \times 5.78 \times 6.55^4 / (384 \times 210 \times 81.0) = 0.0081 \text{ m (span / 809)}$$

In service, the topping has increased I_y for the slab to 224×10^{-6} m^4/m. The additional loading (imposed, partitions and finishes) is 5.4 kN/m^2. The effective span is now 7 m. The deflection is

$$5 \times 5.4 \times 7.0^4 / (384 \times 210 \times 224) = 0.0036 \text{ m}$$

The **total deflection, 11.7 mm**, is span/598. This will be reduced by the initial camber of the slabs, which should be available from the manufacturer. Their subsequent creep deflection is assumed to have been allowed for by the use of modular ratio 12.6 for the cross-section including the topping.

Deflection in service, composite beam

The thickness of the concrete flange is less than 30% of the depth of the section, so it can be assumed to be Type I. From *clause I.4(2)*, discussed above, a cracked section is assumed.

From the values given above, the loading is

$$6.6 \times (5.0 + 6.08) + 2 \times 2.62 + 1.76 = 80.1 \text{ kN/m}$$

The **mid-span deflection** is $5 \times 80.1 \times 8^4/(384 \times 210 \times 776) =$ **0.026 m** (span/308).

This ignores any help from the precast slabs and is certainly conservative. Assuming an uncracked section, with I increased from 776 to 891 m^2 mm^2, the deflection is 23 mm. Thus, even with propped construction for the beam, the total deflection is quite high. This determined some of the dimensions assumed.

Bending of the bottom plate

The elastic bending stress from the slabs supported by the bottom plate affects its design yield strength in the longitudinal direction. The bending moment is calculated at the centre of the fillet weld to the bottom flange of the UC section. For a 10-mm weld, the lever arm is $46.3 + 25 - 5 = 66.3$ mm (Figure 19.3).

How is the vertical shear from the loading after construction shared between the bottom plate and the flange of the UC section? As the flange region is stiffer, it is reasonable to assume that it resists the extra load. Then, the design loading for the plate is that from the slab at the 'wet concrete' stage, which is $9.28 \times 3.3 = 30.6$ kN/m.

The bending moment is $30.6 \times 0.0663 = 2.03$ kN m/m.

From *clause I.4.1.2(2)*, the bending resistance of the plate, $m_{ybt.Rd}$, is

$$355 \times 0.015^2 \times 1000 / 4 = 20.0 \text{ kN m/m}$$

The factor η_m is $2.03/20 = 0.105$. From *Formula I.1*, the reduced yield strength for the plate is $f_{y,red} = 0.99 f_{yd}$. The effect on $M_{pl,Rd}$ is negligible.

Plastic resistances of the composite beam

Resistance to bending

The resistance of the steel cross-section is needed in the design for partial shear connection. Allowing for the slight reduction in the yield strength of the bottom plate, the results are $\boldsymbol{M_{a,pl,Rd}} = \mathbf{849\ kN\ m}$, with neutral axis just within the steel bottom flange at depth 332 mm below the top of the concrete slab.

Where hollow-core slabs are used in composite T-beams, *clause J.4.1(1)* limits the effective width of the concrete flange for structural analysis (ULS) to the gap between the slabs plus the sum of their lengths infilled with concrete. From *clause J.7(2)*, these lengths should be at least 500 mm, and have transverse reinforcement. This is presumably also required where hollow-core slabs are used in shallow floors, though *Annex I* does not refer to it.

For effective width 'for the determination of action effects', *clause I.3.1(2)* refers to *clause 7.4.1.2*, which gives a value about 2 m. No width is given for resistance to bending, so the rule of *clause J.4.1(1)* (above) is followed. Allowing for tolerance in the placing of the slabs, b_{eff} is assumed to be 1.3 m, shown in Figure 19.4 as 650 mm each side of the steel web.

Assuming full shear connection, plastic theory gives the resistance as $\boldsymbol{M_{pl,Rd}} = \mathbf{1466\ kN\ m}$, with neutral axis in the web, at depth $z_{pl} = 139$ mm below the top of the slab. The stress blocks are shown in Figure 19.4. The method assumes that the whole of the steel section is at its yield stress in tension, and that the areas in compression are at a stress of $2f_{yd}$. The 99-kN tensile force shown corrects this double use of the area of the steel top flange, so the compressive force in the concrete is 2982 kN. A similar correction for the web would be negligible. Part of the top 44 mm of the hollow-core slabs is being assumed to resist compression.

Figure 19.4. Stress blocks for calculation of $M_{pl,Rd}$ of composite beam with full shear connection (units: kN and mm; not to scale)

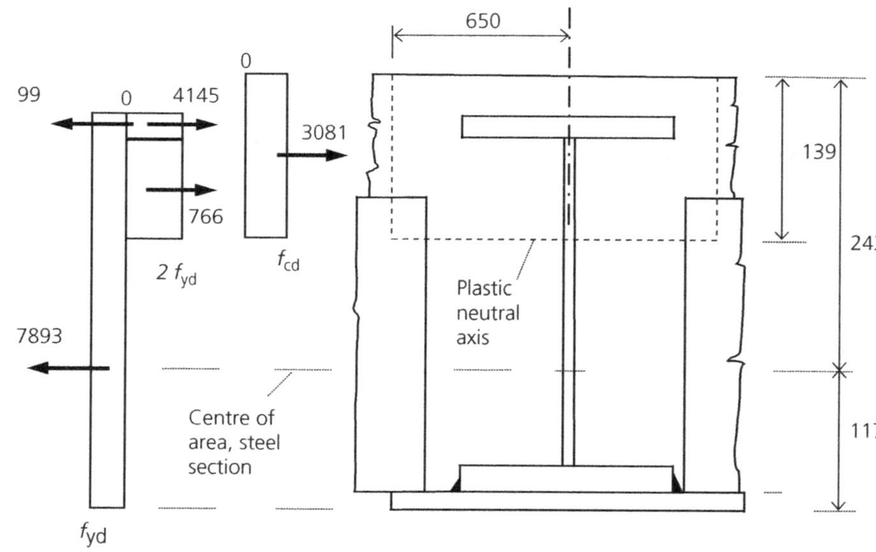

Clause 8.2.1.2 requires resistance $M_{pl,Rd}$ to be reduced to allow for premature crushing of the concrete flange. Its factor β depends on the ratio z_{pl}/h, where h is the overall depth of the composite section, 0.360 m here, giving $z_{pl}/h = 0.39$. From *Figure 8.3*, with S355 steel, $\beta = 0.93$, and $\boldsymbol{M_{Rd}} = \mathbf{1364\ kN\ m}$. M_{Ed} was found, above, to be 907 kN m, or 0.665 M_{Rd}, so partial shear connection is possible.

Vertical shear

From clause 8.2.6(3)(a) of EN 1993-1-1 the shear area of the steel web is 0.0043 m^2. Its plastic resistance to vertical shear is $V_{Rd} = 0.0043 \times 355\,000/(3^{0.5}) = 881$ kN.

As V_{Ed} is only 454 kN, no further check is necessary.

Lateral buckling

During propped construction, there is no major-axis bending of the beam. In service, the top flange is restrained by concrete.

Longitudinal shear, ultimate limit state

Partial shear connection

The simplest design method is often to make checks on a guessed degree of shear connection. Here 0.55 is assumed, based on an earlier trial. As this exceeds 0.5, no allowance need be made for the effect of longitudinal slip on deflections (*clause 9.3.1(4)(b)*).

The new resistance M_{Rd} is found by the interpolation or 'simplified' method shown as line 6 in *Figure 8.6*. From *Key 8* in the figure, the rule given by line 6 is as shown in Figure 19.5.

Using $M_{pl,Rd}$ and $M_{a,pl,Rd}$ as found above, the resistance for $\beta = 0.55$ is $\mathbf{M_{Rd}}$ **(η) = 1188 kN m**, which exceeds M_{Ed} at 907 kN m.

Figure 19.5. Bending resistance, partial shear connection, by the interpolation method (units: kNm)

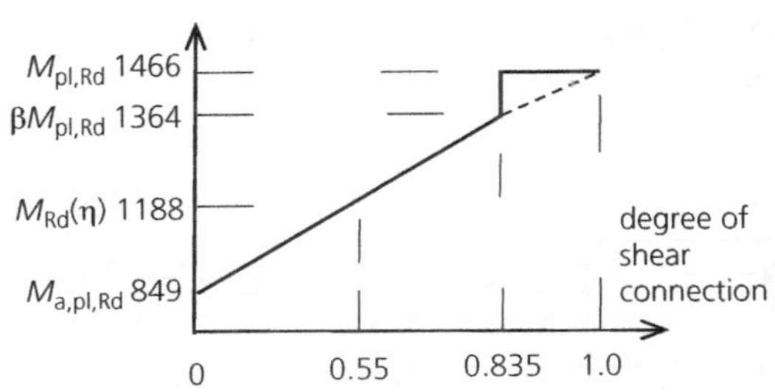

Formula 8.12 gives the lower limit to the degree of shear connection as

$$\eta \geq \eta_0 \rho_m{}^2 k_{up} \geq \eta_{min}$$

Term η_0 depends on the ratio A_{bf}/A_{tf} of the areas of the steel flanges, which for this beam is

$$(0.307 \times 0.019 + 0.5 \times 0.015)/(0.307 \times 0.019) = 2.29$$

Interpolation between *Formulae 8.14* and *8.16* with $L_e = 8.0$ m gives $\eta_0 = 0.70$

For propped construction, $k_{up} = 1.0$

From results above, $\rho_m = 907/(0.95 \times 1132) = 0.843$

From *Formula 8.13*, $\eta \geq 0.7 \times 0.843^2 = 0.50$, so the use of $\eta = 0.55$ is satisfactory.

The compressive force in the slab for $\eta = 1$ is 2982 kN, so shear connection is required for a longitudinal force of $0.55 \times 2982 =$ **1640 kN**. The conditions of *clauses 8.6.3.3(2)* and *I.4.3.2(1)* for partial shear connection are satisfied if the shear connectors are in ductility category D2 or D3. Reinforcing bars through holes in the steel web are likely to be in ductility category D1 (*clause I.4.3.3*). They will be provided to resist longitudinal shear in the concrete slab, but are not available as shear connection. Stud connectors will be used, welded to each side of the steel web. The design shear resistance of a 19-mm stud in C30/37 concrete is 82 kN. Pairs of studs at 0.4 m, uniformly spaced along the shear span of 4.0 m, provide shear resistance of 1640 kN, which is just sufficient.

It may be possible to omit the stud connectors, as CEN/TS 1994-1-102, 'Composite dowels' provides in its clause B.6(3) a method for reinforcing bars to be treated as category D2. It is available, as *clause I.4.3.3(4)* permits an 'advanced method' to be used for the shear resistance of transverse bars. The penalty is that for plastic design, as used here, the shear action effect must be increased by a factor 2.4, so $v_{L,Ed} = 1640 \times 2.4/4 = 984$ kN/m for the 4-m shear span here. Its clause B.5(2) gives $P_{Rk} = 122$ kN per bar, based on concrete failure, as it does not depend on bar diameter (≥ 12 mm). The minimum bar spacing is 125 mm (*clause I.3.3.3(2)*), so the maximum shear resistance $v_{L,Rd}$ is $8 \times 122/1.25 = 781$ kN/m, which is not enough here.

Longitudinal shear in the concrete flange, and transverse reinforcement

The depth of the compressive-stress block in the concrete at $M_{pl,Rd}$, 139 mm, is reduced to 77 mm where $\eta = 0.55$. For a vertical cross-section at an edge of the steel top flange, the longitudinal shear is $1640 \times (7.0 - 0.307)/(2 \times 7 \times 4) = 196$ kN/m.

From clause 8.2.5(1) of EN 1992-1-1, the longitudinal shear stress on a shear plane of thickness h_f (here taken as the thickness of the topping, 95 mm) is

$$\tau_{Ed} = \Delta F_d / (h_f . \Delta x) = 196/95 = 2.06 \text{ N/mm}^2$$

For the 'truss' model of Eurocode 2, angle θ_f is assumed to be 45°. From its Formula 8.69,

$$A_{sf} / s_f \geq \tau_{Ed} h_f f_{yd} = 2.06 \times 0.095 \times 1.15 / 0.500 = 0.450 \text{ mm}$$

For 12-mm bars, $s_f \leq 36\pi/0.45 = 251$ mm.

Figure 19.6. Cross-section of beam during construction with in situ concrete on one side of the web (dimensions in mm)

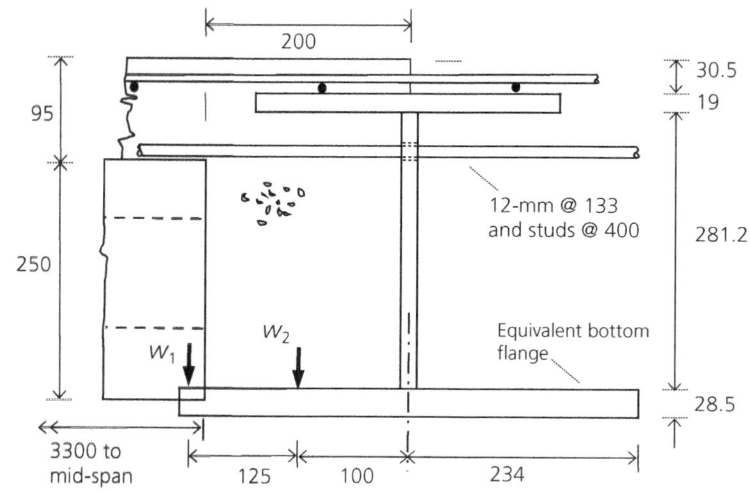

The spacing of the bars, Figure 19.6, must relate to that of the studs (400 mm), as they are at the same level in the web. Hence, bars through 25-mm holes will be provided at 133 mm spacing. The bars can be high enough to avoid the tops of the hollow-core slabs (Figure 19.6), by aligning the top of the holes with the bottom of the web fillets. The compressive stress in the concrete at 45° is well below the limit of Formula 8.70 in EN 1992-1-1.

There will be some longitudinal shear in the 30.5-mm thickness of concrete above the steel flange. This is one reason why it needs reinforcement, shown here as a mesh with 7-mm bars at 200-m spacing.

Torsion of the steel member during construction

It was found above that with propped construction of the beam, and precast slabs and wet concrete topping on one side only of the beam, the total design torque applied to the beam is $\boldsymbol{T_{Ed} = 57.2}$ **kN m**. The ends are assumed to be free to warp but not to twist, so the torsional restraint at each end of the beam is 28.6 kN m.

The torsion at any cross-section is $T = T_{t,Ed} + T_{w,Ed}$, where $T_{t,Ed}$ is the St. Venant torsion, and $T_{w,Ed}$ is the warping torsion. From symmetry, both of these are zero at mid-span, where warping restraint causes longitudinal stress in the steel flanges, to be determined. From clause 8.2.7 of EN 1993-1-1, elastic analysis is used.

EN 1993-1-1 does not give formulae for the warping torsion of an asymmetric I-section, so reference is made here to Publication P385, 'Design of steel beams in torsion' of the Steel Construction Institute, referred to below as 'P385' (Hughes *et al.*, 2011). The formulae here are for I-section beams with unequal flanges of rectangular cross-section.

For calculation of stresses from torsion, the combined bottom flange section here is replaced by a single rectangle, with unchanged total area, minor-axis moment of area and distance 0.305 m between centroids of flanges (symbol h_f), as shown in Figure 19.6.

The torsion constant for the steel section, I_T, is approximately $\sum [(bt^3)/3]$ for the three rectangular components

$$I_T = (0.307 \times 0.019^3 + 0.276 \times 0.012^3 + 0.468 \times 0.0285^3)/3 = 4.47 \times 10^{-6} \text{m}^4$$

This is from pp. 31 and 63 of P385, omitting the minor correction for ends of rectangles.

The distance of the shear centre from the centre of the bottom flange is

$$e_{s,bf} = I_{z,bf}\, h_f / (I_{z,tf} + I_{z,tf}) = 0.0484 \text{ m} \qquad \text{(P385, p. 62)}$$

The warping constant is

$$I_w = I_{z,bf} e_{s,bf} h_f = (0.468^3 \times 0.028/12) \times 0.0484 \times 0.305 = 3.60 \times 10^{-6} \text{m}^6$$

(The extra symbol E in this formula, on p. 67 of P385, is a mis-print.)

The torsional bending constant is $\alpha = (EI_w / GI_T)^{1/2} = 1.447$ m (P385, p. 12)

For a span $L = 8$ m, $L/\alpha = 5.528$

This beam with uniform applied torsion per unit length is Case 4 in Report P385. Its page 73 gives the solution of the differential equation for ϕ, the angle of twist of the beam, with origin of the x-axis at one support,

$$\phi = [Ta^2 / (GI_T L)]\, \{(xL - x^2)/(2a^2) + \cosh(x/a) - \tanh[L/(2a)]\sinh(x/a) - 1\} \qquad \text{(X19.1)}$$

Denoting $d\phi/dx$ by ϕ' and so on, the theory uses values of ϕ', ϕ'' and ϕ''' for the interaction between St. Venant and warping torsion

$$T(x) = GI_T(\phi' - \alpha^2\phi''') \qquad \text{(P385, p. 12)}$$

The following results are obtained by differentiation of Equation X19.1 and substitution.

The angle of twist at mid-span is

$$\phi_{(x = L/2)} = 0.015 \text{ rad}$$

The St. Venant torsional moment at each support is

$$\boldsymbol{T_{t,Ed}} = GI_T\phi'_{(x=0)} = \mathbf{18.2\ kN\ m} \qquad \text{(P385, p. 27)}$$

The warping moment in each flange at mid-span is

$$\boldsymbol{M_w} = EI_{bf}\phi''_{(x=L/2)} e_{s,bf} = \mathbf{41.7\ kN\ m} \qquad \text{(P385, p. 66)}$$

The warping torsion at each end of the span is

$$\boldsymbol{T_w} = EI_{bf}\phi'''_{(x=0)} e_{s,bf} h_f = \mathbf{9.98\ kN\ m} \qquad \text{(P385, p. 66)}$$

At mid-span, the lateral movement of one flange relative to the other is

$$0.015 \times 305 = 4.6 \text{ mm}$$

The longitudinal elastic stress at the tips of the top flange at mid-span is

$$\sigma = 6\, M_w / (b_{tf}^2 t_{tf}) = 6 \times 41.7 / (0.307^2 \times 0.019 \times 1000) = 140 \text{ N/mm}^2$$

The maximum St. Venant shear stress at a support occurs in the thinnest component, the web:

$\tau_{Ed} = T_{t,Ed}\, t_w/I_T = 18.2 \times 0.012/(4.47 \times 10^{-3}) = 49\ N/mm^2$ (P385, p. 9)

The warping torsion at each support is $\boldsymbol{T_{w,Ed}}$ = 28.6 – 18.2 = **10.4 kN m**, which checks the 9.98 kN m above, allowing for rounding errors. It is resisted by equal and opposite transverse shear forces in each of the two flanges of 10.4/0.305 = 34 kN.

None of these results is excessive. Erection of the slabs and concreting on the other side of the web will reduce them, probably substantially. There is no need to consider them further, as this is assumed to be the only loading on the steel beam at this stage. However, the stress 140 N/mm^2, above, is not negligible. It shows the importance of considering torsion at the construction stage, and in edge beams at all stages. Calculations can sometimes be simplified – for example, by checking if the St. Venant torsional resistance is sufficient for the whole torsion.

Roger P Johnson and Stephen J Hicks
ISBN 978-1-83662-921-4
https://doi.org/10.1108/978-1-83662-920-720253023

CHAPTER 20
Annex J (Informative). Other flooring types using precast concrete slabs in buildings

This chapter corresponds to *Annex J* of EN 1994-1-1, which has the following clauses.

- Use of this annex *Clause J.1*
- Scope and field of application *Clause J.2*
- Materials *Clause J.3*
- Structural analysis *Clause J.4*
- Ultimate limit states *Clause J.5*
- Design resistance of shear connectors used with precast floors in buildings *Clause J.6*
- Longitudinal shear in precast concrete slabs *Clause J.7*
- Detailing of the precast slabs *Clause J.8*
- Tests on shear connectors *Clause J.9*

The use of precast concrete slabs was not within the scope of EN 1994-1-1:2004. Clause 8 of EN 1994-2:2005, 'Precast concrete slabs in composite bridges' (a single page) has, in practice, also been used for buildings. Although the new ten-page *Annex J* is limited to buildings, parts of it could be applied for other structures without fatigue loading.

J.2 Scope and field of application

Annex J covers solid slabs with precast concrete permanent formwork, and hollow-core slabs with or without concrete topping. Subjects outside its scope include beams with regions of hogging bending, a non-uniform steel section, lightweight-aggregate concrete, and loading with repetition sufficient to require fatigue verification. Example 19.1 includes the use of hollow-core slabs in a shallow floor structure.

Figure 20.1. A precast beam providing non-uniform support to hollow-core slabs

Clause J.2(7) and its Note 1 refer to the flexibility of the supporting beam. Its curvature, although small, causes vertical shear from a supported hollow-core precast slab to be non-uniform across its width, Figure 20.1. Torsional moment within the hollow-core slab can reduce its resistance to vertical shear at its corners. This a problem for the design of the slab to EN 1992-1-1, so no help is given in *Annex J*. *Clause J.2(7)*

A report from the Steel Construction Institute (Couchman, 2014) gives useful guidance on this subject. Referring to hollow-core slabs

> for unpropped non-composite beams, the influence of support stiffness need not be considered if the design shear force that is applied to the slab is less than $0.35V_{Rd}$, where V_{Rd} is the shear resistance of the hollow core units provided by the manufacturer. For cases when propped construction is used, or when the design shear force applied to the slab is greater than $0.35V_{Rd}$, advice from the manufacturer of the precast units should be sought.

J.4 Structural analysis

Clause J.4.1 From ***clause J.4.1***, the usual rules for effective flange width (*clause 7.4.1.2*) are applicable at serviceability limit states for 'solid composite slabs', which means precast concrete slabs acting as formwork for *in situ* topping. For hollow-core slabs the length L_f of core that is broken out and reinforced (Figure 20.2 and *Figure J.2*) is likely to limit the effective width for ULS (*clause J.4.1(1)*).

Figure 20.2. Composite floor with precast hollow-core slabs (courtesy of the Steel Construction Institute)

J.5 Ultimate limit states

Clause J.5(1) As beams are designed to be simply supported, the reference to 'lateral torsional buckling' in ***clause J.5(1)*** relates to top-flange buckling during construction. The inclusion of resistance to 'transverse forces on webs' does not imply that this is a particular problem for beams of this type.

J.6 Design resistance of shear connectors used with precast floors in buildings

Clause J.6.1 ***Clause J.6.1*** gives signposts to other clauses. *Paragraph (4)* shows that only headed studs and non-preloaded bolt connectors are within the scope of *Annex J.*

Clause J.6.1(5) The rule in ***clause J.6.1(5)*** that relates the gap between the ends of slabs to the size of the studs expresses a principle: there must be enough space for compaction of concrete around shear connectors. For studs and aggregate both of size 19 mm, the nominal (design) gap is at least $3 \times 19 + 5 = 62$ mm, given here as 65 mm. Construction deviations should be assumed to be at least 5 mm for each slab. For post-erection placing of connectors, the designer should then check whether a nominal gap of 75 mm is sufficient.

Clause J.6.2(1) Hollow-core slabs are often preferred to the much heavier solid precast slabs, which are unlikely to be thicker than the 100-mm limit in ***clause J.6.2(1)***. For transverse reinforcement, reference is made to *clause 8.6.10.1(1)* on the minimum projection of connectors above bottom reinforcement. This '30-mm rule' would require the topping to be at least 70 mm thick.

The limiting dimensions in ***clauses J.6.2*** and ***J.6.3*** are likely to be based on current practice and the scope of relevant research. The use of other dimensions may need supporting evidence. *Clause J.6.2* *Clause J.6.3*

Clause J.6.3(4) gives a reduction factor for the resistance of stud connectors, with values from 0.6 to 1.0. The width of gaps between slabs are unlikely to be much less than 70 mm, and then the factor depends on the diameter of the reinforcement within the cores, and is about 0.75 for 10-mm bars. Similar to *clause 8.6.9.2*, *Formula J.1* was developed by assuming that all the variability is accounted for within the *Formulae 8.23* and *8.24*. It is presumed that this assumption was made due to the magnitude of the tolerances on the bearing lengths for the precast units. *Clause J.6.3(4)*

Shear failure of a bolt connector depends on its interaction with the surrounding concrete. This is simplified in ***clause J.6.4.2(1)*** by taking the resistance as the lower of the values given for steel failure, *Formula J.2*, and concrete failure, *Formula J.3*. Guidance on which formula to use is given in Figure 20.3 for 8:8 bolts of diameter d. Each curve is for bolts of a particular height, h_{sc}, and gives the characteristic concrete strength at which the two resistances are equal. For data that plots above the relevant curve, *Formula J.2* gives the lower resistance. The limit '≤ 1.0' in *Formula J.5* causes the change in slope of the curves at $d \approx 20$ mm. *Clause J.6.4.2(1)*

Figure 20.3. Concrete strength for the shear resistances from *Formulae J.2* and *J.3* to be equal, assuming grade 8:8 bolts

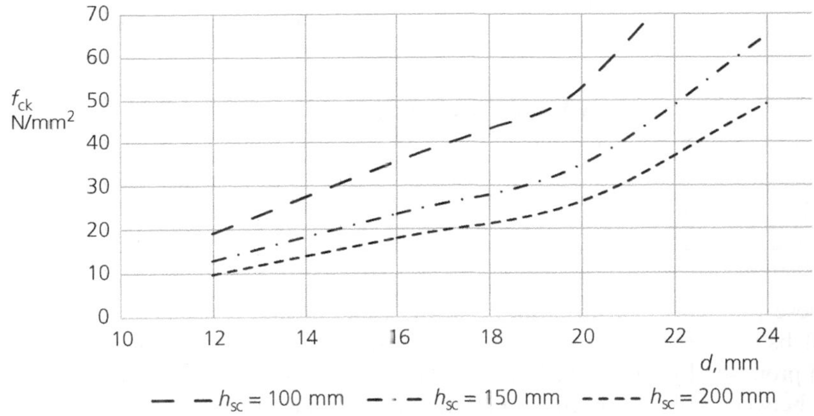

In ***clause J.6.4.3***, the condition given by *Formula J.6* for a non-preloaded bolt to be in ductility category D.2 (*clause J.5.4.1(5)*) is $P_{b,Rd}/P_{c,Rd} > 0.97$. In the formula, e is the base of natural logarithms. The formula is based on a parametric study of load-slip behaviour of bolt shear connectors. *Clause J.6.4.3*

J.7 Longitudinal shear in precast concrete slabs

Clause J.7(2) says that transverse reinforcement should be placed 'in each open core and between adjacent units if necessary to comply with the minimum bar spacing'. Reinforcement of this type is shown in *Figure J.5(a)*.

J.8 Detailing of the precast slabs

Clause J.8.1(2) gives the minimum nominal bearing length for a precast slab on a steel beam as 55 mm. Where there are slabs on both sides of the beam, the rules for minimum gaps between slabs also apply (about 70 mm, see above), plus tolerances. Hence, the minimum width for the steel top flange is about $70 + 2 \times 55 + 20 = 200$ mm. *Clause J.8.1(2)*

J.9 Tests on shear connectors

The principle here is that test specimens should follow the materials and details of the intended construction as closely as possible – the rule for chamfers in *clause J.9.1(2)e)* is an example.

Clause J.9.1 ***Clause J.9.1*** gives the dimensions of the parts of each hollow core slab to be used for push-test specimens, based on the specific push tests of *Annex B*.

Clause J.9.1(2)a) In ***clause J.9.1(2)a)***, width b_e corresponds to dimension L in *Figure B.2*. The first diagram there shows test slabs with three troughs of profiled sheeting on each side, and studs between the troughs. A hollow-core specimen would be similar, with the cores replacing the troughs. Longer specimens could be used, and would be needed to represent a beam with studs in alternate troughs.

Clause J.9.1(2)b) *Figure B.2* also shows the width of the slabs as b_c, defined for hollow-core slabs in ***clause J.9.1(2)b)***. This is b_g plus 'the length of the in-situ infill to either side of the beam'. It is clear from the shaded length in *Figure J.(2)(a)* that this means the width $2L_f + b_g$, over which the concrete infill should be complete. Construction of such test specimens is not straightforward. After casting of the slabs attached to one steel flange, the specimen would need to be inverted to reproduce the correct direction of pouring. Diamond sawing of slabs would also be needed.

Figure 20.4. Test specimen for headed stud connectors with hollow-core slabs

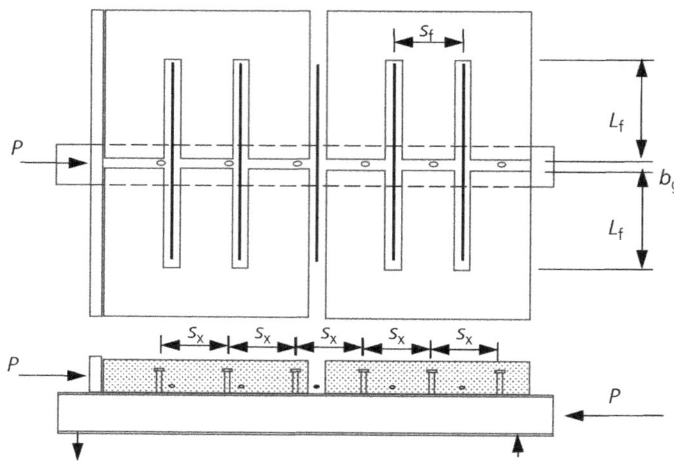

To remedy these problems, and to ensure that the experimental resistance of shear connectors in hollow-core slabs is determined consistently, an alternative one-sided arrangement for specific push tests has been proposed by Hicks *et al.* (2023). This arrangement is shown in Figure 20.4, in which concrete infill between slabs has been omitted, for clarity. It has been successfully used by Lam *et al.* (1998). Results from tests using this type of specimen formed the basis of *Formula J.1*.

emerald PUBLISHING ice

Roger P Johnson and Stephen J Hicks
ISBN 978-1-83662-921-4
https://doi.org/10.1108/978-1-83662-920-720253024

APPENDIX A
Lateral-torsional buckling of composite beams for buildings

This Appendix supplements the comments on *clause 8.4*.

Simplified expression for 'cracked' flexural stiffness of a composite slab

It is assumed that the composite slab spans between the relevant beams. The 'cracked' stiffness per unit width of a composite slab, $(EI)_2$, is defined in *clause 8.4.2(6)* as the lower of the values at mid-span and at a support. The latter usually governs, because the profiled sheeting may be discontinuous at a support. It is now determined for the cross-section shown in Figure A.1 with the sheeting neglected.

Figure A.1. Model for stiffness of a composite slab in hogging bending

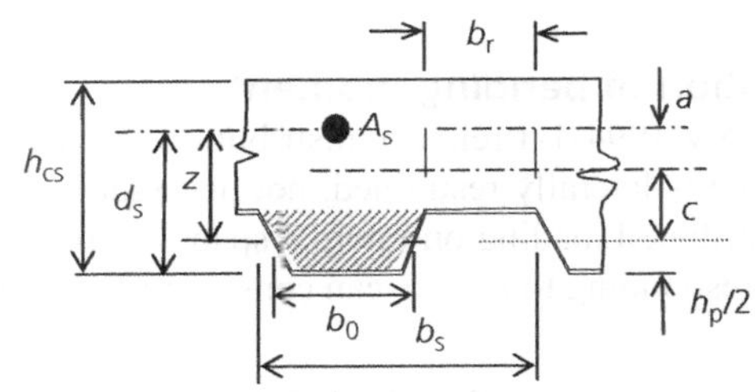

The elastic neutral axis may be close to the top surface of the sheeting, and accurate calculation becomes complex. For simplicity, it is assumed that the compressed region is the concrete within the net depth of the sheeting, h_p. Its transformed area per unit width of slab in 'steel' units is

$$A_e = b_0 h_p / n b_s \tag{a}$$

where n is the modular ratio. The position of the elastic neutral axis is defined here by the dimensions a and c, so that

$$A_e c = A_s a \quad \text{and} \quad a + c = z \tag{b}$$

where A_s is the area of top reinforcement per unit width of slab, and

$$z = d_s - h_p/2 \tag{c}$$

Assuming that each trough is rectangular, the second moment of area per unit width is

$$I = A_s a^2 + A_e (c^2 + h_p^2/12) \tag{d}$$

Using equations (b) to (d), the flexural stiffness per unit width of slab is

$$(EI)_2 = E_a [A_s A_e z^2 / (A_s + A_e) + A_e h_p^2 / 12] \tag{XA.1}$$

This result is used in Example 8.6.

Flexural stiffness of beam with encased web

For a beam with a web encased by concrete of width b_c, the model used for the derivation of *Formula 8.10* for the flexural stiffness k_2 is as shown in Figure A.2. A lateral force F applied to the steel bottom flange causes displacement δ. The rotation of line AB is

$\phi = \delta/h_s$. It is caused by a bending moment Fh_s acting about A. The stiffness is

$$k_2 = M/\phi = Fh_s^2/\delta$$

The force F is assumed to be resisted by vertical tension in the steel web and compression in a concrete strut BC, of width $b_c/4$. Elastic analysis, with the approximation that lengths AB and CB are equal, leads to *Formula 8.10.*

Figure A.2. Flexural stiffness of an encased web in an inverted U-frame

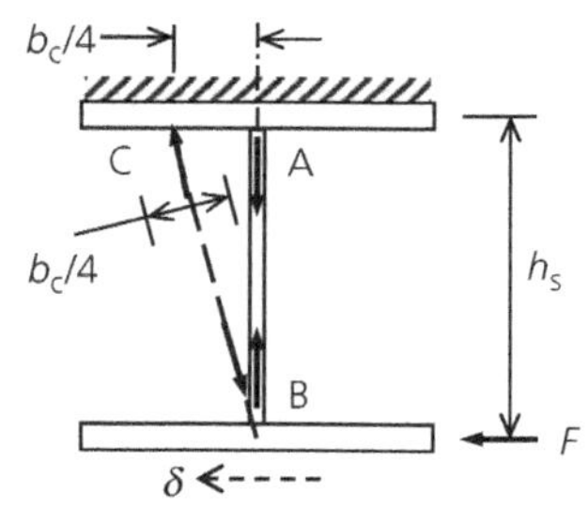

Factor C_4 for the distribution of bending moment

The tables that were given in ENV 1994-1-1 relate to distributions of bending moment between points at which the steel bottom flange is laterally restrained, not necessarily to complete spans. The more commonly used values for distributed loading on internal spans are plotted in Figure A.3. For values of ψ exceeding 3.0, values corresponding to $\psi \to \infty$ can conservatively be used. These are also shown.

Figure A.3. Values of the factor C_4 for uniformly distributed and centre-point loading

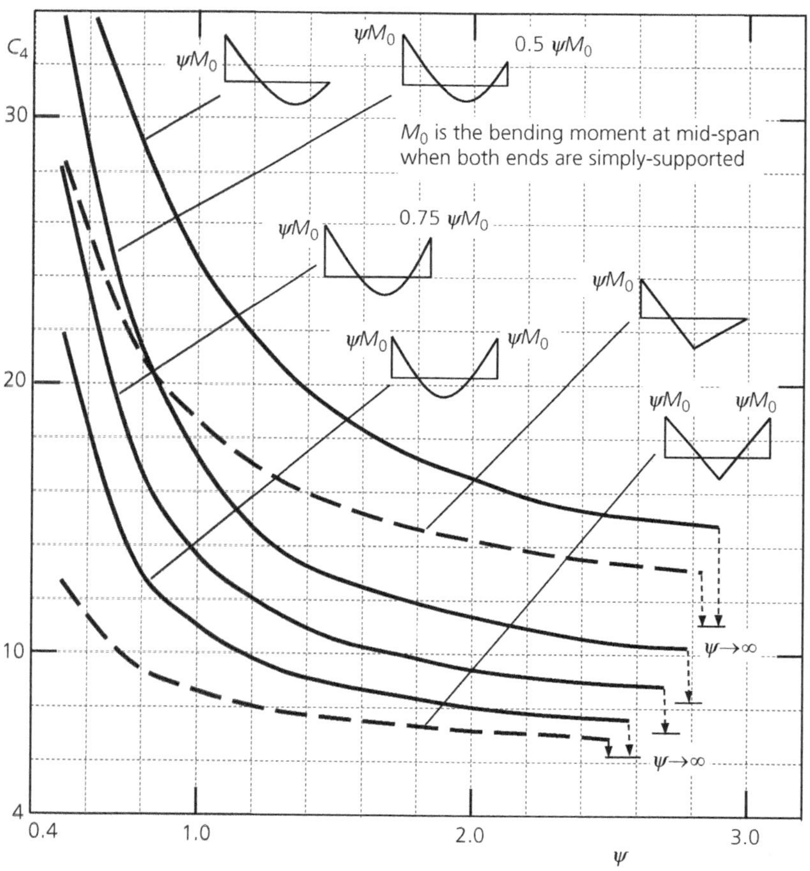

The dashed lines in Figure A.3 are for point loads at mid-span. Two other sets of values are plotted in Figure A.4. The solid lines apply for lateral buckling of a cantilever of length L_c, where both it and the adjacent span of length L have the same intensity of distributed loading. The dashed lines are for an unloaded span with one or both ends continuous.

Figure A.4. Values of the factor C_4 at end supports, for uniformly loaded spans with a cantilever extension, and for spans without transverse loading

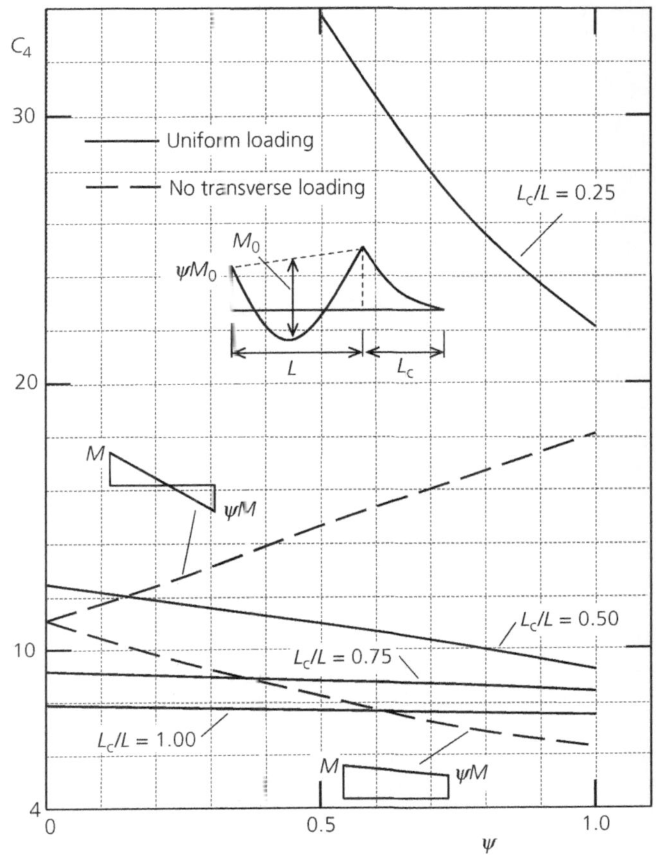

Roger P Johnson and Stephen J Hicks
ISBN 978-1-83662-921-4
https://doi.org/10.1108/978-1-83662-920-720253027

APPENDIX B
Simplified calculation method for the interaction curve for resistance of composite column cross-sections to compression and uniaxial bending

Scope and method

Equations are given for the coordinates of points B, C and D in *Figure 8.28*, also shown in Figure B.1. They are applicable to cross-sections of columns where the structural steel, concrete and reinforcement are all doubly symmetric. The steel section should be an I- or H-section or a rectangular or circular hollow section. Examples are shown in *Figure 8.24*.

Plastic analysis is used, with rectangular stress blocks for structural steel, reinforcement and concrete in accordance with *clauses 8.8.3.2(2)* to *(6)*. For concrete-filled tubes of circular section, the stresses f_{cd} and f_{yd} used below should be replaced by the modified values given by *Formula 8.40* and *Table 8.5*, to allow for confinement effects. These depend on the slenderness of the column, the ratio $M_{Ed}/(N_{Ed}\,d)$ and other parameters.

In this Annex, the compressive stress in concrete in a rectangular stress block is denoted f_{cd}, where generally, $f_{cd} = 0.85\, f_{ck}/\gamma_C$. However, for concrete-filled steel sections, the coefficient 0.85 may be replaced by 1.0, following the Note to *clause 5.1(3)*. These values can be altered by national annexes.

Figure B.1. Polygonal interaction curve

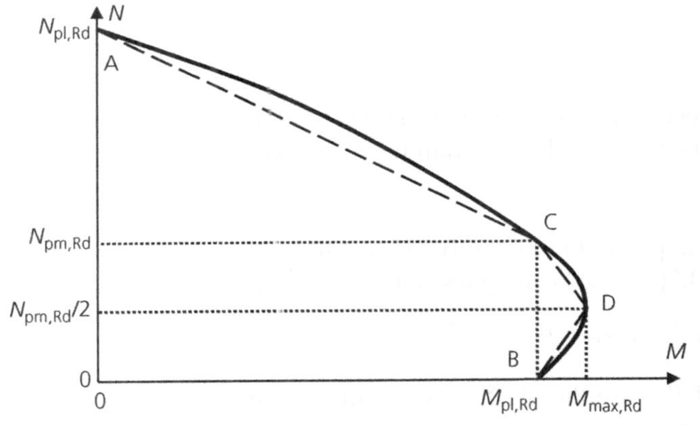

Resistance to compression

The resistance $N_{pm,Rd}$ in Figure B.1 is calculated as follows. Figure B.2(a) represents a generalised cross-section of the structural steel and reinforcement (shaded area), and of the concrete, symmetrical about two axes through its centre of area G. For bending only (point B in Figure B.1) the neutral axis is line B–B which defines region (1) of the cross-section, within which concrete is in compression. The line C–C at the same distance h_n on the other side of G is the neutral axis for point C in Figure B.1. This is because the areas of structural steel, concrete and reinforcement in region (2) are all symmetrical about G, so that the changes of stress when the axis moves from B–B to C–C add up to the resistance $N_{pm.Rd}$ and the bending resistance is unchanged. Using subscripts 1 to 3 to indicate regions (1) to (3),

$$N_{pm.Rd} = R_{c2} + 2|R_{a2}| \quad \text{(XB.1)}$$

where R_{c2} is the resistance of the concrete in region (2), and R_{a2} is the resistance of the steel in region (2).

In the notation of *clause 8.8.3.2(1)*, $R_{c2} = A_{c2} f_{cd}$ and $R_{a2} = A_{a2} f_{yd} + A_{s2} f_{sd}$ where compressive forces and strengths of materials are taken as positive.

From symmetry, $R_{a1} = |R_{a3}|$ and $R_{c1} = R_{c3}$ (XB.2)

When the neutral axis is at B–B, $N = 0$, so that $R_{a1} + R_{c1} = |R_{a2}| + |R_{a3}|$ (XB.3)

From Equations XB.2 and XB.3, $|R_{a2}| = R_{c1} = R_{c3}$. Substituting in Equation XB.1,

$$N_{pm,Rd} = R_{c2} + R_{c1} + R_{c3} = R_c \quad \text{(XB.4)}$$

where R_c is the compressive resistance of the whole area of concrete, which is easily calculated.

Figure B.2. Composite cross-sections symmetrical about two axes

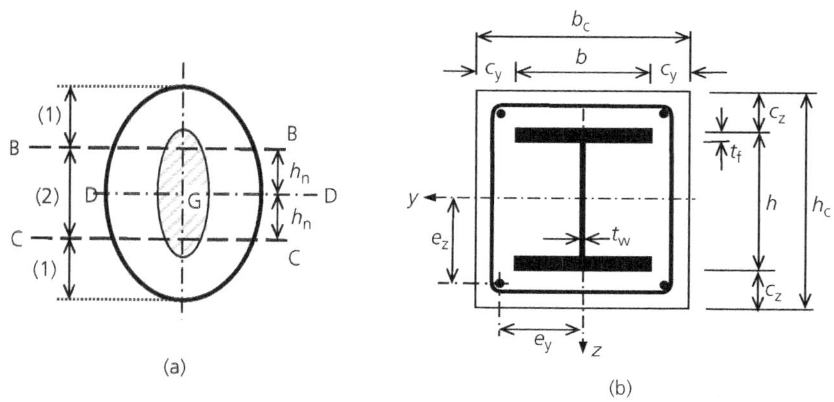

Position of neutral axis

The value of h_n depends on the axis of bending, the type of cross-section and the cross-section properties. The equations can be derived from Equations XB.1 and XB.4.

Bending resistances

The axial resistance at point D in Figure B.1 is half that at point C, so the neutral axis for point D is line D–D in Figure B.2(a). The bending resistance at point D is

$$M_{max,Rd} = W_{pa} f_{yd} + W_{ps} f_{sd} + W_{pc} f_{cd}/2 \quad \text{(XB.5)}$$

where W_{pa}, W_{ps} and W_{pc} are the plastic section moduli for the structural steel, the reinforcement and the concrete part of the section (for the calculation of W_{pc} the concrete is assumed to be uncracked), and f_{yd}, f_{sd} and f_{cd} are the design strengths for the structural steel, the reinforcement and the concrete.

The bending resistance at point B is

$$M_{pl,Rd} = M_{max,Rd} - M_{n,Rd} \quad \text{(XB.6)}$$

with

$$M_{n,Rd} = W_{pa,n} f_{yd} + W_{ps,n} f_{sd} + W_{pc,n} f_{cc}/2 \quad \text{(XB.7)}$$

where $W_{pa,n}$, $W_{ps,n}$ and $W_{pc,n}$ are the plastic section moduli for the structural steel, the reinforcement and the concrete parts of the section within region (2) of Figure B.2(a). Equations for these plastic section moduli of some cross-sections are given below.

Interaction with transverse shear

If the shear force to be resisted by the structural steel is considered according to *clause 8.8.3.2(4)* the appropriate areas of steel should be assumed to resist shear alone. The method given here can be applied using the remaining areas.

Neutral axes and plastic section moduli of some cross-sections

General

The compressive resistance of the whole area of concrete is

$$N_{pm,Rd} = A_c f_{cd} \quad \text{(XB.8)}$$

The value of the plastic section modulus of the total reinforcement is given by

$$W_{ps} = \sum_{i=1}^{n} |A_{s,i} e_i| \quad \text{(XB.9)}$$

where e_i are the distances of the reinforcement bars of area $A_{s,i}$ to the relevant middle line (y-axis or z-axis).

The equations for the position of the neutral axis h_n are given for selected positions in the cross-sections. The resulting value h_n should lie within the limits of the assumed region.

Major-axis bending of encased I-sections

The notation is given in Figure B.2(b). The plastic section modulus of the structural steel may be taken from tables or be calculated from

$$W_{pa} = \frac{(h - 2t_f)^2 t_w}{4} + bt_f(h - t_f) \quad \text{(XB.10)}$$

and

$$W_{pc} = \frac{b_c h_c^2}{4} - W_{pa} - W_{ps} \quad \text{(XB.11)}$$

For the different positions of the neutral axes, h_n and $W_{pa,n}$ are given by

(*a*) Neutral axis in the web, $h_n \leq h/2 - t_f$

$$h_n = \frac{N_{pm,Rd} - A_{sn}(2f_{sd} - f_{cd})}{2b_c f_{cd} + 2t_w(2f_{yd} - f_{cd})} \quad \text{(XB.12)}$$

$$W_{pa,n} = t_w h_n^2 \quad \text{(XB.13)}$$

where A_{sn} is the sum of the area of reinforcing bars within the region of depth $2h_n$.

(*b*) Neutral axis in the flange, $h/2 - t_f < h_n < h/2$,

$$h_n = \frac{N_{pm,Rd} - A_{sn}(2f_{sd} - f_{cd}) + (b - t_w)(h - 2t_f)(2f_{yd} - f_{cd})}{2b_c f_{cd} + 2b(2f_{yd} - f_{cd})} \quad \text{(XB.14)}$$

$$W_{pa,n} = bh_n^2 - \frac{(b - t_w)(h - 2t_f)^2}{4} \quad \text{(XB.15)}$$

(*c*) Neutral axis outside the steel section, $h/2 \leq h_n \leq h_c/2$,

$$h_n = \frac{N_{pm,Rd} - A_{sn}(2f_{sd} - f_{cd}) - A_a(2f_{yd} - f_{cd})}{2b_c f_{cd}} \quad \text{(XB.16)}$$

$$W_{pa,n} = W_{pa} \quad \text{(XB.17)}$$

The plastic modulus of the concrete in the region of depth from $2h_n$ then results from

$$W_{pc,n} = b_c h_n^2 - W_{pa,n} - W_{ps,n} \quad \text{(XB.18)}$$

$$\text{with } W_{ps,n} = \sum_{i=1}^{n} |A_{sn,i} e_{z,i}| \quad \text{(XB.19)}$$

where $A_{sn,i}$ are the areas of reinforcing bars within the region of depth $2h_n$, and $e_{z,i}$ are the distances from the middle line.

Minor-axis bending of encased I-sections

The plastic section modulus of the structural steel may be taken from tables or be calculated from

$$W_{pa} = \frac{(h - 2t_f)t_w^2}{4} + \frac{2t_f b^2}{4} \quad \text{(XB.20)}$$

$$\text{and } W_{pc} = \frac{h_c b_c^2}{4} - W_{pa} - W_{ps} \quad \text{(XB.21)}$$

For the different positions of the neutral axes, h_n and $W_{pa,n}$ are given by

(*a*) Neutral axis in the web, $h_n \leq t_w/2$,

$$h_n = \frac{N_{pm,Rd} - A_{sn}(2f_{sd} - f_{cd})}{2h_c f_{cd} + 2h(2f_{yd} - f_{cd})} \quad \text{(XB.22)}$$

$$W_{pa,n} = hh_n^2 \quad \text{(XB.23)}$$

(*b*) Neutral axis in the flanges, $t_w/2 < h_n < b/2$,

$$h_n = \frac{N_{pm,Rd} - A_{sn}(2f_{sd} - f_{cd}) + t_w(2t_f - h)(2f_{yd} - f_{cd})}{2h_c f_{cd} + 4t_f(2f_{yd} - f_{cd})} \quad \text{(XB.24)}$$

$$W_{pa,n} = 2t_f h_n^2 + \frac{(h - 2t_f)t_w^2}{4} \quad \text{(XB.25)}$$

(*c*) Neutral axis outside the steel section, $b/2 \leq h_n \leq b_c/2$,

$$h_n = \frac{N_{pm,Rd} - A_{sn}(2f_{sd} - f_{cc}) - A_a(2f_{yd} - f_{cc})}{2h_c f_{cc}} \quad \text{(XB.26)}$$

$$W_{pa,n} = W_{pa} \quad \text{(XB.27)}$$

The plastic modulus of the concrete in the region of depth $2h_n$ then results from

$$W_{pc,n} = h_c h_n^2 - W_{pa,n} - W_{ps,n} \geq 0 \qquad \text{(XB.28)}$$

with $W_{ps,n}$, according to Equation XB.19, changing subscript z to y. For a web-encased section (*Figure 8.24(b)*) the neutral axis is likely to be within the web, and then $W_{pc,n}$ may be found to be negative.

Figure B.3. Concrete-filled (a) rectangular and (b) circular hollow sections, with notation

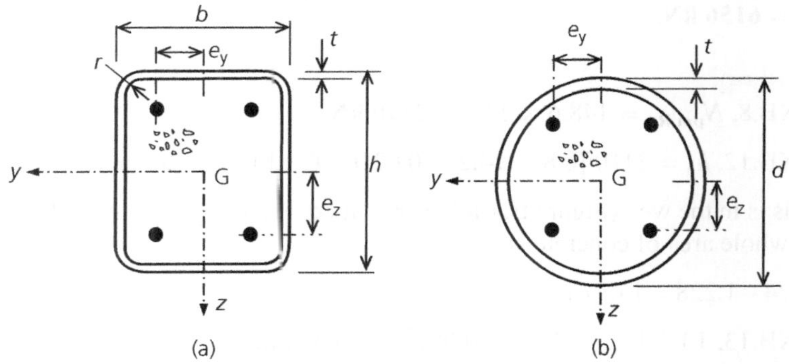

Concrete-filled rectangular and circular hollow sections

The following equations are derived for rectangular hollow sections with bending about the y-axis of the section (see Figure B.3). For bending about the z-axis the dimensions h and b are to be interchanged as well as the subscripts z and y. Equations XB.29 to XB.33 may be used for circular hollow sections with good approximation by substituting

$$h = b = d \text{ and } r = d/2 - t$$

$$W_{pc} = \frac{(b-2t)(h-2t)^2}{4} - \frac{2}{3}r^3 - r^2(4-\pi)(0.5\,h - t - r) - W_{ps} \qquad \text{(XB.29)}$$

with W_{ps} according to Equation XB.9.

W_{pa} may be taken from tables or be calculated from

$$W_{pa} = \frac{(bh^2)}{4} - \frac{2}{3}(r+t)^3 - (r+t)^2(4-\pi)(0.5\,h - t - r) - W_{pc} - W_{ps} \qquad \text{(XB.30)}$$

$$h_n = \frac{N_{pm,Rd} - A_{sn}(2f_{sd} - f_{cd})}{2bf_{cd} + 4t(2f_{yd} - f_{cd})} \qquad \text{(XB.31)}$$

$$W_{pc,n} = (b-2t)h_n^2 - W_{ps,n} \qquad \text{(XB.32)}$$

$$W_{pa,n} = bh_n^2 - W_{pc,n} - W_{ps,n} \qquad \text{(XB.33)}$$

with $W_{ps,n}$ according to Equation XB.19.

Example B.1: *N–M* interaction polygon for a column cross-section

The method of Appendix B is used to obtain the interaction polygon given in Figure 8.39 for the concrete-encased H-section shown in Figure 8.38. The small area of longitudinal reinforcement is neglected. The data and symbols are as in Example 8.7 and Figures 8.38, B.1 and B.2.

- Design strengths of the materials: $f_{yd} = 355$ N/mm^2; $f_{cc} = f_{cd} = 14.2$ N/mm^2
- Other data: $A_a = 11\,400$ mm^2; $A_c = 148\,600$ mm^2; $t_f = 17.3$ mm; $t_w = 10.5$ mm; $b_c = h_c = 400$ mm; $b = 256$ mm; $h = 260$ mm; $10^{-6}\ W_{pa,y} = 1.228$ mm^3; $10^{-6}\ W_{pa,z} = 0.575$ mm^3; $\boldsymbol{N_{pl,Rd}}$ **= 6156 kN**.

Major-axis bending

From Equation XB.8, $\boldsymbol{N_{pm,Rd}} = 148.6 \times 14.2 =$ **2110 kN**

From Equation XB.12, $h_n = 2110/[0.8 \times 14.2 + 0.021(710 - 14.2)] = 81$ mm

so the neutral axis is in the web (Figure B.4(a)), as assumed. From Equation XB.11, the plastic section modulus for the whole area of concrete is

$$10^{-6} W_{pc} = 4^3/4 - 1.228 = 14.77 \text{ mm}^3$$

From Equation XB.13, $10^{-6}W_{pa,n} = 10.5 \times 0.081^2 = 0.069$ mm^3

From Equation XB.18, $10^{-6}\ W_{pc,n} = 400 \times 0.081^2 - 0.069 = 2.555$ mm^3

From Equation XB.5, $\boldsymbol{M_{max,Rd}} = 1.228 \times 355 + 14.77 \times 14.2/2 =$ **541 kN m**

From Equations XB.6 and XB.7, $\boldsymbol{M_{pl,Rd}} = 541 - (0.081 \times 355 + 2.555 \times 14.2/2) =$ **494 kN m**

The results shown above in bold type are plotted on Figure 8.39.

Minor-axis bending

From Equation XB.4, $N_{pm,Rd}$ is the same for both axes of bending. Thus,

$$\boldsymbol{N_{pm,Rd}} = \mathbf{2110\ kN}$$

Assuming that the neutral axis B–B intersects the flanges, from Equation XB.24,

$$h_n = [2110 + 0.0105(34.6 - 260)(710 - 14.2)]/[0.8 \times 14.2 + 0.0692(710 - 14.2)] = 7.8 \text{ mm}$$

so axis B–B does intersect the flanges, Figure B.4(b). From Equation XB.21,

$$10^{-6} W_{pc} = 4^3/4 - 0.575 = 15.42 \text{ mm}^3$$

Figure B.4. Neutral axes at points B, C and D on the interaction polygons

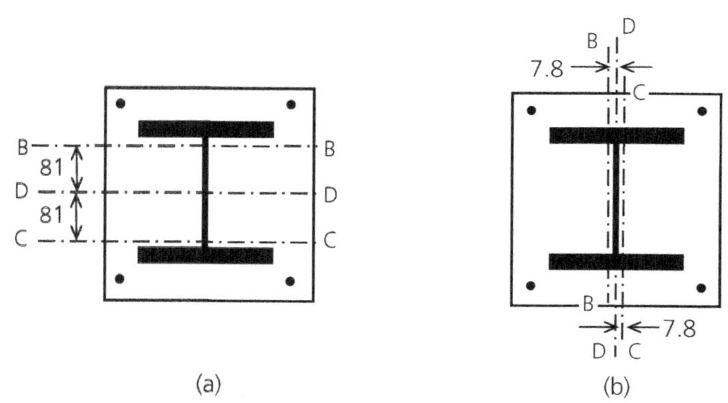

From Equation XB.25,

$$10^{-6} W_{pa,n} = 34.6 \times 0.0078^2 + 0.0105^2 (260 - 34.6)/4 = 0.0083 \text{ mm}^3$$

From Equation XB.28, $10^{-6}\ W_{pc,n} = 400 \times 0.0078^2 - 0.0083 = 0.0160 \text{ mm}^3$

From Equation XB.5, $\boldsymbol{M_{max,Rd}} = 0.575 \times 355 + 15.42 \times 14.2/2 =$ **314 kN m**

From Equation XB.7, $M_{n,Rd} = 0.0083 \times 355 + 0.0160 \times 14.2/2 = 3.1$ kN m

From Equation XB.6, $\boldsymbol{M_{pl,Rd}} = 314 - 3 =$ **311 kN m**

These results are plotted on Figure 8.39 and used in Example 8.7.

Roger P Johnson and Stephen J Hicks
ISBN 978-1-83662-921-4
https://doi.org/10.1108/978-1-83662-920-720253028

APPENDIX C
Composite dowel shear connectors

The new CEN/TS 1994-1-102, 'Composite dowels' covers connectors developed from the tooth-shaped steel plates known as Perfobond-Leiste that were first used in the 1980s. They are welded in long strips to steel top flanges of beams, or used as teeth cut into the top of a steel web of a steel inverted T-section (Johnson *et al.*, 2025).

The scope of Part 1-102 extends over several shapes of dowel, each defined by a spacing e_x, which can range from 0.15 m to 0.5 m. Figure C.1 shows one wavelength of shape MCL, which is the subject here. All dimensions except thickness are proportional to e_x, so a wide range of sizes is available. The larger ones can replace multiple rows of stud connectors. They have been used in about twenty bridges, mostly in Poland. Part 1-102 also includes an annex on shear connection by reinforcement passing through holes in steel webs, as shown in Figure 19.6. It provides a less conservative design method than the one in *Annex I* of EN 1994-1-1.

Figure C.1. Details of composite dowel shear connector

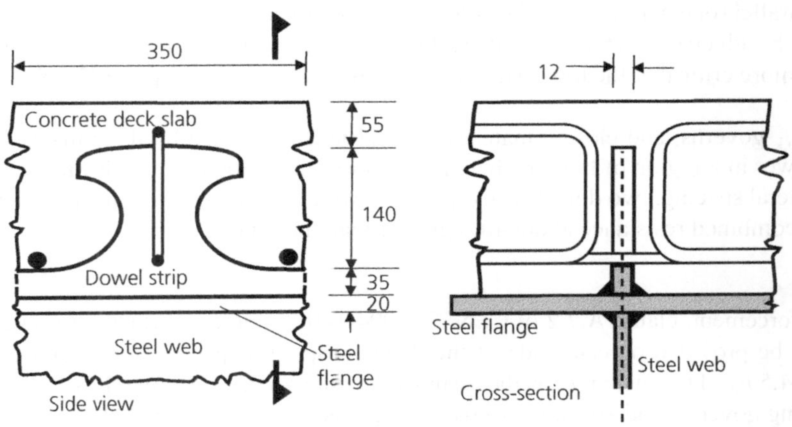

Toothed connectors usually require more local reinforcement than studs, but are well suited to the steel fabrication process. An early use was in a region where stud-welding equipment was not available. According to *clause 8.6.1(6)* of EN 1994-1-1, if dowels are used in combination with headed stud connectors within the same span of the beam, account should be taken of the difference in load-slip properties. The load-slip performance of composite dowels can vary between the curves shown in *Figure 5.1(d)* to *(f)* of EN 1994-1-1.

In the following example, a dowel strip is used to replace two rows of studs in a length of a bridge beam in sagging bending. It is not possible here to include the details of the design rules in Part 1-102, which are mostly based on tests and numerical modelling. It is found that the shear resistance is governed by pry-out of a concrete cone at the top surface of the slab. It would be difficult, using these materials and dimensions, to design a single dowel strip for a higher resistance than that found here.

Example C.1: comparison between stud and dowel shear connectors

Shear connection in a sagging moment region of a bridge beam is provided by two rows of 22-mm welded studs at 260 mm spacing. The studs will be replaced by a dowel strip of shape MCL designed for the same longitudinal shear. Part 1-102 includes rules for fatigue, which is not included in this design.

The strengths of the materials are: $f_{ck} = 30$ MPa; $f_{sk} = 500$ MPa; $f_y = 355$ MPa.

Part 1-102 uses characteristic resistances. Design of the studs to EN 1994-1-1 gives the design shear resistance as 846 kN/m, so $\boldsymbol{V_{L,Rk,stud}} = 1.25 \times 846 =$ **1058 kN/m**.

Dowel strips can have a thickness between 4 mm and 40 mm. Another type of cross-section for a beam, not shown here, is to use a steel inverted T-section, with teeth cut into the top of its web, and embedded in concrete either in a flange or in the web of a concrete T-section.

After some trials, these dimensions are chosen: $e_x = 350$ mm; $t_w = 12$ mm. This design is for a characteristic resistance close to $0.35 \times 1058 =$ **370 kN/dowel**. For type MCL, the height of the dowel is $0.4e_x$, which is 140 mm. From Figure C.1, the top and bottom covers to the dowel are $c_{D,o} = 55$ mm and $c_{D,u} = 35$ mm.

The characteristic shear resistance per dowel $P_{s,Rk}$ is the lowest of resistances found for four failure modes.

- $P_{pl,k}$, yield of steel in the dowel, which depends on e_x, t_w and f_{yk} and is **386 kN**.
- $P_{sh,k}$, shearing of the concrete, which depends on $e_x, f_{ck}, E_a/E_{cm}$ and the area A_b of transverse reinforcement that passes between the teeth of the dowel. The required area is 346 mm^2. Two 15-mm bars per dowel (353 mm^2) are used. The result is $\boldsymbol{P_{sh,k}} =$ **772 kN**.
- $P_{poe,Rk}$, pry-out of a concrete cone at an edge of the slab; not applicable here.
- $P_{po,k}$, pry-out of a concrete cone at a surface of the slab. This depends on the number of dowel strips in parallel (one here), e_x, f_{ck}, the concrete covers given above, and a factor ψ_{crack} (≤ 1) that allows for the adverse effect of flexural cracking in a region of hogging bending. Here, $\psi_{crack} = 1$. Pry-out is more critical at the top surface of the slab. The result is $\boldsymbol{P_{po,k}} =$ **372 kN**.

Resistance $P_{po,k}$ governs, and closely matches the resistance provided by the pairs of 22-mm studs. It would be lower in a region of hogging bending, where calculation of ψ_{crack} is quite complex. From the rules on lateral spacing, two dowel strips spaced as widely as possible on this flange 400 mm wide would have a combined resistance about 40% greater than that of one strip.

Reinforcement

For local reinforcement, clause A.7.2 of CEN/TS 1994-1-102 requires vertical stirrups at least 8 mm in diameter to be provided on each side of the dowel strip, at a spacing not greater than the lesser of 300 mm or 4.5 h_{po}. Dimension h_{po} is the assumed height of a pry-out cone. It is 79.5 mm here. The 300-mm spacing governs, and 10-mm stirrups are used, as shown in Figure C.1.

Only the dowel reinforcement is shown in Figure C.1. The stirrups cannot be combined with top reinforcement needed for hogging transverse bending. Their legs must be fully anchored. The 15-mm bars can be combined with any other bottom reinforcement that may be required.

Roger P Johnson and Stephen J Hicks
ISBN 978-1-83662-921-4
https://doi.org/10.1108/978-1-83662-920-720253026

References

Abdullah JA and Sumei Z (2009) Prediction of lateral deflection and moment of partially-encased composite beam-columns. In *Steel Concrete and Composite and Hybrid Structures* (Lam D (ed.)), Research Publishing Services, Singapore, pp. 649–662.

Al-Deen S, Ranzi G and Uy B (2015) Non-uniform shrinkage in simply-supported composite steel-concrete slabs. *Steel and Composite Structures* **18(2)**: 375–394.

Allison RW, Johnson RP and May IM (1982) Tension-field action in composite plate girders. *Proceedings of the Institution of Civil Engineers, Part 2* **73**: 255–276.

Anderson D and Aribert JM (2000) Design rotation capacity of composite joints. *Structural Engineer* **78(6)**: 25–29.

Aribert J-M (1990) Design of composite beams with a partial shear connection. *Symposium on Mixed Structures, including New Materials, Brussels.* IABSE, Zurich, Switzerland, Reports 60, pp. 215–220 (in French).

Aribert J-M (1997) Analyse et formulation pratique de l'influence de la nuance de l'acier du profilé sur le degré minimum de connexion partielle d'une poutre mixte. *Construction Métallique* **3**: 39–55.

Aribert J-M (1999) Theoretical solutions relating to partial shear connection of steel-concrete composite beams and joints. In *Proceedings of the International Conference on Steel and Composite Structures.* TNO Building and Construction Research, Delft, Netherlands, pp. 7.1–7.16.

Aribert J-M (2010) Influence éventuelle du degré de connexion sur la flèche de poutres mixtes de bâtiment. *Construction Metallique* **47(2)**: 27–41.

AS/NZS 2327:2017 (2017) *Composite Steel-Concrete Construction for Buildings*. Standards Australia/Standards New Zealand, Sydney/Wellington.

Banfi M (2006) Slip in composite beams using typical material curves and the effect of changes in beam layout and loading. In *Composite Construction in Steel and Concrete V* (Leon RT and Lange J (eds)), American Society of Civil Engineers, New York, NY, USA, pp. 356–368.

BCSA (British Constructional Steelwork Association) (2021) *UK Structural Steelwork: 2050 Decarbonisation Roadmap*. BCSA, London, UK.

Bergmann R and Hanswille G (2006) New design method for composite columns including high strength steel. In *Composite Construction in Steel and Concrete V* (Leon RT and Lange J (eds)), American Society of Civil Engineers, New York, NY, USA, pp. 381–389.

Bode H and Sauerborn I (1991) Partial shear connection design of composite slabs. *Proceedings of the 3rd International Conference.* Asscciation for International Co-operation and Research in Steel-Concrete Composite Structures, Sydney, Australia, pp. 467–472.

Bose B and Hughes AF (1995) Verifying the performance of standard ductile connections for semi-continuous steel frames. *Proceedings of the Institution of Civil Engineers: Structures and Buildings* **110(4)**: 441–457.

Brown D and Iles D (2013) *Joints in Steel Construction: Moment-Resisting Joints to Eurocode 3.* Publication P398, Steel Construction Institute, Ascot, UK.

BS 5950-3.1 + A-1:2010 Structural use of steelwork in buildings. Design in composite construction. Code of practice for design of simple and continuous composite beams. BSI, London, UK.

BS 5950-4:1994 Code of practice for design of floors with profiled steel sheeting. BSI, London, UK.

BS 6472-1:2008 Guide to evaluation of human exposure to vibration in buildings – Part 1: Vibration sources other than blasting. BSI, London, UK.

BS 8110-1:1997 Structural use of concrete. Code of practice for design and construction. BSI, London, UK.

BS EN 206:2013+A2:2021 Concrete. Specification, performance, production and conformity. BSI, London, UK.

BS EN 1090-2:2018+A1:2024 Execution of steel structures and aluminium structures. Part 2: Technical requirements for steel structures. BSI, London, UK.

BS EN 1090-4:2018 Execution of steel structures and aluminium structures. Part 4: Technical requirements for cold-formed structural steel elements and cold-formed structures for roof, ceiling, floor and wall applications. BSI, London, UK.

BS EN 1990:2023 Eurocode – Basis of structural and geotechnical design. BSI, London, UK.

BS EN 1991-1-1:2025 Actions on structures. Part 1-1: General actions. BSI, London, UK.

BS EN 1991-1-5:2025 Actions on structures. Part 1-5: Thermal actions. BSI, London, UK.

BS EN 1992-1-1:2023 Design of concrete structures. Part 1-1: General rules and rules for buildings. BSI, London, UK.

BS EN 1993-1-1:2022 Design of steel structures. Part 1-1: General rules and rules for buildings. BSI, London, UK.

BS EN 1993-1-3:2024 Design of steel structures. Part 1-3: Cold formed members and sheeting. BSI, London, UK.

BS EN 1993-1-5:2024 Design of steel structures. Part 1-5: Plated structural elements. BSI, London, UK.

BS EN 1993-1-8:2024 Design of steel structures. Part 1-8: Joints. BSI, London, UK.

BS EN 1993-1-13:2024 Design of steel structures. Part 1-13: Beams with large web openings. BSI, London, UK.

BS EN 1993-1-14:2025 Design of steel structures. Part 1-14: Design assisted by finite-element analysis. BSI, London, UK.

BS EN 1994-1-1:2026 Design of composite steel and concrete structures – Part 1-1: general rules and rules for buildings. BSI, London, UK.

BS EN 1998-1:2004+A1:2013 Design of structures for earthquake resistance. Part 1, General rules, seismic actions and rules for buildings. BSI, London, UK.

BS EN ISO 6892-1: 2019 Metallic materials – tensile testing; Part 1: Method of test at room temperature. BSI, London, UK.

BS EN 13670:2009 Execution of concrete structures. BSI, London, UK.

BS EN ISO 13918:2018+A1:2021 Welding – studs and ceramic ferrules for arc stud welding. BSI, London, UK.

BS EN ISO 14555:2017 Welding – arc stud welding of metallic materials. BSI, London, UK.

BSI (1994) DD ENV 1994-1-1:1994 Design of composite steel and concrete structures – Part 1-1: General rules and rules for buildings. BSI, London, UK.

BSI (2005) NA to BS EN 1990:2002+A1:2005 UK National Annex for Eurocode Basis of structural design. BSI, London, UK.

CEN/TS 1994-1-102:2026 Design of composite steel and concrete structures – Part 1-102: Composite dowels. CEN, Brussels, Belgium.

Couchman G (2014) *Design of Composite Beams Using Precast Concrete Slabs in Accordance with Eurocode 4*. Publication P401, Steel Construction Institute, Ascot, UK.

Couchman G (2015) *Minimum Degree of Shear Connection Rules for UK Construction to Eurocode 4*. Publication P405, Steel Construction Institute, Ascot, UK.

Couchman G and Way A (1998) *Joints in Steel Construction – Composite Connections*. Publication P213. Steel Construction Institute, Ascot, UK.

Degtyarev VV and Hicks SJ (2023) Shear resistance of welded studs in deck slab ribs transverse to beams. *Engineering Structures* **294**: 116709.

Dougherty LA (2023) *Composite Slabs and Beams Using Steel Decking: Best Practice for Design and Construction* (3rd ed.). SCI Publication P300, Metal Cladding & Roofing Manufacturers Association and Steel Construction Institute, Ascot, UK, pp. 119.

Dunant C, Drewniok M, Eleftheriadis S, Cullen J and Allwood JM (2018). Regularity and optimisation practice in steel structural frames in real design cases. *Resources, Conservation and Recycling* **134**: 294–302.

ECCS/IPHA (European Convention for Constructional Steelwork and International Prestressed Hollowcore Association) (1998) *Guidelines for Application of Prestressed Hollow Core Slabs Supported on Built-In Beams*. ECCS & IPHA, Report No. 103, Brussels.

ECCS (1999) *Design of Composite Joints for Buildings*. ECCS, Brussels, Belgium, Reports, 109.

Elliott JS and Nethercot D (1991) *Non-Composite Flexural and Shear Tests on CF70 Decking*. Department of Civil Engineering, University of Nottingham, Nottingham, UK, Report SR 91033.

Gardner L (2011) *Stability of Beams and Columns*. Publication P360, Steel Construction Institute, Ascot, UK.

Girão Coelho AM, Lawson RM, Lam D and Yang J (2020) *Guidance on Demountable Composite Construction Systems for UK Practice*. Steel Construction Institute, Ascot, UK, Report P428.

Grant JA, Fisher JW and Slutter RG (1977) Composite beams with formed metal deck. *Engineering Journal of the American Institute of Steel Construction* **14(1)**: 24–43.

Guo S and Bailey C (2007). Experimental and numerical research on multi-span composite slab. In *Steel Concrete and Composite and Hybrid Structures* (Lam D (ed.)). Research Publishing Services, Singapore, pp. 339–344.

Haensel J (1975) *Effects of Creep and Shrinkage in Composite Construction*. Institute for Structural Engineering, Ruhr-Universität, Bochum, Germany, Report 75-12.

Hanswille G (1997) Cracking of concrete – mechanical models of the design rules in Eurocode 4. In *Composite Construction in Steel and Concrete III* (Buckner CD and Sharooz BM (eds)), American Society of Civil Engineers, New York, pp. 420–433.

Hanswille G (2002) Lateral torsional buckling of composite beams. Comparison of more accurate methods with Eurocode 4. In *Composite Construction in Steel and Concrete IV* (Hajjar JF, Hosain M, Easterling WS and Shahrooz BM (eds)). American Society of Civil Engineers, New York, NY, USA, pp. 105–116.

Hanswille G, Porsch M and Ustundag C (2007) Resistance of headed studs subjected to fatigue loading Part II: analytical study. *Journal of Constructional Steel Research* **63(4)**: 485–493.

Hanswille G, Bergmann M and Bergmann R (2017) Design of composite columns with cross-sections not covered by Eurocode 4. *Steel Construction* **10**: 10–16.

Henderson JR (2017) *The Design of Cast-In Plates*. Publication P416. Steel Construction Institute, Ascot, UK.

Hendy CR and Johnson RP (2006) *Designers' Guide to EN 1994-2. Eurocode 4: Design of Composite Steel and Concrete Structures. Part 2: General Rules and Rules for Bridges*. Thomas Telford, London, UK.

Hicks SJ (2017) Design shear resistance of headed studs embedded in solid slabs and encasements. *Journal of Constructional Steel Research* **139**: 339–352.

Hicks SJ and Smith AL (2014) Stud shear connectors in composite beams that support slabs with profiled steel sheeting. *Structural Engineering International* **24(2)**: 246–253.

Hicks SJ, Pennington AF and Jones AS (2014) Longitudinal shear resistance of composite slabs. *Proceedings of the Institution of Civil Engineers Structures & Buildings* **167(10)**: 610–617.

Hicks SJ, Ciutina A and Odenbreit C (2020) Development of a new push test for Eurocode 4. In *Composite Construction in Steel and Concrete VIII* (Rassati GA, Hajjar JF and Leon RT (eds)), American Institute of Steel Construction, Chicago, IL, USA, pp. 62–73.

Hicks SJ, Braun M, Markovic Z and Way J (2023) New Eurocode 4 design rules for composite beams with precast concrete slabs. In *Composite Construction in Steel and Concrete IX* (Knobloch M, Kuhlmann U, Kurz W and Schäfer M (eds)). Ernst & Sohn, Berlin, Germany, pp. 40–49.

Huber G (2000) *Non-linear Calculations of Composite Sections and Semi-continuous Joints*. Ernst & Sohn, Berlin, Germany.

Huber G (2001) Semi-continuous beam-to-column joints at the Millennium Tower in Vienna, Austria. *Steel and Composite Structures* **1(2)**: 159–170.

Hughes AF, Iles DC and Malik AS (2011) *Design of Steel Beams in Torsion*. Steel Construction Institute, Ascot, UK, Report P385. https://www.steelconstruction.info/images/6/6f/Sci p385.pdf

ISO 10137:2017 (2017) Bases for design of structures – Serviceability of buildings and walkways against vibrations. International Organization for Standardization, Geneva, Switzerland.

Jähring A (2008) *Zum Tragverhalten von Kopfbolzendübeln in hochfestem Beton*. Thesis presented at Technische Universität München (in German). http://mediatum.ub.tum.de/doc/620984/620984.pdf

Johnson RP (1987) Shrinkage-induced curvature in cracked composite flanges of composite beams. *Structural Engineer* **65B(4)**: 72–77.

Johnson RP (2003) Cracking in concrete tension flanges of composite T-beams – tests and Eurocode 4. *Structural Engineer* **81(4)**: 29–34.

Johnson RP (2004) The m–k and partial-interaction models for shear resistance of composite slabs, and the use of non-standard test data. In *Composite Construction in Steel and Concrete V* (Leon RT and Lange J (eds)). American Society of Civil Engineers, New York, NY, USA, pp. 157–165.

Johnson RP (2011) Control of cracking in composite bridge decks in regions of biaxial tension. *Stahlbau* **80(12)**: 880–884.

Johnson RP (2012) *Designers' Guide to Eurocode 4: Design of Composite Steel and Concrete Structures*, 2nd edn. ICE Publishing, London, UK.

Johnson RP (2019) *Composite Structures of Steel and Concrete*. Wiley Blackwell, Oxford, UK.

Johnson RP and Allison RW (1983) Cracking in concrete tension flanges of composite T-beams. *Structural Engineer* **61B(1)**: 9–16.

Johnson RP and Buckby RJ (1986) *Composite Structures of Steel and Concrete*, 2nd edn, Vol. 2, Bridges. Collins, London, UK.

Johnson RP and Chen S (1991) Local buckling and moment redistribution in Class 2 composite beams. *Structural Engineering International* **1(4)**: 27–34.

Johnson RP and Fan CKR (1988) Strength of continuous beams designed to Eurocode 4. *Proceedings of the IABSE P-125/88*. IABSE, Zurich, Switzerland, pp. 33–44.

Johnson RP and Fan CKR (1991) Distortional lateral buckling of continuous composite beams. *Proceedings of the Institution of Civil Engineers, Part 2* **91(1)**: 131–161.

Johnson RP and Hanswille G (1998) Analyses for creep of continuous steel and composite bridge beams, according to EC4: Part 2. *Structural Engineer* **76(15)**: 294–298.

Johnson RP and Huang DJ (1995) Composite bridge beams of mixed-class cross-section. *Structural Engineering International* **5(2)**: 96–101.

Johnson RP and Ivanov RI (2001) Local effects of concentrated longitudinal shear in composite bridge beams. *Structural Engineer* **79(5)** 19–23.

Johnson RP and Molenstra N (1990) Strength and stiffness of shear connections for discrete U-frame action in composite plate girders. *Structural Engineer* **68(19)**: 386–392.

Johnson RP and Molenstra N (1991) Partial shear connection in composite beams for buildings. *Proceedings of the Institution of Civil Engineers, Part 2* **91(4)**: 679–704.

Johnson RP and Oehlers DJ (1981) Analysis and design for longitudinal shear in composite T-beams. *Proceedings of the Institution of Civil Engineers, Part 2* **71(4)**: 989–1021.

Johnson RP and Oehlers DJ (1982) Design for longitudinal shear in composite L-beams. *Proceedings of the Institution of Civil Engineers, Part 2* **73(1)**: 147–170.

Johnson RP and Shepherd AJ (2013) Resistance to longitudinal shear of composite slabs with longitudinal reinforcement. *Journal of Constructional Steel Research* **82**: 190–194.

Johnson RP and Yuan H (1998) Models and design rules for stud shear connectors in troughs of profiled sheeting. *Proceedings of the Institution of Civil Engineers Structures & Buildings* **128(3)**: 252–263.

Johnson RP, Lorenc W and Kurz W (2025) A new Eurocode, EN 1994-1-102, for composite dowel shear connectors. *Structural Engineer* **103(1)**: 24–26.

Kemp AR (2006) Avoiding fracture of slab reinforcement and enhancing ductility in composite beams. In *Composite Construction in Steel and Concrete V* (Leon RT and Lange J (eds)). American Society of Civil Engineers, New York, NY, USA, pp. 369–380.

Kilpatrick A and Rangan V (1999) Tests on high-strength concrete-filled tubular steel columns. *ACI Structural Journal* **96-S29**: 268–274.

Konrad M, Eggert F, Kuhlmann U and Schorr J (2020) New approach for the design shear resistance of headed studs in profiled steel sheeting with ribs transverse to supporting beam. *Steel Construction* **13**: 252–263.

Kuhlmann U and Breuninger U (2002) Behaviour of horizontally lying studs with longitudinal shear force. In *Composite Construction in Steel and Concrete IV* (Hajjar JF, Hosain M, Easterling WS and Shahrooz BM (eds)). American Society of Civil Engineers, New York, NY, USA, pp. 438–449.

Kuhlmann U and Hauf G (2015) Deformation calculation methods for slim floors. *Steel Construction* **8**: 96–101.

Kuhlmann U and Kürschner K (2006) Structural behaviour of horizontally lying shear studs. In *Composite Construction in Steel and Concrete V* (Leon RT and Lange J (eds)). American Society of Civil Engineers, New York, NY, USA, pp. 534–543.

Kuhlmann U, Wald F, Hofmann J *et al.* (2014) *Design of Steel to Concrete Joints, Design Manual I and II*. European Convention for Constructional Steelwork, Brussels, Belgium.

Lam D and Fu F (2006) Behaviour of semi-rigid beam-column connections with steel beams and precast hollow core slabs. In *Composite Construction in Steel and Concrete V* (Leon RT and Lange J (eds)). American Society of Civil Engineers, New York, NY, USA, pp. 443–454.

Lam D, Elliott KS and Nethercot DA (1998) Push-off tests on shear studs with hollow-cored floor slabs. *Structural Engineer* **76(9)**: 167–174.

Lam D, Elliott KS and Nethercot DA (2000) Designing composite steel beams with precast concrete hollow core slabs. *Proceedings of the Institution of Civil Engineers: Structures and Buildings* **140(2)**: 139–149.

Lawson RM and Hicks SJ (2011) *Design of Composite Beams with Large Web Openings*. Publication P355, Steel Construction Institute, Ascot, UK.

Lawson RM and Rackham JW (1989) *Design of Haunched Composite Beams in Buildings*. Publication 060. Steel Construction Institute, Ascot, UK.

Lee PG, Shim CS and Chang CS (2005) Static and fatigue behaviour of large stud shear connectors for steel-concrete composite bridges. *Journal of Constructional Steel Research* **61(9)**: 1270–1285.

Leskelä MV (2017) *Shear Connections in Composite Flexural Members of Steel and Concrete*. European Convention for Constructional Steelwork (ECCS), Report No. 138, Brussels.

Liew JYR, Xiong M and Lai B (2021) *Design of Steel-Concrete Composite Structures Using High-Strength Materials*. Elsevier, New York, NY, USA.

Lindner J and Budassis N (2002) Lateral distortional buckling of partially encased composite beams without concrete slab. In *Composite Construction in Steel and Concrete IV* (Hajjar JF, Hosain M, Easterling WS and Shahrooz BM (eds)). American Society of Civil Engineers, New York, NY, USA, pp. 117–128.

Moreno EN (2014) *Joints in Steel Construction: Simple Joints to Eurocode 3 (2014)*. Publication P358. Steel Construction Institute, Ascot, UK.

Piel W and Hanswille G (2006) Composite shear head systems for improved punching shear resistance of flat slabs. In *Composite Construction in Steel and Concrete V* (Leon RT and Lange J (eds)). American Society of Civil Engineers, New York, NY, USA, pp. 226–235.

Plumier A and Degée H (2023) *Design of Hybrid Structures – Where Steel Profiles Meet Concrete*. Routledge, Abingdon-on-Thames, UK.

Porsch M and Hanswille G (2006) Load introduction in composite columns with concrete filled hollow sections. In *Composite Construction in Steel and Concrete V* (Leon RT and Lange J (eds)). American Society of Civil Engineers, New York, NY, USA, pp. 402–411.

Rackham JW, Hicks SJ and Newman GM (2006) *Design of Asymmetric Slimflor Beams with Precast Concrete Slabs*. Publication P342. Steel Construction Institute, Ascot, UK.

Randl E and Johnson RP (1982) Widths of initial cracks in concrete tension flanges of composite beams. *Proceedings of the IABSE*, P-54/82, 69–80.

Roik K and Bergmann R (1990) Design methods for composite columns with unsymmetrical cross-sections. *Journal of Constructional Steelwork Research* **15(1-2)**: 153–168.

Roik K and Bergmann R (1992) Composite columns. In *Constructional Steel Design – an International Guide* (Dowling PJ, Harding JE and Bjorhovde R (eds)). CRC Press, FL, USA, pp. 443–469.

Roik K, Hanswille G, Cunze O and Lanna A (1989) *Report EC4/4/88, Eurocode 4, Clause 5.3, Cracking of Concrete*. University of Bochum, Bochum, Germany.

Roik K, Hanswille G and Kina J (1990) Solution for the lateral torsional buckling problem of composite beams. *Stahlbau* **59**: 327–332 (in German).

Sandberg J and Hendy CR (2010) The change from BS 5400 to Eurocodes – implications, benefits and challenges. In *Codes in Structural Engineering* (Hirt MA, Radic J and Mandic A (eds)). International Association for Bridge and Structural Engineering, Zurich, Switzerland, pp. 259–266.

Schäfer M, Braun M and Hauf G (2018) Flachdecken in Verbundbauweise – Bemessung und Konstruktion von Slim-Floor-Trägern. In *Stahlbau-Kalender 2018 – Verbundbau – Fertigung* (Kuhlmann U (ed.)), pp. 631–680 (in German).

Schäfer M, Zhang Q, Braun M and Banfi M (2021) Limitations of plastic bending resistance for composite beams deviated from strain-limitation. *Journal of Constructional Steel Research* **180**: 106562.

Schmeckebier N and Kurz W (2023) Engineering model for the vertical shear capacity of composite slabs with additional reinforcing steel. In *Composite Construction in Steel and Concrete IX* (Knobloch M, Kuhlmann U, Kurz W and Schäfer M (eds)). Ernst & Sohn, Berlin, Germany, pp. 3–14.

Simms WI and Hughes AF (2011) *Composite Design of Steel Framed Buildings*. Publication P359, Steel Construction Institute, Ascot, UK.

Smith AL and Couchman GH (2010) Strength and ductility of headed stud shear connectors in profiled steel sheeting. *Journal of Constructional Steel Research* **66(6)**: 748–754.

Smith AL, Hicks SJ and Devine PJ (2009) *Design of Floors for Vibration: A New Approach* (Revised edn.). Publication P354, Steel Construction Institute, Ascot, UK.

Stark JWB and Brekelmans JWPM (1990) Plastic design of continuous composite slabs. *Journal of Constructional Steel Research* **15(1-2)**: 23–47.

Stark JWB and van Hove BWEM (1990) *The Midspan Deflection of Composite Steel-and-Concrete Beams Under Static Loading at Serviceability Limit State*. TNO Building and Construction Research, Delft, Netherlands, Report BI-90-033.

Stark JWB and van Hove BWEM (1991) *Statistical Analysis of Pushout Tests on Stud Connectors in Composite Steel and Concrete Structures*. TNO Building and Construction Research, Delft, Netherlands, Report BI-91-163.

Steel Construction Institute (2011) Composite floors – Wheel loads from forklift trucks. Advisory Desk Note AD 150. *New Steel Construction* **1**.

Tan E and Uy B (2011) Nonlinear analysis of composite beams subjected to combined flexure and torsion. *Journal of Constructional Steel Research* **67(5)**: 790–799.

Thai H-T, Thai S, Ngo T *et al.*(2021) Reliability considerations of modern design codes for CFST columns. *Journal of Constructional Steel Research* **177**: 106482.

Trahair NS, Bradford MA, Nethercot DA and Gardner L (2008) *The Behaviour and Design of Steel Structures to EC3*, 4th edn. Taylor & Francis, Oxford, UK.

Umamaheswari N, Jayachandran SA and Lakshmipathy M (2007) Effect of concrete confinement on axial strength of circular concrete-filled steel tubes. In *Steel Concrete and Composite and Hybrid Structures* (Lam D (ed.)). Research Publishing Services, Singapore, pp. 271–278.

Uy B (2003) High strength steel-concrete composite columns for buildings. *Proceedings of the Institution of Civil Engineers – Structures and Buildings* **SB1: 156(1)**: 3–14.

van Hove BWEM (1991) *Experimental Research on the CF70/0.9 Composite Slab*. TNO Building and Construction Research, Delft, Netherlands, Report BI-91-106.

Vasdravellis G and Uy B (2014) Shear strength and moment-shear interaction in steel-concrete composite beams. *Journal of Structural Engineering* **140(11)**.

Vasdravellis G, Uy B, Tan EL and Kirkland B (2015) Behaviour and design of composite beams subjected to sagging bending and axial compression. *Journal of Constructional Steel Research* **110**: 29–39.

Vigneri V, Odenbreit C and Romero A (2021) Numerical study on design rules for minimum degree of shear connection in propped steel–concrete composite beams. *Engineering Structures* **241**: 112466.

Vigneri V, Hicks SJ, Taras A and Odenbreit C (2022) Design models for predicting the resistance of headed studs in profiled sheeting. *Steel and Composite Structures* **42(5)**: 633–647.

Wheeler AT and Bridge RQ (2002) Thin-walled steel tubes filled with high strength concrete in bending. In *Composite Construction in Steel and Concrete IV* (Hajjar JF, Hosain M, Easterling WS and Shahrooz BM (eds)). American Society of Civil Engineers, NY, USA, pp. 584–595.

Zhong S-T and Goode CD (2001) Composite construction for columns in high-rise buildings in China. *Proceedings of the Institution of Civil Engineers, Structures and Buildings* **146(4)**: 333–340.

Roger P Johnson and Stephen J Hicks
ISBN 978-1-83662-921-4
https://doi.org/10.1108/978-1-83662-920-720253029

Index

Notes: references to 'beams' and to 'columns' are to composite members
many entries (eg, 'annex') refer to EN 1994-1-1

Printed and bound by CPI Group (UK) Ltd, Croydon, CR0 4YY
07/07/2026
14916048-0003